THE GREENHOOD ORCHIDS OF THE SOUTH MAITLAND COALFIELDS.

1, Pterostylis rufa. 2, P. Mitchellii. 3, P. nutans. 4, P. curta. 5, P. parviflora. 6, P. truncata. 7, P. concinna. 8, P. longifolia. 9, P. mutica. 10, P. pedunculata. 11, P. furcillata. 12, P. ophioglossa. 13, P. acuminata. 14, P. reflexa. 15, P. revoluta. 16, P. pusilla var. prominens. 17, P. obtusa. 18, P. grandiflora.

THE ORCHID MAN

'The Orchid Man' as visitors to the Herbarium in Sydney's Botanic Gardens knew him in the early 1940s. (By courtesy of Mrs Rachel Cox)

THE ORCHID MAN

*The Life, Work and Memoirs of
The Rev. H. M. R. Rupp,
1872–1956*

Lionel Gilbert

Kangaroo Press

Gilbert, Lionel A. (Lionel Arthur), 1924– .
  The orchid man : the life, work and memoirs of the Rev. H.M.R.
  Rupp, 1872–1956.

  Bibliography.
  Includes index.
  ISBN 0 86417 415 2.

  1. Rupp, H.M.R. (Herman Montague Rucker), 1872–1956. 2. Church
  of England – Clergy – Biography. 3. Botanists – Australia –
  Biography. 4. Orchids – Australia – Identification. 5. Clergy –
  Australia – Biography. I. Rupp, H.M.R. (Herman Montague Rucker),
  1872–1956. Recollections of an amateur botanist. II. Rupp, H.M.R.
  (Herman Montague Rucker), 1872–1956. Retrospect. III. Title.
  IV. Title: Recollections of an amateur botanist. V. Title: Retrospect.

584.15092

First published in 1992 by Kangaroo Press Pty Ltd
3 Whitehall Road (P.O. Box 75) Kenthurst NSW 2156
Typeset by G. T. Setters Pty Limited
Printed in Singapore by Kyodo Printing Co (S'pore) Pte Ltd

ISBN 0 86417 415 2

Contents

Foreword

It is fitting that there should be an ample record of the life of the Rev. H. M. R. Rupp, for he represented a period in scientific research when direct personal observation was a requirement, and the assessment of technical features was the rule. That methodology contrasts with the present-day approach of using highly sophisticated instruments and recording results on smaller and yet smaller storage facilities such as discs and wires, although there still remains the demand for individual human attention in determining elemental qualities.

H. M. R. Rupp lived at a time when it was still a recognized objective for otherwise qualified persons, such as medical practitioners, parsons (for example the Rev. William Woolls, who is commemorated by the genus *Woollsia*) and other educated people to undertake studies in Botany, or in specific aspects of this discipline, mainly as a hobby. One of these was the Rev. H. M. R. Rupp, who by reason of his movement by posting to various parishes, and his service with the Australian Board of Missions, was able to travel and to observe widely in the south-eastern part of Australia.

Rupp's special interest was in the identification of species of the family Orchidaceae, which in Australia contains many species of terrestrial orchids, but much fewer epiphytes. Over thirty years, Rupp was responsible for describing many new species and for establishing new genera to accommodate these or the transfer of other species already described. Of particular interest is *Cryptanthemis slateri* (now *Rhizanthella slateri*) a saprophytic orchid which develops and matures its flowers below the surface of the soil, emerging later to distribute its seeds. Mr Rupp described this species from specimens collected by Mr Ernest Slater on Alum Mountain at Bulahdelah, N.S.W., where he had served a term as rector.

Mr Rupp's botanical work is recorded in a host of scientific papers, and in two reference books, *Guide to the Orchids of New South Wales* (1930) and *The Orchids of New South Wales* (1943) which was reissued in 1969 with a comprehensive supplement by D. J.

McGillivray of the National Herbarium of New South Wales.

The subject of this work must also be remembered as a great letter-writer, this skill being necessary for adequate communication in the times when he was making his researches. Consequently there is a vast collection of the correspondence kept by many of his fellow investigators including W. H. Nicholls and J. H. Willis of Victoria, T. E. Hunt of Queensland and F. Fordham of New South Wales.

I knew Mr Rupp from 1945, after he had retired from the ministry and was working on the orchid collection in the National Herbarium of New South Wales in the Royal Botanic Gardens, Sydney. For this he received an honorarium, small even for those times, until the year of his death in 1956. He was the epitome of 'Nature's Gentleman', and while retaining some mid-Victorian attitudes, was a most lovable character. It was a signal honour when he began to address me by my surname only, although even with the closest associate, he was always called '*Mr* Rupp'.

In addition to his family, the Rev. H. M. R. Rupp had two great loves—his Church and the native orchids of Australia. In presenting this biography, along with Mr Rupp's memoirs, to a wider public, Dr Lionel Gilbert has provided a biographical study in which these major aspects of Rupp's interests receive due attention. He has travelled widely over many of the scenes of Rupp's labours and sifted through masses of correspondence in seeking to produce for posterity an accurate record of this remarkable man's life and work.

Further, my old friend Dr Jim Willis has provided a bibliography, painstakingly listing for the first time Rupp's astonishing output of botanical books, papers and articles published during nearly seventy years.

Knowles Mair
Former Director and Chief Botanist,
Royal Botanic Gardens
Sydney

Preface

Some forty-five years ago, I presented myself rather formally (for there was then no official receptionist, public reference collection or visitor centre) at the magnificent old cedar counter at the office of the National Herbarium of New South Wales in Sydney's (pre-Royal) Botanic Gardens. The purpose of this visit was to have identified a rather precious specimen of an epiphytic orchid. It was considered 'precious' because I had but one specimen which had been painstakingly reassembled from an unpromising mass of pressed stems, leaves and flowers, and there seemed to be no prospect of ever obtaining another.

Other specimens, for example, of *Acacia, Grevillea, Melaleuca* and even of delicate *Nymphaea* (waterlilies) and small *Utricularia* (bladderworts) gathered in the vicinity of a radar station near Darwin during the last days of the Second World War had proved much more hardy, and had dried well, despite the high humidity. On the other hand, the orchid, collected with some difficulty from a large paperbark tea-tree, had been something of a disarticulated disaster until meticulously mounted on a sheet of double foolscap paper.

My enquiry, and display of the prized specimen on the counter, drew the response: 'Ah . . . you had better see our Orchid Man about that one'. Thereupon I was privileged to be ushered behind and beyond the counter into a labyrinth of passageways, galleries and alcoves, shelved to their ornate ceiling cornices, and packed with herbarium boxes of white cardboard, measuring in those pre-metric times, eighteen inches by twelve, and either three or six inches deep. Visitors to Sydney's 'old Herbarium'— that is, before the new Herbarium Complex was opened in 1982—will recall this rather dim yet fascinating place, with its distinctive naphthalenic atmosphere and grey- or white-coated staff members who lurked and worked amid huge piles of boxes and herbarium sheets. One could only hope that the Department of Agriculture hade made some special arrangement with the Board of Fire Commissioners to ensure the continued existence of the institution.

At a specimen-laden table in one cramped alcove sat a benign-looking, faintly-smiling, bespectacled gentleman in a dark suit. I was then introduced to the Rev. H. M. R. Rupp (clearly 'our Orchid Man') as 'someone with a query'. We exchanged some introductory pleasantries, and he asked a few direct questions about the orchid's place of origin and habitat, and the colour of the living flowers. Looking penetratingly at the specimen, he quietly remarked, 'It's probably *Dendrobium dicuphum*'. Then he paused, still looking intently at the specimen sheet, both with and without the aid of a small hand lens, all the while drawing upon a well-stoked pipe. Then with a sudden movement of conviction, he swept off his glasses, peered again at the specimen, and confirmed with smiling enthusiasm, 'Oh, it *is dicuphum*—no doubt about it!'

Thus I made my first personal contact with the genial 'Orchid Man' of the National Herbarium. Had I but known it, this affable gentleman, elderly yet youthful, could have recounted his childhood reaction to the capture of the notorious Ned Kelly and his personal meeting with the celebrated Baron von Mueller!

We met only once or twice again, but became good botanical penfriends between June 1947 and June 1956, when shortly before his death, he sadly advised, 'I am afraid my orchid days are over. I cannot get about much, and anything in the way of bush excursions is impossible.' He wished me well in future botanical adventures and concluded in his customarily courteous manner, 'with kind regards and all good wishes'. We had exchanged thirty-two letters—a very modest score I later discovered when examining his wider, prodigious correspondence with others of longer standing and of greater erudition.

Unfortunately, during the next thirty years, my pursuit of orchidological knowledge lost much of its former vigour, partly because of the lack of encouragement through the mail, but chiefly because vocational commitments and further studies allowed little time for field work beyond that associated with some delving into Australian botanical history. After a disastrous fire destroyed the Belshaw Block of the

University of New England, Armidale, in February 1958, I presented my herbarium, including the *Dendrobium dicuphum* (now known as *D. affine*), to the Botany Department to help compensate for its loss, thereby unconsciously emulating an action of my late penfriend.

Work in botanical history ultimately led to interesting developments. Early in 1982, Mr Don Blaxell, Assistant Director of the Royal Botanic Gardens, Sydney, asked whether he could borrow for copying any letters I had received from Rupp. Returning the letters in April, he advised that copies had been lodged in the Herbarium Library 'with all the others' he had been able to borrow for copying or otherwise to acquire from some of Rupp's numerous correspondents. He also expressed the wish that 'some day someone will write a biography of Rupp from a botanico-historical point of view' and that I would perhaps consider the matter.

Mr Blaxell's wish came closer to realization when, in August 1985, there came a request to write about 500 words on H. M. R. Rupp for the *Australian Dictionary of Biography*. This rather forced the issue, and it was then that the initiative to assemble originals or copies of Rupp's letters in one place was fully appreciated. Rupp was a fluent and frequent letter-writer whose voluminous correspondence exhibited a delightful literary style and an infectious enthusiasm for botanical investigation. As he told one long-standing correspondent, Fred Fordham, in June 1933, 'one of life's pleasures for me is corresponding with orchidy folks'. Thereafter they corresponded for another twenty-three years.

When searching for personal details of Rupp's life and work, I had the pleasure of making the acquaintance, first of Rupp's elder daughter, Mrs Rachel Cox of Armidale, then of his younger daughter, the late Mrs Eileen Cox of Noosa Heads, and subsequently of his son, the late Mr Arthur Rupp of Willoughby. Having had Mrs Rachel Cox's kind assistance to produce the brief article for the *Dictionary*, it seemed proper to ask her to read the contribution before it was sent to Canberra. She graciously agreed to do this, and events took a rather exciting course.

In preparing the article, I had the great advantage of being able to use the splendid obituary written by one of Rupp's most valued penfriends, Dr James H. Willis (former Assistant Government Botanist of Victoria) for one of Rupp's favourite journals, the *Victorian Naturalist* (November 1956). Mrs Rachel Cox had assisted in the preparation of this by making available some 'autobiographical notes' prepared by

her father. In a quarto page of careful typing, Rupp had recorded in a pithy direct manner, his life story in about 550 words.

Such a terse record was valuable enough, but Mrs Cox produced two further items from her father's papers—one in manuscript, comprising about 17,000 words, entitled 'My Australian Hobby' or 'The Recollections of an Amateur Botanist', written December 1932–January 1933, and the other in typescript, comprising about 35,000 words, entitled 'Retrospect', written November 1948–January 1949, with amendments made in the early 1950s. Thus it seemed that any prospective biographer had been virtually pipped to the post. Rupp had fascinatingly recorded his life story, twice—once rather briefly with a botanical emphasis, and once in a more extended, general way.

Naturally, the accounts overlap at certain periods, but hardly enough to warrant any attempt to combine them into a single narrative. They were obviously written at different times, from different points of view and for readers with different interests. It is hoped that gentle and judicious editing has reduced the inevitable repetition to an acceptable minimum. Like Rupp's 'Gardens of Nature'—the 'recollections of over half a century of bush ramblings'—neither autobiograpical work was published.

While nothing would have been gained from attempting to produce a composite paraphrase from these two accounts, they so teem with geographical, historical, botanical, ecclesiastical and personal allusions that some annotation was clearly necessary to complement the texts without disturbing them. Hence there are extensive references, and these, together with Dr Willis's admirable bibliography, compiled over thirty years ago, will, we trust, help readers to enjoy and appreciate the memoirs of one who lived a long and fruitful life, who was never wealthy, always perceptive, and who did not have the opportunity to travel widely beyond south-eastern Australia. Despite enormous changes at the Herbarium since Rupp's death, he is still remembered there as 'a charming man, with a chubby, cheery face and twinkling eyes, an enjoyment of life, and without any enemies'—an appropriately worthy tribute.

As Mr Knowles Mair has indicated in his Foreword, the Rev. H. M. R. Rupp belonged to 'the old school' of botanists, whose approach to their subject was principally morphological and taxonomic. Although sometimes gently critical of the conclusions he found in the work of Robert Brown,

George Bentham, Baron von Mueller and Robert David FitzGerald, Rupp had profound respect for these and other botanists who worked on the Australian flora during the nineteenth century. His methodology was similar to theirs—explore, discover, examine, dissect, compare and determine—or, if it seemed appropriate, name, illustrate and describe. He suggested that other amateur investigators could do likewise and in June 1934, for example, advised readers of the *Australian Woman's Mirror* that in order to study orchids, 'the only requisites are sharp eyes, a little patience, a small magnifying glass, and some knowledge of the parts of the orchid flower'.

In thus encouraging others, he could make it all sound rather too easy, but a popular magazine was no place for suggesting that long experience, a keenly perceptive mind, finely tuned to discern possible affinities and subtle differences, and a clear scientific insight—in short, the qualities which Rupp himself saw in those to whom he attributed possession of 'the true orchid eye'—were similarly desirable. In addition, he possessed a remarkable memory not only for the pertinent literature, but also for localities and habitats, scents and seasons, and the idiosyncrasies likely to be exhibited by particular species living in certain places under unusual conditions. In today's terms, Rupp would be described as a 'natural', with a 'back-to-basics' approach to his field of study. It was characteristic of him to publish a short, well-illustrated article, 'What is an Orchid' in the *Australian Orchid Review* in September 1945, nearly nine years after the journal first appeared. He began:

This seems a stupid question for the title of an article in the official journal of an Orchid Society. Yet I am assured that it is not what it seems; and that many members of our Society would welcome a simple exposition of the botanical characters which differentiate the Orchid from all other flowering plants.

The editor immediately received 'a number of congratulatory messages', and Rupp had yet more correspondence to answer.

This scientific clergyman's firm faith was not disturbed by palaeontology, Darwinism or other evolutionary theory. Today's refined phylogenetic systematics and ingenious cladograms would have intrigued him; the use of the electron microscope and the application of computers to taxonomy would have astounded him. Yet the most recent revisions of the family Orchidaceae in Australia indicate that many of the botanical judgements and determinations of

the man whose most sophisticated aid was the binocular optical microscope have been vindicated. His name is still associated with the naming of a significant number of the 660 or so species of native orchids presently recognized.

Some of the background to Rupp's life, and aspects which were overlooked or only briefly noticed in his reminiscences, have been recorded in the biographical study. This was compiled 'from the outside', as it were, with the aid of diverse sources, especially Rupp's extensive correspondence. The Orchid Man told his own story 'from the inside' in his memoirs, which have been deliberately separated from the remainder of the work to stand in their own right as their author intended.

It will be evident that this work would not have had much substance had not Mrs Rachel Cox of Armidale so generously offered me her father's memoirs to read, and then given permission for them to be copied for detailed study. Her kindness—and patience—in answering a host of queries and in lending family papers must also be warmly acknowledged.

Similarly her sister, the late Mrs Eileen Cox of Noosa Heads, readily provided answers to queries, photographs, and many 'personal touches' to enliven the picture of a personality known to me chiefly through correspondence and other writings. The late Mr Arthur Rupp and his daughter Miss Elizabeth Rupp received me graciously at Willoughby, where I was treated to a feast of documentary material, in the form of books, photographs, certificates and other papers. Such ready cooperation is gratefully recorded.

Thanks are also due to Mr Don Blaxell, Assistant Director, Royal Botanic Gardens, Sydney, for his forethought in assembling much of Rupp's correspondence; to Dr James H. Willis of Brighton, Victoria, for answering many enquiries, for graciously agreeing to the amendment and publication of his bibliography, and for generously adding 284 letters to the Sydney collection of Rupp papers; to Professor Carrick Chambers, Director of the Royal Botanic Gardens, Sydney, for providing me with a working copy of those letters, for locating and providing copies of still more letters, and for constant encouragement.

I am grateful also to Ms Anna Hallett, Librarian of the Royal Botanic Gardens, Sydney, for access to the Rupp papers and much other material in her care; to Ms Helen Cohn, Librarian of the Royal Botanic Gardens, Melbourne, for locating some additional Rupp letters; to the Librarian of the Needham Memorial Library of the Australian Board of Missions, Sydney; to Dr Gwenda Davis (formerly Associate Professor of Botany at the University of New England) of Port Macquarie, for answers to queries and for locating further Rupp letters; to Mrs Jean Uhl of Blackburn, Victoria, for both her kind interest and a generous supply of information gleaned during her own extensive research; to Dr Peter Thomson of Blackburn, for seeking Rupp correspondence in the papers of his late grandmother, Mrs Edith Coleman; to Mr L. Scott Rogers of Buderim, Queensland, and to Ms Joyce Gibberd of St Peters, S.A., for seeking Rupp letters in the papers of the late Dr R. S. Rogers of Adelaide; to Dr Pat Flecker of Townsville for advice about the papers of his late father, Dr Hugo Flecker of Cairns; to Mrs Betty Mathieson for information about her husband's brother-in-law, the late Trevor E. Hunt of Ipswich; to Mrs Lorna Scammell of Neutral Bay for material relating to her late husband, George Vance Scammell; to Mrs Jean Trotter of Port Macquarie and Mrs Helen Faint of Hillgrove for family records relating to their father, the late Fred Fordham; to Mr C. Keith Ingram of Mt Tomah for material relating to H. M. R. Rupp and G. W. Althofer; to Mr Ronald Kerr, former Editor of *Australian Orchid Review*, for the loan of a collection of Rupp's journal articles; and to Mr Ernest Slater of Glenbrook, N.S.W., for an interview relating to his discovery of *Cryptanthemis*.

Thanks are also due to my good friend, Dr Geoffrey Burkhardt of Canberra, for direction into the unknown (to me) field of German immigration resources and to Mr Robert Wuchatsch of Melbourne and Mr David Thiele of Walla Walla, N.S.W., for similar help; to Mr Michael D. de B. Collins Persse, Keeper of the Archives at Geelong Grammar School for constant interest and generous contributions of documentary material; to the Reverend Dr Arthur de Q. Robin for kind assistance while working in the Diocesan Archives, Melbourne; to Miss Jean Waller, Librarian of the Mollison Library and Mr Nikolaos Sakellaropoulos, Honorary Archivist, of Trinity College, Melbourne, for access to essential sources relating to Rupp's student days; to Mr Frank Baines of the Registrar's Office, University of Melbourne, for answers to queries, and to Mr Ian Clarke, formerly Technical Officer in the Botany Department, University of Melbourne, for a warm welcome to the Department's Herbarium and for access to specimens, records and photographic material; to Mr Denis Rowe, Archives Officer, and Mr Tim Mawson, Diocesan Secretary, for ready access to Diocesan records in the Auchmuty Library, University of Newcastle; to Mr John Hansen, Registrar, for similar access to the records of the Diocese of Armidale; to Mr E. D. Hatch of Auckland, for information about his association with Rupp; to Mrs Gwen Artlett of Mt Irvine and Mrs Jean Harslett of Amiens, Queensland, for readily supplying reminiscences and kindly donating their collections of original letters to the Library of the Royal Botanic Gardens, Sydney; to Norman and Marjory Loader of Dural for interesting reminiscences and generous access to documentary material; to Mrs Susan Cantrell for important information relating to the early naturalists of the Mount Tamborine area of Queensland; to Professor John Pearn of the University of Queensland, Dr Robert W. Johnson, Director of the Queensland Herbarium, to Ms Joan Cribb and Mr H. S. Curtis of Brisbane; to Professor Ken Cable of Sydney and to Fr Bede Lowery of Parramatta for prompt replies to requests for specific information. Mrs Valerie Jones, Honorary Phycologist at the National Herbarium of New South Wales, will notice her welcome contribution to this Preface.

The help received from members of the staffs of institutions not previously noticed is thankfully acknowledged: the State Library of New South Wales, the State Library of Victoria, the Australian National Library, the Australian Archives, the State Archives of Tasmania, the Geelong Historical Records Centre, the Library of the Australian Museum, the Blue Mountains City Library, Springwood, and the Library of the Royal Botanic Gardens, Kew, Surrey.

Thanks are also due to H. M. R. Rupp's good friend, and mine, Mr Knowles Mair, former Director of the Royal Botanic Gardens, Sydney, for his encouragement, his hospitality and his Foreword; to Mr Edwin Wilson, Liaison Officer at the Royal Botanic Gardens, Sydney, for permission to reproduce his poems; to Mrs Dianne Hill, Mrs Wendy Thomson and Ms Mary Tudhope for their patience and skill in producing an acceptable script from mysterious machines made long after that on which the original copy was typed; and to the many amiable and cooperative people met at Port Fairy and at several other places during the long pilgrimages to localities associated with the story that Montague Rupp has told.

The friendly encouragement and practical advice received from Mr David Rosenberg, director of Kangaroo Press, and from his editor, Mr Carl

Harrison-Ford, are also warmly acknowledged, as is the financial assistance so kindly given by the Australian Orchid Foundation and by the Sydney Royal Botanic Gardens and Domain Trust which enabled production of the book to proceed.

Lionel Gilbert
Armidale, N.S.W.
October 1991

A Note on Nomenclature

Orchid taxonomy is a difficult subject, and a sound knowledge of our native orchids may be obtained only by long and close study of the plants and of the literature relating to them.

—Donald J. McGillivray
in his 'Supplement' to H.M.R. Rupp,
The Orchids of New South Wales,
Facsimile Edition, 1969

Chiefly working alone, or occasionally in collaboration with other enthusiasts, such as C. T. White, T. E. Hunt, or W. H. Nicholls, H. M. R. Rupp was associated with describing, naming and publishing about 120 taxa of Australian orchids. He was fiercely patriotic in this activity, declaring that the appropriate medium through which a new Australian species should be made known was an Australian journal. His bibliography clearly reflects this view.

However, just as Rupp did not always agree with the interpretations and conclusions of his predecessors, for example, R. D. FitzGerald (or indeed with those of some of his contemporaries, for example, W. H. Nicholls), so have Rupp's successors not necessariy agreed with some of his findings and determinations. Many of the taxa he carefully described in the papers listed in the Bibliography are not currently recognized.

During the three decades and more since Rupp's death, there has been further intensive fieldwork despite the destruction of vast areas of bushland which once provided ideal habitats for orchids; keen-eyed investigators have rigorously re-examined type specimens in Australian, British, Continental and other herbaria, and reassessed the relevant literature.

The methodology of classification has also been developed and refined, and the variability of certain taxa has become more evident with the discovery of ranges of intermediate forms.

Accordingly, the most recent revision of Australian orchid nomenclature (Mark Clements, 'Catalogue of Australian Orchidaceae', in *Australian Orchid Research*, Vol.1, August 1989, published by the Australian Orchid Foundation) has, for various reasons, relegated about seventy of Rupp's classifications to the status of synonyms. On the other hand, a few of his varietal determinations have been raised to specific rank.

About fifty taxa currently considered valid carry Rupp's name as the primary or secondary author. He would be delighted that a dozen of his beloved 'Prassies' still bear the names he bestowed; he would be saddened to note that *Prasophyllum bowdenae, P. hopsonii, Chiloglottis dockrillii* and *Saccolabium loaderianum*, which honoured his esteemed friends and co-workers, have now been declared synonyms. He would no doubt steadfastly defend his genus *Cryptanthemis* against its supersession by Dr R. S. Rogers's *Rhizanthella*.

Yet one suspects that, overall, Rupp would regard the progress made in the investigation of his favourite plants with as much admiration as astonishment. He would surely enjoy the gratification of knowing that he had made a significant contribution according to the procedures and expectations of his time. He would be quick to appreciate that advances in scientific technology and knowledge had been matched by advances in book production, thereby preserving the new information in clear and attractive ways. As one who had been nurtured on the rather stark botanical keys and prose of George Bentham and Ferdinand von Mueller, Rupp must have taken great pleasure in the black-and-white lithographs which softened some of Mueller's works, and characterized (together with photographs) most of J. H. Maiden's remarkable output.

The hand-coloured lithographs of the great folios of R. D. FitzGerald's *Australian Orchids* must have appeared as the ultimate pictorial record of these fascinating plants before W. H. Nicholls began his similarly ambitious project. How Rupp would have delighted in Thomas Nelson's 1969 production of *Orchids of Australia* by D. L. Jones and T. B. Muir, who provided the text for Nicholls's superb water-colours. Unhappily, he would have looked in vain for any acknowledgement from his old friend and sparring-partner in the Preface, written in July 1950; but he would have heartily admired the well-reproduced illustrations, notwithstanding any disapproval of certain aspects of morphology and colouring!

As a keen photographer himself, Rupp would have taken tremendous delight in David Lloyd Jones's *Native Orchids of Australia*, published by Reed Books in 1988, and whatever he may have thought about the disappearance of some of his nomenclature, the inclusion of *The Orchids of New South Wales* in the Bibliography would have brought a quiet knowing smile.

To avoid anachronistic as well as taxonomic problems, the nomenclature Rupp used in his memoirs has been left virtually unchanged. The opportunity to make 'corrections' (a term not used in an absolute sense!) has been taken in the notes and in the Botanical Index.

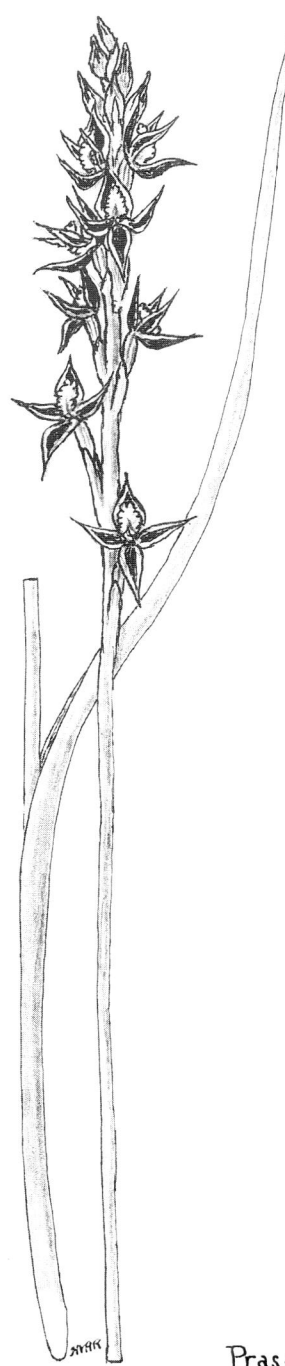

Prasophyllum australe R.Br.
Mt. Irvine.

11/41.

PART I

THE ORCHID MAN

Say, botanist! within whose province fall
The cedar and the hyssop on the wall,
Of all that deck the lanes, the fields, the bow'rs,
What parts the kindred tribes of weeds and flow'rs?
Sweet scent, or lovely form, or both combin'd,
Distinguish ev'ry cultivated kind,
The want of both denotes a meaner breed,
And Chloe from her garland picks the weed.
Thus hopes of every sort, whatever sect
Esteem them, sow them, rear them, and protect;
If wild in nature, and not duly found
Gethsemane! in thy dear, hallowed ground,
That cannot bear the blaze of scripture light,
Nor cheer the spirit, nor refresh the sight,
Nor animate the soul to Christian deeds,
Oh cast them from thee! are weeds, arrant weeds.

—from *Hope* (1781) by William Cowper,
(1731–1800)

Australian orchids fetch no fancy prices. Most of
them are very humble members of the great family
to which they belong... But orchids are orchids,
whether they belong to the flamboyant battalions of
cattleyas, cymbidiums, oncidiums, and their
associates, or to the modestly clad rank and file of
the terrestrials that inhabit our Australian
heathlands and open forests. It is the unique
character of the orchid flower, irrespective of its size
or colouring, that has always held me spell-bound.

H. M. R. Rupp, *Journal of the New York
Botanical Garden*, July 1946.

1 Despair and Deliverance

When I made my first Home trip, in 1847, I resolved to
open, if I possibly could, German immigration to Port Phillip.

—William Westgarth, 1888

On the evening of Tuesday, 6 March 1849, the German ship *Wappaus*,[1] 304 tons, from Hamburg, commanded by Captain Peterson, ended its long and 'boisterous passage' in Hobson's Bay, near the burgeoning metropolis of Melbourne. On board was a relatively light cargo of sundry merchandise, including a case of watches, a cask of bluestone and fifteen packages of passengers' luggage. The passengers themselves, all steerage, numbered 191—61 women, 69 men, 32 girls and 29 boys. Their precise trades and professions proved somewhat difficult to ascertain,[2] but they included agricultural labourers and shepherds, cartwrights, tailors and masons, shoemakers and cabinetmakers, a soapboiler, a lithographer, a schoolmaster, one or two missionaries—and a vine-dresser.

The newcomers were considered 'a very respectable looking set' and Superintendent Charles Joseph La Trobe 'placed the Immigrant Barracks at their disposal' until they found employment.[3] Some quickly 'met with engagements much at the same rates of wages as those given to English mechanics and labourers',[4] and it was hoped, indeed assumed, that most, if not all, would choose to remain in the Port Phillip District. However, when the *Wappaus* sailed for Adelaide on 15 March, fifty-two of the Hamburg passengers, together with eight English emigrants, had decided on South Australia.[5]

The Melbourne *Argus* proposed, in vain as it happened, to provide its readers with 'some information' about the new arrivals in order to 'keep up the public attention towards them, and facilitate their engagement and settlement amongst us'. An *Argus* reporter 'unfortunately encountered Mr Charles Hotson Ebden' at the barracks, and 'was favoured with a prolonged and elaborate essay upon the state of the Port Phillip press', more especially 'the degraded paper ... he was doomed to represent'. Accordingly, *The Argus* remained 'in a state of Cimmerian darkness as to the doings of the poor Germans'.[6]

Other doings were more readily brought into the light, and the recent immigrants, depending on their access to comprehensible sources of gossip and reported news, may have learned something of local opinion and events. For example, there were, in addition to notices and results of sundry race meetings at Melbourne, Geelong, Portland and Albury, indignant reports of house robbery and forgery, and alarming reports of the 'dreadful bush fire' at Mt Eliza; there was great concern over increasing instances of 'Bushranging and bailing-up' in the Portland District, and relief that three suspected miscreants had been apprehended at Belfast (later Port Fairy) in connection with the Warrnambool Mail Robbery. The superintendent of bridges, Mr David Lennox, was calling tenders for new work on dams, jetties and bridges; Mr Thomas Ham, engraver and publisher, was promoting his new maps of Australia Felix for the benefit of travellers and prospective settlers, and Captain Peterson may have learned, with relief, of proposals to increase the number of shipping beacons in Port Phillip Bay.

Of much greater moment was the imminent arrival of His Excellency the Governor, Sir Charles Augustus FitzRoy, expected 'hourly' from Sydney on HMS *Havannah*. The governor would arrive at a time of great excitement and anger following vigorous protest meetings over the risk of 'pollution' from other new arrivals, not free migrants from Germany, but ticket-of-leave 'exiles' from England. Such an acceptance of convicts would surely mean 'the stoppage of the present redundant supply of virtuous and industrious Immigrants', warned the *Melbourne Morning Herald*.

Perhaps more in keeping with the German experience was the fact that Melbourne's citizens seemed fiercely opposed to certain decrees, emanating not from a Prussian king, nor even directly from Queen Victoria, but from an apparently unsympathetic and rather despised Sydney legislature, condemned for its 'ungrateful and unprincipled parsimony ... towards Port Phillip'. While 'bilious animosity' divided factions of bickering aldermen of the recently proclaimed cathedral city of Melbourne, 'Separation' seemed generally acknowledged as the inevitable—and only—panacea for the ills and problems of all colonists, whether of long standing or of recent arrival.

Meanwhile, at a more modest level, Henry Clegg enthusiastically offered at the attractive rate of six shillings a bucketful, or 'about threepence per dozen', the fresh Sydney oysters he imported by the steamer *Shamrock*. Those unpatriotic enough to over-indulge could, perchance, seek relief from their discomfort by having recourse to Parr's Life Pills (which 'in their operation ... go direct to the disease') or to Dr Thomas Holloway's much-advertised pills, for there were ample testimonies of their efficacy vividly composed by grateful erstwhile sufferers from (it seemed) most maladies known to mankind.[7]

The promotion of such miracle medicines may well have prompted some of the new residents of the Immigration Barracks to reflect on their good fortune in surviving not only the voyage, but also the pre-embarkation period, for many other relatives and friends, now absent, had been with them at Hamburg. Despite the attentions of the surgeon-superintendent, Frederick Asschenfeldt, four adults and nine children had died before the captain of the *Wappaus* 'was obliged to put into Bahia, in the Brazils, for a supply of water', and after thirteen days there, the vessel lost another adult and two more children before gaining its destination.[8]

The nineteen-week voyage had been especially tragic for two of the passengers, who arrived as sad and bewildered orphans in a strange land full of strangers. They were Carl Ludwig Herman Rupp, aged ten, and his five-year old sister, Augusta Adelaide Johanna, who must have wondered why they had been brought to the other side of the world and what would become of them.

With ample justification, the year 1848 has been dubbed 'the year of Continental Revolution and Reaction'.[9] By the end of June there had been revolutions in Sicily (which had declared its independence), in Paris, Venice, Parma and Milan, and in Vienna, Berlin and Prague; Austria and Sardinia were at war; the Prussians had suppressed a Polish insurrection in Warsaw before invading Danish territory over the 'problem' of the provinces of Schleswig and Holstein, then Danish, but with considerable German populations. By the end of the year Naples had recovered Sicily, Hungary had a proclaimed dictator, Louis Napoleon was President of the French Republic and Francis Joseph I had been installed on the Hapsburg throne. Sundry treaties and truces had been arranged throughout Europe, but hopes for general peace and stable government remained uncertain.

Elsewhere there were less bellicose, if no less significant, events during 1848—gold was discovered in California; Karl Marx and Friedrich Engels published their *Communist Manifesto* and William Makepeace Thackeray completed his *Vanity Fair*; William Holman Hunt, Dante Gabriel Rossetti and John Everett Millais founded the Pre-Raphaelite Brotherhood in Rome, while in England the Rev. Frederick Denison Maurice's Christian Socialists were organizing themselves, and, inspired by Edwin Chadwick (and by an outbreak of cholera), the House of Commons passed its first Public Health Act.

In far-off Australia, Mr Surveyor Edmund Besley Court Kennedy led his disastrous expedition into the north of Cape York Peninsula in 1848; a young chemist, Dr Ferdinand Mueller, just arrived in Adelaide from Schleswig-Holstein, devoted every spare minute to studying the flora of his newly adopted country; young Mr Charles Moore assumed directorship of Sydney's Botanic Gardens, while 'Dr' Ludwig Leichhardt, the Prussian naturalist who had aspired to that position, led his last exploratory expedition into oblivion somewhere in the remote interior—and some brave souls even proposed linking Sydney and Melbourne by a railway.

Like some other European countries, Germany experienced a vast population increase during the nineteenth century, notwithstanding wars, famines, floods and other disasters. Its population 'more than doubled from 24.5 million to 56.4 million' and its resources were sorely strained.[10] It has been estimated that over 5,000,000 people (about 10 per cent of the population) emigrated from Germany between 1815 and 1885, the majority leaving between 1845 and 1860.[11] About 95 per cent went to North America, and some of the remainder came to Australia.

The German Lutherans who came to South Australia, 1838–42, with Pastor A.L.C. Kavel, Captain Dirk Hahn and Pastor G.D. Fritzsche, were largely motivated by the religious persecution resulting from attempts by Frederick William III, King of Prussia, 1797–1840, to unite his Lutheran and Calvinist subjects by decree.[12] After the King's death, however, this issue was not so significant, and a recent study has concluded that:

although there is no question that religious, political, military and social considerations did influence some emigrants, the myth prevails in Australia, that most, if not all, nineteenth-century German immigrants, were escaping problems such as religious persecution, political unrest or military service. In reality, they left to escape the lean times.[13]

These 'lean times', intensified by crop failures and rising prices, provided a most appropriate occasion for the launching of any emigration scheme.

It so happened that William Westgarth (1815–89), who left his native Scotland in 1840, had by 1847 become a successful Melbourne import merchant and shipping agent. He returned to Britain on a business trip in January 1847 determined 'to open ... German immigration to Port Phillip', for in South Australia the Germans had demonstrated qualities dear to his heart—'industry, frugality, sobriety and general good conduct' which, as one might expect, 'had made them excellent colonists'.[14]

Believing he had the assurance of a government subsidy to bring German vine-dressers, shepherds and agricultural labourers to Port Phillip, Westgarth went to Hamburg. Armed with a letter from Earl Grey, secretary of state for the colonies, he called on John Caesar Godeffroy & Son, 'at that time the chief shipowners of the city', and with the aid of an agent, Edward Delius of Bremen, recruitment of approp-riate emigrants began. Although chiefly based in Scotland, Westgarth revisited Hamburg at the end of September 1848 'to see the first party away' on the *Godeffroy*, shortly after.

The departure was not without its problems, as the emigrants

were in a good deal of trouble, for most of them in spite of all advice, clung to old family lumber . . . But, worst of all, the cholera was then raging in Hamburg, and it attacked several of the party . . . while they waited, under such shelter as they could improvise until the ship could take them.[15]

Dejected first by the general upheaval associated with emigration, then by the loss of 'a number of the young men' through recruitment in the militia during the Danish crisis, and finally by the cholera outbreak, the prospective emigrants in Hamburg probably had mixed feelings about Westgarth and Delius when the pair visited 'to cheer them with the near prospect of the sunshine and plenty in Australia'.[16] When Westgarth left Hamburg on 9 October 1848, the *Godeffroy* had sailed,[17] and the *Wappaus* 'was just completing her finishings and about to receive the second detachment' of emigrants.[18] She did not sail, however, until the twentieth.[19]

Clearly Earl Grey and the Colonial Land and Emigrant Office considered that both Westgarth and Delius had rather too liberally interpreted the terms under which a subsidy might be allowed for the passages of approved emigrants. There were several conditions, one of which was the submission of carefully compiled lists, indicating names, ages and occupations of intending emigrants, for official approval. Great concern was expressed over 'certain irregularities' which could well invalidate any agreement to pay a bounty of £2 10s. 0d. per head on 400 emigrants. For his part, Edward Delius had tremendous difficulty in compiling a final list, with the details required. Then the cholera epidemic began, and it became imperative that the emigrants 'should immediately quit Hamburg, where the malady was raging'.[20] Lists were hurriedly prepared, the ships were hastily loaded, and to this day there remain problems of determining who sailed on which ship. Years later, after the protracted and difficult negotiations over the subsidy to be paid, Westgarth drily commented, 'I fear they were not all vine-dressers'.[21]

It is clear, however, that among those who embarked or intended to embark on the *Wappaus* was a family of five—Carl Ludwig Rupp, a schoolmaster, his wife Caroline, née Freyer, and their three

children, Carl, Augusta and Paul, an infant. The
elder boy was born on 2 July 1838 at Frankfort-on-
Oder, where presumably the father was then
teaching.[22] Whether the father, mother and baby
Paul succumbed to cholera at Hamburg, or died on
the voyage from some shipboard malady such as
dysentery, is not known.[23]

It was later noted that, fortunately for the
children, 'there were family friends on board',[24]
greatly concerned about their plight. Perhaps it was
a coincidence that the name Thiele appears on the
hastily compiled passenger list of the *Wappaus*, but
a joiner of that name was one of Carl Ludwig
Herman Rupp's five godparents when he was
baptized in St George's Church at Frankfort-on-
Oder.[25] In any case, someone brought the situation
of the orphans to the notice of W.F.A. Rucker, later
credited with having been 'beyond doubt the first
Melbourne banker',[26] and 'a pioneer Melbourne
should remember with gratitude'.[27] Certainly Carl
and Augusta Rupp remembered him with gratitude,
for he took them into his household and, in effect,
adopted them as his own children.

2 New Kith and Kin

. . . a shrewd and wide awake man of his generation

—Edmund Finn, 1888

William Frederic Augustus Rucker, who had such a profound influence upon the fortunes of the two survivors of this particular branch of the Rupp family, was himself an interesting character. The son of Frederic Augustus Rucker, an officer in the French Army, and his wife Maria, he was born in 1806 or 1807 in Dresden, Saxony. On 31 October 1835 he arrived in Hobart from London on the ship *Augustus Caesar* with his wife, Rebecca (née Greaves, whom he had married in London) and daughter, Rosetta Wilhelmina, then aged about five.[1]

Shortly after his arrival, Rucker composed a press notice advertising that he would open business as 'a Wholesale Wine, Spirit and Porter Merchant' in Trafalgar Place, Macquarie Street, Hobart. From 'a long and intimate connection with the different wine producing countries of Europe', he was assured of commanding 'a regular supply of wines of every description', and of meeting 'the views of his friends, both as regards quality and price'. Being 'content with a fair and moderate profit', he hoped 'to merit extensive patronage and encouragement from a discerning public'.[2]

By mid-1837, Rucker had decided to move to Port Phillip. Accompanied by a servant, the family left Van Diemen's Land on 26 July 1837 on the schooner *Hetty*, bound for Melbourne.[3] Here again, Rucker quickly asserted himself in local commerce. In September 1837 he signed a memorial from the 'proprietors of Stock depasturing in the Western District of Port Phillip' beseeching Governor Sir Richard Bourke to station a magistrate and a troop of mounted police in the Geelong area.[4] At the second Melbourne Land Sale on 1 November 1837, he purchased for £91 a block of land in Collins Street adjacent to J.P. Fawkner's allotment on the Market Street corner.[5] By December 1837 Rucker was well-established in a store in Queen Street, acting as an agent for local and overseas shipping, and offering for sale an astonishing and doubtless welcome array of goods—wines, spirits, ale and porter; flour, rice, raisins and cheese; tea, sugar, pork and potatoes; tobacco, paint, turpentine and sheep wash; timber, glass, nails and garden implements; clothing, saddlery, gunpowder, wool packs, needles and twine.[6]

Rucker took a further, and major, step into Melbourne's brisk young business world when, on 8 February 1838, he opened an agency of the Derwent Bank Company of Hobart Town in his ever-accommodating store.[7] Duly supplied with an 'Iron Chest' containing instructions, record books, stationery and 'a Supply of Notes and Specie to the Amt. of One Thousand & Five Pounds', Rucker assured his superior in Hobart, Charles Swanston,[8] that he would attend to the bank's affairs in 'Matters of regularity in the Management' of the agency, with 'Scrupulous adherence' to the directions received. One and a half centuries later, the records of the agency's brief existence still attest to Rucker's declared scrupulousness.[9]

Melbourne's co-founders, John Batman (1801–39) and John Pascoe Fawkner (1792–1869), and other leading colonists patronized the Derwent

William Frederic Augustus Rucker (1808?–82), pioneer merchant and banker who adopted the Rupp orphans, Charles Ludwig Herman and his sister Augusta, when they arrived in Melbourne in 1848. *(Review of Reviews)*

Bank Agency before it lost its identity. Fawkner's innovative, if quaint, manuscript newspaper, the *Melbourne Advertiser*, which first appeared on 1 January 1838, was immediately well-supplied with Rucker's notices relating to his vigorous retailing, shipping and banking activities.

Nor did the enterprising Mr Rucker confine his business interests to Melbourne. At the beginning of 1838 he had Frederic Champion open a branch store in 'several tents on the beach' for the benefit of Geelong settlers, and he acted as agent for 'the well-known schooner *Lapwing*' which plied between Geelong and Melbourne with passengers and cargo, leaving the principal settlement 'every tenth day'.[10] By March 1838 the tents at Geelong had been replaced by a timber building.[11] In October 1838, when Melbourne boasted some 1,300 citizens housed in about 350 dwellings including many tents,[12] Rucker was continuing to advertise a wide range of wines, spirits and general merchandise 'at his new Store', shortly referred to as 'The Melbourne Store'.[13]

Energetic and irrepressible, the prosperous Mr Rucker, as was fitting, 'used to dine every day at Fawkner's public table' at one stage in his career.[14] When his bank agency was taken over by the Union Bank of Australia in 1839, he promptly became a director of the enlarged institution. The following year he was elected an Anglican church warden, a director of the Melbourne Fire and Marine Insurance Company, and a member (the thirty-first on the roll, and the first merchant) of the rather exclusive Melbourne Club.[15]

Yet even successful merchants are apt to have their times of uncertainty. He fell out with George Arden, the fiery young editor of the *Port Phillip Gazette*, whose pages had so frequently featured Rucker's advertisements. Following an argument, Arden published some less than complimentary remarks about 'Mr Rucker, dealer in slops and spirits'. The merchant promptly took the issue to court where, in Port Phillip's first libel conviction in May 1839, the editor 'was sentenced to twenty-four hours' imprisonment and a fine of £50'.[16] Clearly Melbourne's pioneer merchant and banker was not to be trifled with, and the well-bred George Arden was left to 'bitterly regret having come into collision with Rucker', something which should have been avoided by a gentleman 'of birth, education, rank and character'.[17]

Rucker compliantly accepted additional recognition. In 1839 he was nominated to the Melbourne Mechanics Institute Committee for 1840; he was also elected a director of the Pastoral and Agricultural Society in the same year,[18] and he subscribed his name to the complimentary address presented to the Police Magistrate, William Lonsdale.[19] Less happily, on Queen Victoria's birthday, 24 May 1839, he won the dubious distinction of being expelled from the select fraternity of the Melbourne Club following a dispute over the dissemination of rumours, the threat of a duel, and the holding of a court of honour.[20]

In June 1840 William Rucker added significantly to his land holdings by purchasing for £2,266 10s. 0d. over 260 acres in the Northcote area, including 'the most prominent geographical feature...the steep, flat-topped hill' which was given its purchaser's name.[21] In the following year he bought an adjoining block, and on the hill near the corner of High and Bayview Streets built an impressive bluestone mansion.[22] For Rucker, as for many others, this proved a most inopportune time at which to embark on new financial ventures. From accounts of dealings between Rucker and some of his business associates, such as Jonathan Binns Were

(1809–85), it is clear that he had over expanded his commercial activities, allegedly owning by 1841, 'several houses and areas of land in and around Melbourne, of a nominal value of £40,000'[23]—a huge sum for the time.

When depressed conditions began to threaten the economic boom previously enjoyed in the rapidly developing city, Rucker enlisted the aid of eleven other businessmen, including J.P. Fawkner and J.B. Were, to form an association (dubbed 'The Twelve Apostles', or more sinisterly, 'The Ring', with Rucker as 'the Arch-Priest of the sanctified circle'[24]) which would enable him to retain his estate through the crisis, despite an overdraft. A change in the management of the Union Bank, and Rucker's additional obligations to the Bank of Australasia, produced a difficult, embarrassing, and for some a potentially disastrous situation which was later neatly described as 'the Rucker *imbroglio*'.[25] In February 1845 he was publicly declared insolvent.[26]

However, like others, Rucker survived, bruised but not beaten, although he did not emerge from insolvency still owning the fine house on Rucker's Hill. The Rucker property on Merri Creek, comprising 'the most extensive country house in the district, with stables, coach-house, sheds, &c., and a garden in a high state of cultivation' (sourly called 'Rucker's Folly') was advertised for sale in February 1843.[27] The sale was presumably forestalled by falling prices, and the Union Bank from which Rucker had retired as a director during his transitory affluence in 1841 became the new owner.[28]

News of Rucker's reduced state travelled slowly, for in May 1842 at least one Scottish firm still believed that its 'friend, W.F.A. Rucker, Esqr. a Bank Director' was 'the most influential man in Port Phillip'.[29] The resilient Rucker re-established himself as a wine and spirit merchant in 1847 in Roach's Store in Elizabeth Street,[30] so that by the time the *Wappaus* arrived in March 1849 he was once again on the way towards being a man of substance. In September 1849 he became a British subject, the fourth to take such a step under new legal provisions.[31] By the end of the year, Rucker was a member of a Committee elected at William Westgarth's suggestion during a public meeting in the Royal Hotel, called to consider ways of providing assistance to immigrants due shortly from Hamburg and Rotterdam.[32]

Rucker proceeded to lend money for short terms, to sell goods on commission, to collect debts for creditors, and to enter into transactions for the milling of 'fine flour' and for the production of 'sound, good and merchantable bricks' from clay he still owned in a paddock near Merri Creek—and in March 1850 he bought the appropriately named schooner, *Enterprise*, fifty-eight tons, from John S. Spotswood for £400.[33]

Tragedy struck the newly expanded Rucker household at St Kilda soon after the Rupp children were welcomed to it, for Rebecca Rucker died on Wednesday, 8 May 1850, 'at half-past 7 o'clock' in the morning.[34] Disaster of a less personal nature, the Black Thursday bushfires of 6 February 1851, shortly brought more concern, for Rucker had been made a director of the newly inaugurated Victorian Fire and Marine Insurance Company,[35] and soon after he was elected to 'a Committee of shrewd, hard-headed business men' to investigate the feasibility of a jetty-and-railway scheme proposed for Sandridge and Melbourne.[36]

On 2 August 1851, at St Peter's Church, Eastern Hill, Rucker married Susan Emily, the teenage daughter of a fellow merchant, Thomas Foster and Eliza, née Watson.[37] News of a gold-strike at Buninyong, near Ballarat, was about to break, thereby causing even greater euphoria among the already rejoicing citizens of Victoria, now at last a separately-constituted Colony. On 1 July 1852, the first anniversary of the celebrated Separation, William Sigismund Rucker was born.[38] The Rupp children now had a foster-brother as well as their foster-sister, Rosetta, by then a young lady of twenty-two. Four years later she married William Philpott in St Andrew's, Brighton.

A contemporary journalist considered that W.F.A. Rucker was 'a shrewd and wide awake man of his generation'.[39] Happily for the Rupp children, his shrewdness and business acumen were combined with practical compassion.

3 Country and Calling

. . .a man of singular sweetness of disposition and breadth of outlook.

—*Church Standard*, 5 October 1917

Despite William Rucker's history of fluctuating fortune, his family, including the two foster-children, probably lived quite comfortably. It was later noted that young Carl, to be known as Charles, attended 'a Mr Brookfield's school in Melbourne'.[1] Charles himself recalled in old age the time 'when Richmond was "out in the bush" ', Emerald Hill was the site of Aboriginal camps, 'stumps were common in Collins Street, and Elizabeth Street was a gully'. He also stated that 'as a lad' he had 'served as a clerk to his foster-father' and on occasions had been obliged to clamber 'to a look-out among the trees on the lower Yarra, to watch for Mr Rucker's trading-schooner coming up the bay'—presumably the *Enterprise*. Charles was even taken to Launceston on this schooner, the voyage being memorable for its duration; it was becalmed in Bass Strait for a week and spent two weeks working up the Tamar River.[2]

For a time, it seems, Rucker encouraged his new charges to retain links with their (and his) German compatriots. The German Union, which was formed 'quietly' in June 1850, sponsored a Christmas Celebration in St Patrick's Hall on the evening of Christmas Day, 1850. Charles was billed to deliver a Christmas poem in German as the sixth of a sixteen-item programme.[3] Performing before a capacity audience, 'Master Rupp, a German orphan, twelve years of age', won special comment for reciting his piece 'in an exceedingly happy style, which elicited a burst of applause'.[4] It was further reported that he alluded 'to the fact of their having found a new home here', and expressed his 'wish that

the colony might flourish and prosper'.[5] Mr Rucker must have felt very gratified.

It was 'apparently at an early age' that Charles 'felt a call to the Ministry'.[6] By 1861 he was serving as a lay reader at St Peter's Church, Eastern Hill,[7] then as now a centre of Anglican worship 'in the Tractarian tradition'. There being no theological college in Melbourne, Rucker apparently arranged for his foster-son to attend Moore College, then at Liverpool, near Sydney, for two terms during 1862. Thereafter, the young student was guided, instructed and examined on behalf of the college by the bishop of Melbourne's examining chaplain, the Rev. George Goodman, MA (1821–1908) of Christ Church, Geelong, and the bishop's personal chaplain, the Rev. Septimus Lloyd Chase, MA (1818–95) of St James's Cathedral, Little Collins Street.[8]

The examinable subjects of theological training of the time offered daunting fare for any twenty-four-year old: Old and New Testament (both 'Memory' and 'Interpretation'), Greek Testament, Ecclesiastical History, Pastoral Theology, Doctrine and Evidences, the Book of Common Prayer and the Thirty-nine Articles of Religion, as well as Exposition, Public Reading and Sermon. Overall, Charles's results ranged from 'fair' and 'moderate' to 'good' for History and Bible studies.[9] Accordingly, in St James's Cathedral, on Sunday, 21 December 1862, Charles Ludwig Herman Rupp (who for some reason declared himself to be Louis Herman Rupp[10]) was presented by the Rev. S.L. Chase to the first

The Rev. Charles Ludwig Hermann Rupp (1838–1917) about the time of his ordination in 1867. From a carte-de-visite by Wm. Davies & Co., Melbourne. (From the late Mr Arthur Rupp, Willoughby)

Bishop of Melbourne, the Rt Rev. Charles Perry. The bishop, 'well assured' that the candidate was 'of...virtuous and pious life and conversation and competent learning and knowledge in the Holy Scriptures', admitted him as a deacon of the Church.

The occasional sermon was preached by the Rev. D. Seddon of Christ Church, St Kilda, on the text, 'Therefore, seeing we have this ministry, as we have received mercy, we faint not'.[11] Thus inspired, the Rev. Charles Rupp was immediately appointed curate of St John's, Belfast (Port Fairy), as assistant to the celebrated cleric, teacher and author, Dr Thomas Henry Braim (1814–91), Archdeacon of Portland, and instigator of the building of the handsome Belfast Church.[12]

On 21 January 1865, at All Saints' Church, Geelong, the curate of Belfast married Marie Ann Catherine Rowcroft.[13] The bride was a native of Longford, Tasmania, the sixth of nine children of Horatio (Horace) Nelson Rowcroft (1806–78), a London-born Crimean veteran, and Mary Anne Catherine, née O'Donahoo. The Rowcrofts had literary leanings. Horatio's brother Charles (1798–1865) was a novelist,[14] and Horatio himself contributed to the *Geelong Advertiser* under the name of 'John Barleycorn' before joining the staff, first as a commercial reporter and then, from January 1866 to September 1869, as editor. Subsequently he became editor and part proprietor of the *Geelong Express*, and established some association with the *Geelong Times*.[15] The O'Donahoos claimed descent from 'a famous Irish chieftain' of that name.[16]

When Dr Braim left for England in March 1865, the newly married curate, who had joined such bodies as the Belfast Hospital and Benevolent Asylum Committee,[17] assumed charge of the parish until the end of August when he was appointed first incumbent of Cranbourne and Berwick. Here he demonstrated his characteristic pastoral energy, travelling widely to conduct services in outlying centres. On 8 April 1866 the Rupps' first child was born at Cranbourne and named Florence Emily Marie.

All the while, Charles Rupp continued studying for priest's orders, and twice during 1867 he faced his examiners to be tested in the subject areas already indicated. Again scoring assessments ranging from 'fair' and 'very fair' to 'good' for Old Testament, Church History, Pastoral Theology, Exposition and Sermon, he was 'approved for ordination'. In St James's Cathedral he was ordained priest by Bishop Perry on Sunday, 22 December 1867, and promptly appointed to the District of Dandenong, Berwick and The Clyde,[18] where he remained until May 1870.

Charles Rupp was to serve his entire ministry in Victoria, in the dioceses of Melbourne and (after 1875) Ballarat. He returned to St John's, Belfast, in July 1870, and other appointments followed: Learmonth, January 1873 to December 1876; Koroit, December 1876 to November 1883; Coleraine, from January 1884 for some ten years until going to Warrnambool as *locum tenens* for about eighteen months, then to Buninyong, August 1895 to February 1897, and Kingston, February 1897 to October 1898.[19]

While stationed at Kingston, Rupp subscribed to the appropriate oaths—and remitted a postal note for two shillings and sixpence—for Letters of Naturalization, and formally became a dutiful subject of Queen Victoria, 'lawful Sovereign of the United

Kingdom . . . and of this Colony of Victoria' on 10 June 1897.[20] In October the following year Rupp 'received a unique invitation to return to Learmonth, where he had been in charge 25 years before'.[21]

On Christmas Day, 1904, 'unusually large' congregations attended the churches at Learmonth and Waubra where Rupp 'officiated for the last time as incumbent'. In the following February the parishioners presented him with a customary 'purse of sovereigns', and for a short time he lived in Ballarat.[22] He was now nearly sixty-seven and his doctor urged him to leave Ballarat before another winter. Amid ample good wishes for a lengthy retirement, there was concern over the difficulty of replacing him as honorary secretary of the Diocesan Missionary Association.

By now twice widowed, Charles Rupp spent his last years with his daughter and son-in-law, Florence and Arthur Augustus Monypenny at Glen Innes, N.S.W., where he died on 28 September 1917 in his eightieth year. He was described as having had a 'zeal for missionary work', 'a man of singular sweetness of disposition and breadth of outlook, and a faithful parish priest, beloved by his people wherever he served'.[23]

Charles Rupp's sister, Augusta, married a farmer, George Edward Pettett at Stawell in 1872 and their daughter, Mary Caroline, was born the following year. Augusta died in St Kilda in November 1921, aged seventy-seven; her husband, a native of Greenwich, England, was ninety-two when he died at St Kilda in March 1935; their daughter died unmarried at the age of seventy-three in January 1947,[24] just one hundred years after

William Westgarth had returned to Britain, determined 'to open . . . German immigration to Port Phillip'.

The Rev. C. L. H. Rupp in later life, perhaps about 1910. (From a glass negative by H. M. R. Rupp in the possession of Miss E. Rupp, Willoughby)

4 'Tiresome Monkey', but Promising Naturalist

The boy, a tiresome monkey about my own age, was dealt with by the lady herself.
—HENRY HANDEL RICHARDSON, *Myself When Young*, London, 1948

Charles and Marie Rupp's second child was born in the Belfast Vicarage in College Street, on Friday, 27 December 1872. All seemed well until early on Sunday morning, 12 January 1873, when two doctors were urgently summoned to attend the mother, one of them for the second time within an hour. Septicaemia had developed, and then a blood clot apparently caused a fatal heart attack.[1]

Parishioners and other townsfolk were stunned, for the vicar's wife 'was only 31 years old', and had been 'active in deeds of charity . . . a member of the Ladies' Benevolent Society', and 'a hearty and useful worker in the Sabbath School'. St John's Church was 'crowded in every part' on Tuesday, 14 January, for the funeral service. Within the town, 'shutters were on all the windows, and some members of the Church of England closed their places of business altogether' in order to join the procession to the Belfast Old Cemetery where the vicar of the neighbouring parish of Koroit, the Rev. F.H. Smyly, officiated.[2]

On the following Sunday, 19 January, the services in St John's were conducted by the Rev. Dr Peter Teulon Beamish (1824–1914), vicar of Warrnambool. Worshippers found in their pews a farewell message from their distressed pastor, expressing his 'heartfelt gratitude' for their 'touching sympathy' and 'personal kindnesses', and advising them that he now proposed to move to his intended

new post at Learmonth 'almost immediately'; accordingly, he had adopted 'this means' of 'saying farewell'. He subscribed himself, 'With a sorrowful heart . . . your grateful and sincere friend in the Lord Jesus—C.L.H. Rupp.'[3]

The following day Dr Beamish baptised the baby Herman Montague Rucker,[4] who came to be known by his second name. It is at least very probable that either W.F.A. or W.S. Rucker was the boy's godfather, or perhaps both were nominated. Shortly after, Charles Rupp moved to his next parish of Learmonth, near Ballarat, to begin a new phase of his life and ministry.

In the course of his duties, the vicar of Learmonth met Rachel Emma Tillery Kirkpatrick, then a governess at Robertson's Mt Mitchell Station, near Lexton.[5] She was the daughter of James Kirkpatrick and Isabella, née Young, and had some Tasmanian connections, being a niece of Lady Clara Dry, née Meredith, the wife of Tasmania's first knight, Sir Richard Dry (1815–69), a popular politician.[6] There were ecclesiastical associations, too, for Rachel had a brother who was a minister of the Scottish Episcopal Church.

On 4 August 1874 Charles Rupp and Rachel Kirkpatrick were married in All Saints' Church, St Kilda. The bride gave her address as nearby Windsor, where presumably she lived when not tutoring in the country. Rachel Rupp was a

The old vicarage, Port Fairy, where H. M. R. Rupp was born on 27 December 1872, and where his mother died shortly afterwards. The building has subsequently been modified, and is now a private residence. (Photograph: L.A.G., 14 September 1988, with the kind permission of the present owner, Mrs Shirley Roussel)

St John's Church, Port Fairy, where the Rev. C. L. H. Rupp served as Curate, December 1862 to August 1865, and as vicar, July 1870 to January 1873. (Photograph: L.A.G. 14 September 1988)

Monty Rupp, aged about three. Photograph by Wm H. Bardwell, Ballarat. (From Mrs Rachel Cox)

competent artist and singer, a keen gardener and an experienced teacher, and some time after the family moved to Koroit at the end of 1876, she conducted a small, and probably informal, school in the vicarage, then 'a square building of galvanized iron', until the new vicar had weatherboard extensions and a couple of verandahs added.[7] Pupils included not only Florence and Monty Rupp, but also the two daughters of the local postmistress.

At that time, Koroit's postmistress was Mrs Mary Richardson (née Bailey) a native of Leicester. Her husband, a Dublin-born doctor, Walter Lindesay Richardson, died on 1 August 1879 at the age of fifty-three, and on the following day the Rev. Charles Rupp officiated at his burial in Koroit's Tower Hill Cemetery.[8] Their daughters, Ethel Florence Lindesay Richardson, born 3 January 1870, and Ada Lilian Sydney Lindesay Richardson, born 28 April 1871, were known at the vicarage as 'Ettie and Lily'. Charles Rupp baptized the younger girl soon after her father's death.[9]

'Ettie' would achieve fame as the celebrated Australian novelist, Henry Handel Richardson, whose father would be immortalized, however fictitiously, as 'Richard Mahony', while curiously, the vicar and his wife would be transmogrified as the 'Ruckers'.[10] Virtually an expatriate since her study-trip to Europe in 1887,[11] Ethel Richardson (as 'H.H.R.') recorded some impressions of Koroit and its people, including the residents of the vicarage and of the police station in her last and unfinished book, which was published posthumously in 1948.[12]

The vicar, with 'a German-sounding name', was recalled (at first) as 'a kindly, blond-bearded little man, in manner and appearance...a typical German Pastor', who actually stimulated Ethel's musical interests by introducing her to some of Mendelssohn's work 'and by singing lengthy Chorals...in his mother-tongue'. His wife was 'a tall, rather elegant lady' who 'painted pictures on an easel'.[13] There, sadly, the complimentary remarks ended, for then followed dark pictures of the vicar 'mercilessly belabouring his fourteen year-old daughter about the head and ears' for some undisclosed breach of discipline, and of his wife's pursuit, riding-whip in hand, of the agile Monty, whom Ethel considered 'a tiresome monkey' deserving of such chastisement. There was also the assertion that the stepmother hated 'both children as much as they hated her'. Yet, continued H.H.R., 'except for this boy and his sister, the one so cheeky, the other so buttoned-up, Lil and I were thrown on ourselves for company'.[14] However, 'the policeman who lived next door to us had a girl; but she went to the State School and we weren't supposed to know her'.[15]

By the time these allegations were published, the 'tiresome monkey' (who good-humouredly agreed that he 'probably was'[16]) had attained the age of seventy-five, but like his sister, then eighty-two, had clear recollections of their time at Koroit vicarage. Deeply hurt, and disappointed in their childhood friend, Florence and Montague considered H.H.R.'s 'stories of the rectory...a gross libel on our parents'.[17] Rachel Rupp 'could not have lavished more love and devotion' on the children had they been her own. Florence declared that 'our stepmother, far from hating us (or we her) and chasing my brother round the room with a riding-whip, was devoted to him from babyhood to manhood, and was rather inclined to spoil him than otherwise (only that he was unspoilable!)'[18] Montague himself confessed that his stepmother 'idolized me...to such an extent that I was justly dubbed a spoilt brat, often escaping deserved punishments because she would not have me touched'. The 'sheer fiction' of the situation as described led him to 'wonder how much of the rest of the autobiography is true'.[19]

H.H.R.'s geographical memories of Koroit were similarly erroneous, and Montague could only conclude that the esteemed authoress, whose 'literary gifts' he warmly acknowledged, 'for some reason must have hated Koroit very badly'.[20] Others also conceded that 'the description of Koroit...naturally dissatisfied the residents';[21] but the erstwhile 'tiresome monkey' found 'the Koroit story', with its omission of any reference to local natural beauty and its inclusion of 'all that rubbish about our parents', to be 'incomprehensible'.[22] Doubtless H.H.R.'s impressions of the town were largely determined by painful memories of her father's mental decline and death, and of her mother's heroic endeavours in the little post office to help maintain the family's solvency.

Montague Rupp's Koroit years, from late 1876 to late 1883, were extremely significant for him. One vivid memory was of the news of the final stand of Ned Kelly and his three associates at Glenrowan in June 1880, and another, less dramatic but of infinitely greater personal consequence, was of 'a thrilling serial in the *Boys' Own Paper* about Orchid-hunting in Borneo'.[23] There were many local stimuli to aid an observant lad in developing a fondness for the bush and its plant life. There were the botanic gardens, where 'a gardener named Prior...grew beautiful flowers', and 'there was a beautiful private garden opposite the entrance on the road which led down to the business part of the village; the owner, Mr Frank Norman, was a great friend and rival of my stepmother in the matter of raising choice flowers'.[24] Further, 'there was in those days plenty of beautiful bushland quite close at hand; my father used to drive us out when on his rounds, and we revelled in the wildflowers—it was there that I first learnt to love the wild orchids, ''spiders'' and ''doubletails'' and the like'.[25] Sometimes, when on such excursions with his father, Monty found himself 'left...outside the cemetery gates', and he retained 'a vague recollection' of Dr Richardson's funeral.[26]

Mrs Richardson, 'of whom our parents were very fond, was often at the vicarage' and Montague remembered her 'very well indeed—almost more clearly than the girls'.[27] Others also recalled the close link between the post office and the vicarage.[28] The link was retained when Mrs Richardson was succeeded by Mrs MacDougall, for her son Graeme, who is mentioned later, became Montague's new and inseparable playmate.[29] Florence Rupp went away to boarding school, and apparently after the preliminary tuition at the vicarage (*c.* 1878–79), Montague went briefly to a small school associated with the Presbyterian Manse (*c.* 1880–1), then for a couple of years, probably 1882–3, to the Koroit State School.[30]

By this time, 'Mrs Rucker used to pay us long visits', for on 2 March 1882 the doughty old merchant (then accountant) of Melbourne, W.F.A. Rucker, died at his home in Darling Street, South Yarra.[31] Charles Rupp travelled from Koroit to officiate at his foster-father's funeral, held on 4 March in the Melbourne Old Cemetery, now the site of the Victoria Markets. On Rucker's monument, relocated at Fawkner, may still be read the text which Charles and Augusta Rupp doubtless considered appropriate: 'Inasmuch as ye have done it unto one of the least of these my brethren, ye have done it unto me'.[32]

As a country parson in his mid-thirties, with a wife and two children, Charles Rupp was doubtless comforted by the security offered under the terms of his foster-father's will. A lengthy document, prepared just a month before Rucker's death, it provided for a sixth of the income from certain trust funds to be paid to Charles 'by equal quarterly instalments' during Susan Emily Rucker's lifetime, and the payment of £2,000 after her death.[33] The principal beneficiaries were Rucker's widow, son and daughter, but there was no mention of Augusta, who, now married for ten years, was apparently considered to be well provided for. Once liabilities of about £3,500 had been met, Rucker's estate both real (including the South Yarra residence, sold for £1,000) and personal (including sundry promissory notes and shares in various mining and other companies) was valued at nearly £17,000.[34]

Five years later, Susan Rucker married the prominent politician–pastoralist, John Rout Hopkins of Wormbete, near Winchelsea.[35] Tragically, she shortly developed cancer, and towards the end of her life was taken from Winchelsea to Rannock House, Aberdeen Street, Geelong, where she died on 12 November 1889, aged fifty-seven.[36] Her son, William Sigismund Rucker,[37] died relatively young on 7 June 1901, and 'the Rev. C.L.H. Rupp, vicar of Learmonth, the deceased's foster brother' officiated at the funeral in Booroondara Cemetery, Kew.[38]

After Susan Hopkins's death, Charles Rupp presumably received the remainder of his inheritance. In March 1890 he advised his Coleraine parishioners that he had 'somewhat suddenly determined on taking a trip to Europe with Mrs Rupp'.[39] Accordingly, after the marriage of Florence Rupp and Arthur Augustus Monypenny on Easter Tuesday, 8 April 1890 ('the wettest day of the season'), the vicar and his wife left for Europe via Colombo and Suez. They travelled to the Holy Land, the Continent and Britain, where they visited the Kirkpatricks, then living near Liverpool.[40]

Resuming duty at Coleraine about a year later, the vicar delivered lectures and wrote articles on his long journey,[41] extolling the many scenic, architectural and ecclesiastical glories which had so impressed him—more especially those in England and Scotland. Perhaps significantly, the German part of the itinerary consisted merely of one day in Strasburg. The orphan boy from Hamburg had long since became an Australian Anglophile.

The handsome bluestone building which Rupp knew so well, housed the Geelong Grammar School from 1858 to 1914. (Photograph, 1864, from the School's jubilee history, 1907)

5 Geelong Days

*Memories of tramps before dawn, of the splash of the oars and the
rattle of rowlocks, of the lifting of the boat in the chill morning air . . .of
the unforgettable sound of the rising of the swans . . .of the scents of
wattle and clematis in the springtime . . .and the salt tang of the sea . . .*

—H. M. R. Rupp, *The Corian*, May 1941

When the Rev. Charles Rupp, on becoming vicar of Coleraine in January 1884, sent Montague to the Junior Grammar School in Moorabool Street, Geelong, he felt assured that his son would be in good hands, for the headmaster, Alfred P. Rowcroft, was the boy's uncle.[1] After a year, Montague proceeded across the road to the 'big' school, Geelong Grammar, of which, somewhat remarkably, the headmaster was another uncle (by marriage)—the celebrated naturalist John Bracebridge Wilson, MA, FLS.[2]

On 10 February 1885 the headmaster himself enrolled Montague as a boarder,[3] placing him in Third Form for Latin, French and Arithmetic, and in Fourth Form for English. For the next seven years, the name of Rupp figured frequently and prominently in the school's award lists.

At the end of his first year, Rupp gave some indication of his future career by coming second in Church of England Catechism. He also came first in Fifth Form English and won the Crosthwaite Prize 'for passing in the greatest number of subjects'. In the following year, 1886, he topped the Bible Class, came first in Lower Sixth English and in Third Form Arithmetic and won the Crosthwaite Prize again. Never quite at ease in the field of mathematics, Rupp must have made a great effort to win the Head-master's Prize for Arithmetic, which took the rather unlikely form of *The Poetical Works of John Milton*, by

no means the recipient's favourite poet. Not surprisingly, for Rupp developed a distinctive bold but pleasant hand, first prize for Writing in the Lower Sixth followed in 1887, again presented, and doubtless selected by the headmaster, this time a more welcome volume by Henry Walter Bates: *The Naturalist on the River Amazons* (London 1884). In addition, he once more topped the Lower Sixth in English, and came third in Fifth Form French and Greek.

Further success followed in 1888, when Montague won first place in Middle Sixth Latin and English, the latter earning him the Headmaster's Prize in the form of two volumes of Macaulay's *History of England* (London 1886), which probably proved palatable enough. Next year he came second in Upper Sixth Greek Testament and won the School Council's second prize (a volume of Shakespeare) for Upper Sixth Classics and English. His name also appeared in the Matriculation Pass List, University of Melbourne, for October Term 1888, and in the Matriculation Honours List in English and History for October 1889 (3rd Class), October 1890 (2nd Class) and October 1891 (1st Class). These achievements were well supported by success in school examinations, with prizes for top places in Upper Sixth English and Greek Testament in both 1890 and 1891.[4]

The promising scholar found that his School

Alfred P. Rowcroft's old Junior School, at the corner of Moorabul and Maud Streets, Geelong, is now the Source Restaurant. (Photograph: L.A.G., 12 September 1988)

offered ample and diverse opportunities for expending one's energy beyond the classroom. By 1889 he was a member of the cadet corps which was for long zealously fostered by Lieutenant-Colonel Albert Finchett Garrard (a master and later bursar of the school, 1883–1915) who married Oriana Mary (May) Bracebridge Wilson, the headmaster's daughter and Rupp's first cousin. The new recruit applied his literary ability to producing for the December 1889 issue of the school's *Quarterly* an article describing the rain-soaked October cadet camp held at Langwarrin, near Frankston. Early next year he was appointed a member of the editorial committee of four, headed by the Classics master, the poet James Lister Cuthbertson. Further contributions included an account of the 1890 cadet camp (also rain-soaked) and an article on the Wannon District embracing Coleraine and the Wannon Falls.[5]

In 1891, Cuthbertson had only Rupp with whom to share the editorial work for the *Quarterly*. The December issue included an article entitled 'Orchids', apparently Rupp's first published botanical paper, the forerunner of over 270 more to be published during the next sixty-four years. In this he drew attention to over twenty terrestrial species, chiefly of *Corybas* (then *Corysanthes*), *Caladenia*, *Thelymitra*, *Diuris*, *Calochilus* and *Pterostylis*, naming the localities around Geelong where they were then to be found. No doubt the headmaster was delighted.

The botanical littérateur came to be recognized and respected in many other ways. He was made a prefect in 1890 and 1891, and although, as he later recalled, he 'had been a very delicate child, and for some time...not allowed to take part in athletic sports',[6] his life at Geelong entered a new and vigorous phase. At the Junior School and in his first year at 'the School proper', the 'feeble and delicate youngster' whom 'doctors had predicted...would probably not survive adolescence' dutifully took things rather quietly, attending to books rather than to bats, balls and boats. In 1886, however, he 'began to make a move' and promptly won 'a gold medal at the School Sports...for the under 14 flat race'.[7]

This remarkable transformation, while no doubt partly attributable to biological factors, was probably very largely due to the 'outdoor' policy of the head-master, J.B. Wilson, whose 'lasting bequest to the school' as a naturalist was, in the view of one biographer, 'the Saturday ramble'.[8] To later generations, these Saturday exercises, performed between the permitted hours of 4 a.m. and 8 p.m., must have seemed more akin to punishing marathons or to unnecessarily extreme endurance tests than to mere 'rambles'. Even earlier departure times were attempted and unsuccessfully requested. Little wonder that Rupp remembered to the end of his long life these action-packed days of walking ('often...more than 40 miles'), rowing, swimming, surfing, 'bird-nesting or flower-hunting or fishing, or just loafing' at any of the favourite—and approved—picnic spots between Geelong and Barwon Heads, twenty-two miles by water; or

John Bracebridge Wilson, his wife Oriana Maria (née Rowcroft) and daughters, Oriana Mary (May) and Maud, at the famous stairway of the old Geelong Grammar School building. (Photograph, *c.* 1870, by courtesy of Mr M. D. de B. Collins Persse, school archivist)

A Geelong Grammar 'Saturday Ramble' group in 1889. *Seated (l. to r):* Laurence Rutherford, E. W. Bagot, James Lister Cuthbertson, J. J. Bowler, H. M. R. Rupp, John Calvert. Those standing are not known. (By courtesy of the archivist, Geelong Grammar School)

perhaps to Spring Creek (now Torquay) about which Rupp wrote a short but graphic article for the *Quarterly* when the sailing vessel *Joseph H. Scammell* was wrecked there on 7 May 1891;[9] or to 'the beautiful Moorabool Valley near Batesford, the Dog Rocks in the same area' or even to 'the distant You Yangs across Corio Bay'.[10]

Such rigorous sorties into the surrounding country enabled Rupp not only to develop his physical stamina, but also to gather harvests of botanical specimens. These were taken to his headmaster-uncle on Sundays for discussion and identification,[11] presumably between Divine Service at Christ Church in the morning and Wilson's acclaimed lectures on the Greek Testament in the afternoon. In his earlier years at the School, Wilson himself often accompanied the boys on the 'rambles' he had instituted,

and for nearly twenty years J.L. Cuthbertson did likewise, taking every opportunity to increase the efficiency of the school's rowers.

According to his botanical nephew, Wilson 'gave generous encouragement to any boy who showed an inclination towards natural history; to him I owe the beginnings of what became a life-long hobby—the study of Australian flowering plants'.[12] Thus, with the aid of some of the publications of his friend the Government Botanist, Baron Ferdinand von Mueller, and with or without the further aid of the school museum's 'powerful microscope', Wilson developed Montague's interest in the study of Nature in a scientific way and complemented the field work begun at Koroit.

More stringently controlled physical activities were pursued with equal enthusiasm. In 1888 Rupp

Geelong Grammar School Cadet Corps, officers and non-commissioned officers, 1891. *Standing* (l. to r.): Sgt V. Wettenhall, Capt. W. Hall, Act.-Lieut. A. W. D. Macartney, L-Cpl C. Cooper, Act.-Lieut. H. Whittingham, Capt. A. F. Garrard, Sgt A. Green, Act.-Lieut. P. Reynolds. *Seated:* Sgt E. James, L-Cpl G. Hillson, Cpl H. W. Raleigh, Cpl G. Greene, Cpl P. Parsons, Sgt T. Clausen, Cpl H. M. R. Rupp. *Front:* L-Cpl E. Robertson, Col.-Sgt A. W. Whitney, L-Cpl T. R. Cunningham. (By courtesy of the archivist, Geelong Grammar School)

and one of the Rutherford brothers 'won the Junior Pairs at the School Regatta' and in the following year he 'was in the winning crew of the Trial Eights'. Early in 1890 Rupp was No. 2 in a new reserve boat crew and in July he participated in the Trial Fours.[13] Always happy 'on the river', he found that 'the standard of rowing was very high indeed in those years' and he was disappointed in his 'ambition to become a member of either First or Second crew'.[14]

A keen cricket fan later in life, Rupp eschewed the game at school, while he enjoyed Australian Rules football. On reflection, he felt that his 'avoidance of cricket and . . . mediocre performance at football, were really due to defective eyesight' not diagnosed until after he had left school.[15] Nevertheless, by 1891 Rupp had made the First Twenty, receiving honourable mention as a goal-kicker in games against Wesley and Scotch Colleges in September. He was described as 'a good forward

man, but not quick enough on the ball. Good mark and splendid long kick.'[16] Later that year Rupp, then a member of the School's Swimming Committee, came second in the 150 yards race for the Senior Boarders' Swimming Cup.

It was in athletics that Rupp really excelled and thereby proved beyond doubt that his gloomy doctors had been unduly pessimistic. After his initial success in 1886, he was in top form by 1890–1. At the Athletics Sports Meeting of 17 November 1890 the headmaster 'as usual . . . was on the ground through-out the day', no doubt taking tremendous pleasure in his nephew's exploits which were shortly described in the *Quarterly*: 'The competition for the School Cup, which goes to the boy who obtains the best position in open unhandicapped events, went to H. Rupp . . . His victory was thoroughly well deserved and popular.' Rupp won the 100 yards (11 seconds), 200 yards (22 seconds) and quarter mile (440 yards, 56 seconds) and came second in the 140-yard hurdles and the long jump. In the following month he represented the School Cadet Corps in the Naval and Military Sports, coming third in the quarter mile.[17]

On the night before the next Annual Athletics Meeting on 6 November 1891, 'heavy rain fell . . . and rendered the Corio turf rather heavy, but a strong south-wester blew the ground pretty dry by noon'. Apparently unperturbed and unaffected by these conditions Rupp again won the 100 yards (10.5 seconds), 200 yards (22.8 seconds) and 440 yards (53.75 seconds, establishing a long-standing record) and came second in the mile, the long jump and the 120-yard hurdles in both the first heat and the final. The day ended rather sensationally, for Rupp and 'Tommy' Clausen tied in the point score and were obliged to run a further 200-yard race to resolve the matter. It was noted that 'Rupp won easily' in 23.8

seconds to secure the Athletics Cup for the second time: 'It was a splendid duel, fought out in the most honourable spirit between winner and loser, who were both heartily cheered by the whole School at the finish'.

Such high excitement, together with the Wilsons' 'excellently managed' afternoon tea and 'capital music' from the school's 'good friends of the Garrison Artillery Band', all in an atmosphere of 'harmonious and generous feeling . . . made the sports of 1891 so jolly a day for all concerned'.[18] For Rupp, about to leave the school, the day must have been infinitely more than 'jolly'—rather a time of sheer exultation as he attained the zenith of the non-academic aspect of his school career after such unsure beginnings.

By this time Rupp had been promoted to corporal in the cadets, and elected honorary secretary of the

Geelong Grammar School Football Team ('Twenty') of 1891. *First row (back, l. to r.):* C. Crooke, W. Sargood, H. Whittingham, T. Cunningham. *Second row:* T. Clausen, H. Hutchinson, G. Eardley-Wilmot, E. James, E. Jenkyn, H. M. R. Rupp. *Third row:* V. Wettenhall, W. Whitney, P. Reynolds, G. Hillson, A. Green, A. Madders, H. McWilliams, L. Gillespie. *Fourth row (front):* E. Robertson, H. Crosthwaite, L. Elder, C. Bailey, A. Hutchings. (By courtesy of the archivist, Geelong Grammar School)

school's General Committee and of the Athletic Sports Committee. Nor had he neglected his literary abilities, for in 1890 he was also secretary of the Senior Debating Society, and there are records of droll debates on the advisability of exterminating the native fauna, whether travel by water or by train was more convenient, the privatization of railways, capital punishment—and the suitability of English dress for Australian conditions: 'M. Rupp thought not. The chief objects of dress were comfort and decency; he did not see much comfort in collars, ties, waistcoats, or starched shirts, especially if one lived in a hot district.'[19]

Appropriately attired of course, Montague enjoyed social outings while at school, not only to attend Divine Service conducted by Canon George Goodman, but also to visit Mrs Susan Emily Hopkins (formerly Mrs Rucker), then in declining health. Then there was a bitter-sweet journey to meet his father and stepmother at Melbourne Railway Station when they arrived from Coleraine *en route* to Europe. Charles Rupp recorded that 'at Port Melbourne' there was 'a fearful crush' in which members of both Rupp and Rucker families were embroiled on that memorable Saturday, 26 April 1890. However, once conveyed from the confusion by the tender *Albatross* to the steamship *Kaiser Wilhelm*, Charles and Rachel Rupp 'with Monty's help' were safely installed in 'Berths W257 and 258'. Although energetically helpful with the luggage and endeavouring to maintain a brave face, seventeen-year-old Monty betrayed his concern at the departure of his parents. The perceptive father noted, 'my boy I am sure was sad by his very movements'.[20]

There had, of course, been other setbacks. When Montague first enrolled at the school, his only previous illness was recorded as measles, although in retrospect he described himself as having been a 'puny, skinny youngster' who had been 'branded with that unpleasant word "delicate" '. The test soon came. In 1887, typhoid fever invaded the school through the milk supply. There were no fatal cases, but Mrs Wilson, one of the masters, several of the domestic staff and about twenty of the boarders, including Montague, succumbed, to the point of being 'not very far from the borderline of life and death'. He further recalled:

It was an anxious time, and a catastrophe for the School; but Mr Wilson's courage never wavered. I am afraid that what distressed us in the sanatorium most at the moment was not the threat to the welfare of the School, but the fact that we all missed the celebrations of Queen Victoria's Jubilee![21]

The school closed early for the midwinter vacation of 1887.[22] Apparently Montague readily recovered, for, as noticed, he performed well in the examinations towards the end of the year.

The memoirs make it clear that Rupp revered the very thought of his Geelong schooldays and the associated friendships. He delighted in looking back

to those years that were mine at the old
    Gray School, firm set upon the wind-swept hill,
    Deep-ivied to the topmost tower of all,
without any futile regrets that they cannot return, but with the memory of them stored in my soul among the treasures of life that cannot be taken away from me.[23]

Such strong feelings of gratitude and affection made Rupp loath to sever all links with the school which had also been his home for so many of his formative years. He continued for a time to subscribe to the *School Quarterly* he had once helped to produce,[24] and to send contributions. A long account of a camping expedition undertaken from late December 1891 to mid-January 1892 appeared as 'A Trip among the Grampians' in April 1892. In December 1894 the *Quarterly* published a poem, 'At Mungadel', which Rupp sent from Hay.[25] Florid in more ways than one, it would nevertheless have pleased his old master, J.L. Cuthbertson. In 1896 Rupp submitted an obituary of the late headmaster's old friend (and his acquaintance) Baron von Mueller. This appeared in the October issue, and two more poems shortly followed, 'Life's Way' and 'Evening'. Another poem, possibly of Rupp's authorship, appeared in October 1897, but for some fifteen years there seems to have been an uncharacteristic literary silence.[26] Then because 'all Geelong "boys", past and present, are river-lovers' there came a full and glowing account of the Clarence River, 'the finest waterway of New South Wales'. Replete with descriptions of magnificent scenery from Yamba to Copmanhurst and beyond, of river islands, luxuriant vegetation, of shipping, fishing, and of course, rowing, the article provided the *Quarterly* of July 1912 with probably its first and last account of Grafton, 'the Queen City of the North'. Unfortunately the enthusiastic author's fine photographs could not be reproduced.

Almost thirty-five years after leaving the school, Rupp submitted a long two-part article for *The Corian*. Like its precursor of 1891, it was simply entitled 'Orchids'. Rupp referred to his earlier 'crude notes' on the orchids of the Geelong district, and hoped that this new account might be 'of interest to the present readers'. He made a rare reference to some attitudes towards his scientific hobby:

There is a strange idea, which dies hard, that a botanist must be 'a bit of an old woman'. I remember the implication quite well in my own school-days—though how it ever got a footing in the life-time of Bracebridge Wilson, a botanist of renown in both Hemispheres, I do not know, for the absurdity of associating anything 'old womanish' with our old chief must have been too obvious.[27]

If an apologia for the study of plants were required, then this could be declared with ease and conviction:

The study of botany is an essential department of the study of Life, the most familiar yet the most wonderful and mysterious thing in the world. What modern agriculture owes to botany would be hard to estimate, and in the deeper realms of science its workers are continually throwing new light on dark problems. Besides which, investigation of plant life is investigation of one of those domains of beauty and wonder which reflect the Mind of their Creator.[28]

Rupp, then rector of Paterson, had come to share the views of some of those nineteenth-century pantheistic clergymen who managed to reconcile their scientific and religious beliefs with the startling tenets of Charles Darwin. Another interesting aspect of this article was Rupp's generous deference to his friend in Adelaide: 'the outstanding authority on Australian Orchids to-day is Dr R. S. Rogers . . whom other workers all regard, I think, as a sort of final Court of Appeal for the solution of all their problems.'[29]

In 1941 Rupp was either moved or prevailed upon to submit some 'rambling memories' of the school in 'Fifty Years Ago',[30] and further reminiscences were published in *The Corian* some five years later.[31] Quotations from other letters were published in 1949 and 1950, and in 1956, the last year of his life, Rupp obligingly produced yet another manuscript of reminiscences as 'School Recollections' in response to a request from the then School Archivist, James Ponder.[32] In this, his enthusiasm for the old school,

his appreciation of, and affection for Wilson, Cuthbertson, Garrard and others, and his fond recollections of Saturday 'rambles', of rowing on the river, and of the names and exploits of schoolmates of sixty-five or seventy years before were as heartfelt as ever.

For its part, the school proudly watched and recorded the achievements of Rupp and other Old Boys as they proceeded through other places of learning and training and into the wider world. In 1905, J. L. Cuthbertson, although no longer a staff member, assisted the historically conscious school to make a survey of the subsequent careers of former pupils by means of a questionnaire. Rupp responded in October from Warialda, adding with some little pride that this was 'the most extensive parish in the Grafton & Armidale Diocese—about 7000 sq. miles, mostly rough hilly country' through which, during the last month, he had 'travelled nearly 450 miles'. He mentioned meeting 'a *very* Old Boy. . .H. M. Pike, now a surveyor at Moree. I fancy he was at the School with Bishop Stretch.' Another was Alfred Brownscombe 'now a farmer & grazier near Tamworth' (and recently best man at Rupp's wedding). The response concluded appropriately on a nostalgic note: 'I hope you are keeping well. I suppose you still help the School crews on the river—I wish I could have a day down at the Lakes or Clematis!'[33]

Excitement over the victories at the Athletic Sports Meeting of November 1891 had barely cooled when Rupp gained his final honour at Geelong Grammar—the award in December of a Mary Armytage Scholarship.[34] Worth £60 a year and tenable for up to three years, the scholarship was intended to assist promising students of the School to proceed to Trinity College.

So it was that Montague Rupp, the country parson's son, possessed of a scholarship, a fond respect for his school and a letter of introduction to Baron von Mueller from his uncle, headed for 'Marvellous Melbourne'.

6 The Season of Trinity

. . .in science, who can equal Darwin, Tyndall, Spencer, Huxley and
Rupp?
—ANDREW EDWARD PEACOCK IN *The Crab Catcher*, 12 June 1897

Well-accomplished in English and history, and well-versed in botany, Montague Rupp enrolled at the University of Melbourne as a student of Trinity College on 21 March 1892. Happily he had the company of two former schoolmates, Philip Parsons and Harry McWilliams, and all three became members of the college's winning boat crew later in the year.[1]

In accordance with his diverse gifts and interests—and the university's regulations—Rupp enrolled in English I, Latin I, Botany I and Deductive Logic. At the October examinations he gained passes in the two languages and third-class honours in Biology (with Laboratory Work) but was required to take a post-examination in Logic, which he duly passed. He won a Cusack Russell Scholarship for 1893,[2] but lost the academic year to his old enemy, typhoid fever. This meant that between Easter and the end of August 1893 Rupp lost 'more than ten weeks' of tuition and study, so that the retention of his scholarship was brought into question, and without it he could hardly have continued.[3] Fortunately there were compensations, for obligatory convalescence provided the stricken student with opportunities for writing letters and collecting plants. One grateful respondent was his uncle:

My dear Rupp,
                              I have to thank you much for the beautiful specimens of Lyperanthus nigricans, which I have never met with here—We have none of the peaty boggy soil in which such orchids delight—You will be glad to learn that I have now got a regular class for Botany. They have three half hours a week with me—I first put them through a good part of 'Thomé' by way of foundation and now we are working up 'Dendy & Lucas'. I hope you are feeling stronger—
    With kind regards from my feminine folk—
    Yours affectionately
    J. Bracebridge Wilson.[4]

Success followed in 1894, with passes in Ancient History, Jurisprudence and Physical Geology and Mineralogy, and honours in History of the British Empire I. There was also considerable success in 1895, with passes in Political Economy, Constitutional and Legal History, and Stratigraphical Geology and Palaeontology, but there was (perhaps understandably) the necessity for a December post-examination, this time in History of the British Empire II, which was passed.

With a dozen year-long courses thus completed, the requirements for a Bachelor of Arts degree had been met by the end of 1895, but the call of Science remained strong. Accordingly, in 1896 and 1897, Rupp took further single subjects in Arts/Science in pursuit of a combined degree—Systematic Botany, Systematic Zoology, Pure Mathematics I, Greek I and Inductive Logic.[5] He was successful, at the first or second attempt, in all of these except Logic, and he was aided by the award, in December 1896, of a Wyselaskie Scholarship in Natural Science.[6]

In addition, the Cusack Russell theological

Dining Hall.

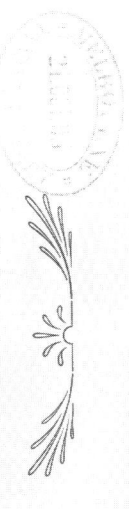

Library.

Corridor Hostel.

Hostel Exterior.

Trinity College as Rupp knew it in the 1890s. *(Alma Mater,* May 1897, Supplement; State Library of Victoria)

scholarship imposed its own obligations on the recipient. The scholarship trustees required a scholar to pass, at the end of each of his second and third years at the university, 'an examination satisfactory to an examiner' appointed by the bishop of Ballarat. Rupp accordingly 'attended lectures in Ecclesiastical History and Biblical Greek' during 1894, but he (and his lecturer) despaired of his chances of passing a Greek examination scheduled for March 1895. There was some earnest correspondence over the matter, for continuance of the scholarship was at stake.[7] Presumably the final passing of Greek I in December 1897 helped to resolve the situation.

Inevitably, there were academic, as well as administrative problems, two of which were Latin and Mathematics. Rupp recalled that the warden of Trinity, Dr Alexander Leeper, a passionate classicist, 'lamented the fact that I lacked the classical mind, which was perfectly true'. Nevertheless, Rupp remained grateful that he had been 'compelled' to study some Latin and Greek, for both were of great significance in his botanical work, as well as in his literary and theological studies.

Rupp further admitted that his Mathematics tutor once declared that 'I was the most hopeless case he'd ever struck. So I got hold of 8 years' exam. papers,

picked out the stock questions, & learnt them off like a parrot... We got 6 of them in the exam'. When the results were announced, the delighted student 'tore over to Trinity' to advise his tutor, Mr Hogg who, 'gazing at me reproachfully...commented, ''The age of miracles, Rupp is *never* past'' '. The immediate cause of this particular miracle was not disclosed.[8]

Unhappily, limitations imposed by time, financial resources and degree regulations thwarted Rupp's full academic plans, and to his sustained regret the Science degree he had so dearly desired remained beyond his reach. On 23 December 1897 Herman Montague Rucker Rupp graduated Bachelor of Arts, albeit with a rather extraordinary pattern of studies. At a later time, the scientific work he ultimately accomplished would have earned him a doctorate.

As at school, the diversity of Rupp's academic programme was matched by that of his extra-curricular activities. He successfully competed in tennis tournaments,[9] and, as noted, in rowing. Over seventy-five years later, it was still appreciated that through J. L. Cuthbertson's efforts at Geelong Grammar,

Trinity reaped a rich harvest of Barwon oarsmen. In 1896 Trinity won following a foul by Ormond, but the

THE CRAB-CATCHER.

THE EDITORS OF "THE CRAB CATCHER."

WE are glad to be able to offer our friends excellent portraits of the Editors of this high-class journal. The photographs were specially taken for us by Messrs. George Grease & Co.

LORD CHIEF JUSTICE VON BELSCHER, K.T.P., M.A., LL.D., Litt.D.

M.A., Trinitikollidge 1868 : LL.D., University of Spitzbergen 1870 : Litt. D. (Literary Devil), South Pole 1872 : Knight Grand Companion of the Order of Tin Pants, 1880 : Baron von Belscher of Belscherdorff, 1887 : Lord Chief Justice of North Greenland, 1892.
Author of works on *Divorce Laws of the South Pole*, *Home Life of the Esquimaux*, *Walrus Poems* and *Whale Verses*, etc.
In 1895 Baron von Belscher was chosen to draw up the terms of the International Arbitration Treaty between Russia and Gabo Island.

HIS EMINENCE CARDINAL O'PEACOCK.

M.A., Trinitikollidge 1801 : B.S. (Back-Slider) Crezwick 1802 – 1850 : D.D. (Doctor of Doggology) Arrarrat 1851 : Father Superior of Arrarrat Asylum for Inebriates 1852 – 60 : Dean of Crezwick 1860 – 5 : Bishop of Boomaanoomoonah West 1865 – 1893 : Cardinal Archbishop of Kangaroo Island, 1893.

HIS GRACE THE LORD ARCHBISHOP OF TRIABUNNA & METROPOLITAN of Van Demen's Land.

M.A. Trinitikollidge 1860 : D.D. Port Arthur 1861.
The Most Rev. Thomas Kerosene Fitz-Pitt was appointed Dean of Boozeborough in 1870 : Assistant-Bishop of the Southern Ocean, 1880 : Bishop of Chinatown (Tas.), 1887 : and Archbishop of Triabunna, 1893.

HIS LORDSHIP THE BISHOP OF TANKTOWN.

M.A., Trinitikollidge 1867 : D.D. University of Yarrahend, 1870.
The Right Rev. Hech Mon Rabid MacRupp was Archdemon of Glupot from 1871 – 1880 ; Dean of Devil's Gully 1880 – 1893 : Bishop of Drunktown 1893 – 5 : translated to Tanktown 1895.

Page depicting the editors of the student paper, *The Crab-Catcher*, of 12 June 1897, executed in the script of H. M. R. Rupp, whose caricature and 'credentials' appear on the right (from a copy in the Leeper Library, Trinity College, Melbourne)

bad steering of the Trinity cox aroused much adverse comment. Among the more prominent rowers were H. M. R. Rupp, the brothers A. H. and H. E. Bullivant, C. F. Belcher, S. D. Green, H. C. Fulford, W. S. Sproule, H. W. Gosse and J. A. Wallace.[10]

Rupp's literary ability and impish sense of humour admirably qualified him for the editorial board of the deliberately boisterous and scurrilous

student paper mentioned in his reminiscences. This was *The Crab-Catcher* ('A high-class journal of Truth and Impartiality') with which he became associated during the year of illness, 1893. He gave the paper good service, for he was still one of the editors four years later.[11] The bumper issue of Saturday, 12 June 1897, marking Queen Victoria's Jubilee, is one of the few surviving examples. Its fourteen pages were obviously duplicated from Rupp's distinctive neat

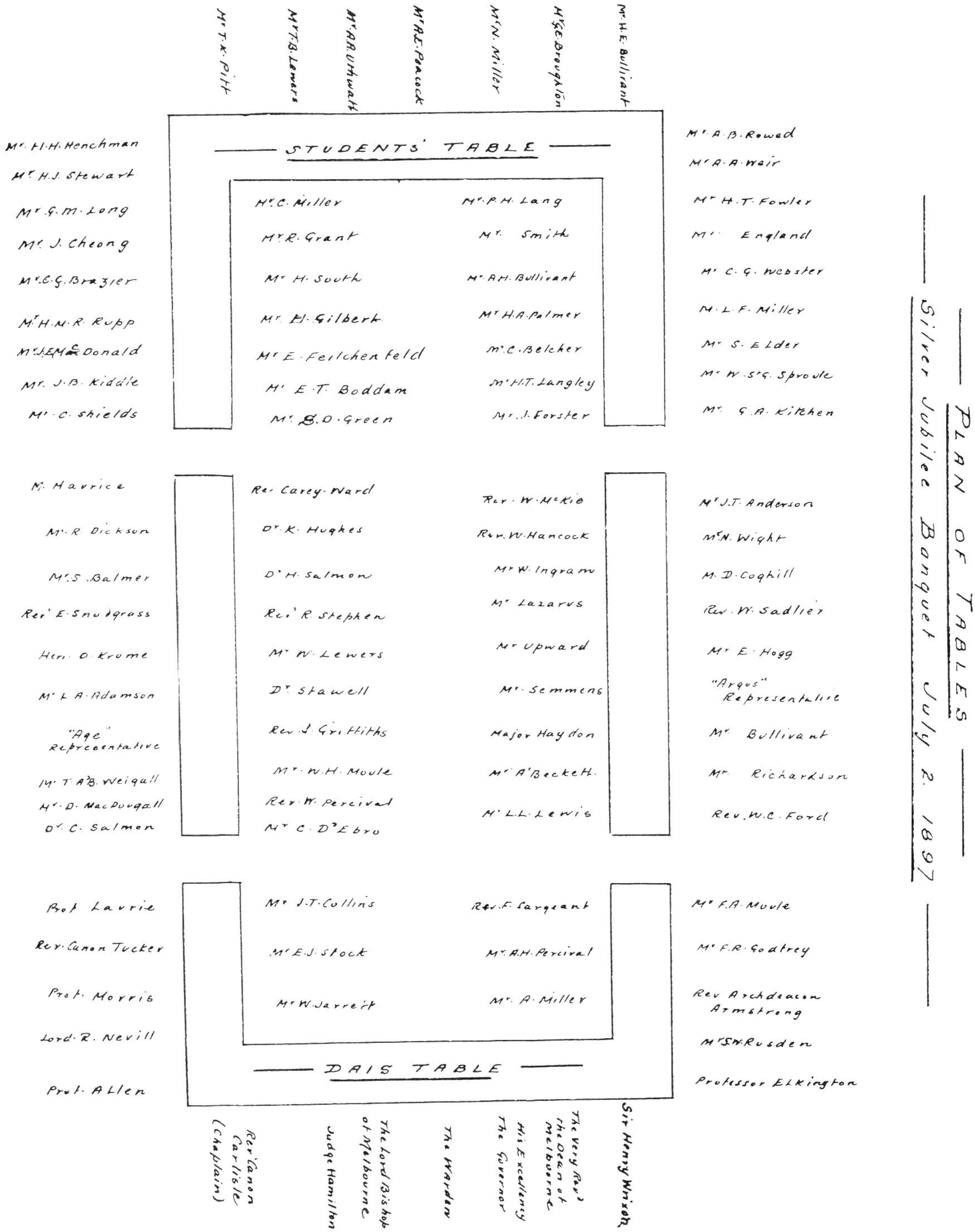

PLAN OF TABLES

Silver Jubilee Banquet July 2, 1897

STUDENTS' TABLE

Top column headers: Mr T.K. Pitt · Mr T.B. Lewers · Mr A.R. Uthwaite · Mr R.E. Peacock · Mr N. Miller · Mr G.E. Broughton · Mr H.E. Bullivant

Left: Mr H.H. Henchman · Mr H.J. Stewart · Mr G.M. Long · Mr J. Cheong · Mr C.G. Brazier · Mr H.N.R. Rupp · Mr J.F. McDonald · Mr J.B. Kiddle · Mr C. Shields

Inner left: Mr C. Miller · Mr R. Grant · Mr H. South · Mr H. Gilbert · Mr E. Feilchenfeld · Mr E.T. Boddam · Mr B.O. Green

Inner right: Mr P.H. Lang · Mr Smith · Mr A.H. Bullivant · Mr H.A. Palmer · Mr C. Belcher · Mr H.T. Langley · Mr J. Forster

Right: Mr A.B. Rowed · Mr A.A. Weir · Mr H.T. Fowler · Mr England · Mr C.G. Webster · Mr L.F. Miller · Mr S. Elder · Mr W.S.G. Sprowle · Mr G.A. Kitchen

Left: Mr Harrice · Mr R. Dickson · Mr S. Balmer · Rev E. Snodgrass · Herr O. Krome · Mr L.A. Adamson · "Age" Representative · Mr T.A'B. Weigall · Mr D. MacDougall · Dr C. Salmon

Inner left: Rev Carey-Ward · Dr K. Hughes · Dr H. Salmon · Rev R. Stephen · Mr W. Lewers · Dr Stawell · Rev J. Griffiths · Mr W.H. Moule · Rev W. Percival · Mr C. D'Ebro

Inner right: Rev W. McKie · Rev W. Hancock · Mr W. Ingram · Mr Lazarus · Mr Upward · Mr Semmens · Major Haydon · Mr A'Beckett · Mr L.L. Lewis

Right: Mr J.T. Anderson · Mr N. Wight · Mr D. Coghill · Rev W. Sadlier · Mr E. Hogg · "Argus" Representative · Mr Bullivant · Mr Richardson · Rev W.C. Ford

Left: Bro Lawrie · Rev Canon Tucker · Prof Morris · Lord R. Nevill · Prof Allen

Inner left: Mr J.T. Collins · Mr E.J. Stock · Mr W. Jarrett

Inner right: Rev F. Sargeant · Mr A.H. Percival · Mr A. Miller

Right: Mr F.A. Moule · Mr F.R. Godfrey · Rev Archdeacon Armstrong · Mr S.H. Rusden · Professor Elkington

DAIS TABLE

Bottom column labels: Rev Canon Carlisle (Chaplain) · Judge Hamilton · The Lord Bishop of Melbourne · The Warden · His Excellency The Governor · The Very Rev'd the Dean of Melbourne · Sir Henry Wrixon

Seating Plan, Trinity College Silver Jubilee Banquet, 2 July 1897. Rupp's position is at the Students' Table, sixth from the corner at the top left of the diagram. (Trinity College Archives)

⬥ TRINITY COLLEGE, ⬥

University of Melbourne.

FRIDAY, 22ND AND SATURDAY, 23RD MAY, 1896.

BROWNING'S ✴ "STRAFFORD"

To be performed by the **STUDENTS OF TRINITY COLLEGE,** *in* **ST. KILDA TOWN HALL,**

Being the first representation given in the Australian colonies of one of Browning's dramas.

DRAMATIS PERSONAE.

King Charles I.	T. SLANEY POOLE
Viscount Wentworth *(afterwards* Earl of Strafford)	D. J. D. BEVAN
John Pym	H. M. R. RUPP
John Hampden	A. E. PEACOCK
Denzil Holles *(Wentworth's brother-in-law)*	W. G. a'BECKETT
Sir Henry Vane *(Secretary of State)*	J. H. FREWIN
Henry Vane *(the younger)*	H. E. BULLIVANT
Benjamin Rudyard	L. S. TOWNSEND
Nathaniel Fiennes	T. K. PITT
Earl of Loudon *(Commissioner from the Scots' Parliament)*	H. H. HENCHMAN
Earl of Holland	H. A. PALMER
Lord Savile	W. L. R. CLARKE
Maxwell *(Usher of the Black Rod)*	G. F. ELKINGTON
Balfour *(Constable of the Tower)*	L. F. MILLER
Strafford's Son *(William)*	L. ELCOATE
Strafford's Daughter *(Anne)*	MISS A. ORR
Queen Henrietta	MISS M. A. BARTROP
Lucy Percy, Countess of Carlisle	MRS. L. L. LEWIS
Bryan	A. H. BULLIVANT
Mainwaring	G. GRICE
Slingsby *(Officers of Strafford's household)*	C. F. G. WEBSTER
Goring	E. FEILCHENFELD
Billingsley	L. S. TOWNSEND
A Puritan Fanatic	C. F. BELCHER
Presbyterians	J. LANG, G. M. LONG, G. A. KITCHEN C. G. BRAZIER, J. REDMOND, R. W. GRANT
Guards	J. E. F. McDONALD, J. B KIDDLE, R. A. O'BRIEN
Executioner	J. B. KIDDLE

Prompter	A. B. ROWED
Musical Director	D. J. COUTTS
Call-boys	W. S. SPROULE, A. A. UTHWATT

The Cast of Robert Browning's *Strafford,* as presented by students of Trinity College in May 1896. (Trinity College Archives) script.[12] He must have relished printing out such declarations as 'I am in the place where I am demanded of conscience to speak libel, and therefore libel will I speak, impugn it whoso list', and the gentler assurance: 'If the literary and artistic masterpieces which appear in this number serve to add a little to the amusement and jollity of our friends, our purpose is fulfilled'. Of further interest is the paper's Official Programme of Jubilee Celebrations in the Town Hall on 21 June, for included in the cast were 'Portia and Shylock... Messrs H. Bullelephant & Mac Rupp'.

Shortly after, on 2 July 1897, the college celebrated its own jubilee with a banquet to mark its first twenty-five years. Doubtless Rupp enjoyed the occasion, for the seating plan shows him within comfortable distance of many of his fellow rowers and actors, and those he so happily lampooned. The plan itself, with its fine array of names, is of more than ordinary interest.

Rupp also gained a reputation for public speaking. In August 1896 he delivered a paper on 'Irreligion' to the Trinity chapter of the Brotherhood of St Andrew. This was 'acknowledged to be one of the most carefully thought-out to which the Chapter has had the opportunity of listening'. In May the following year, Rupp addressed the same group on the 'Problem of Pain',[13] and in August he addressed the Science Club.[14]

However commendable, and commended, such activities were, they faded somewhat in the light of the two performances of Robert Browning's *Strafford* in St Kilda Town Hall in May 1896. Of gargantuan length and difficulty, the play was reduced from its set five hours or so by Dr Leeper, whose 'frequent use' of a blue pencil made the script manageable. From its forty-three members, the college found a cast of twenty-nine, including H. M. R. Rupp as John Pym, D. J. D. Bevan as Strafford, T. Slaney Poole as Charles I and H. E. Bullivant as the younger Vane. In the view of another member of the cast, W. S. Sproule, the most impressive character 'was the headsman, Bert Kiddle, six feet nine inches tall and bearing a ferocious axe'.[15]

It was jocularly recorded that 'The performance of "Strafford" was witnessed by two descendants of the patriot Pym, who were much gratified at the light manner in which Mr Rupp let their distinguished ancestor down'.[16] Rupp himself recorded other views and impressions. Proceeds of the performances went towards the repair and painting of college buildings and the construction of a new fence, for the college's finances were in a sorry state, and even the warden's stipend was in arrears.[17]

There were scientific as well as dramatic and sporting extra-curricular challenges. Maintaining the best Bracebridge Wilson tradition, Rupp vigorously pursued his botanical field work during his years at Trinity. He made weekend excursions into the bush in and around the outer suburbs, especially along Merri Creek in the vicinity of North Coburg, and by 1896 had compiled a catalogue of his discoveries.[18]

A more formal scientific and personal mission was to present Bracebridge Wilson's letter of introduction to Baron von Mueller. Despite some diffidence at the time, Rupp duly made his pilgrimage to the modest cottage at 28 Arnold Street, South Yarra, probably late in 1892, and treasured the memory of the visit ever after. He saw the old botanist again in October 1895 at Bracebridge Wilson's funeral in the Eastern Cemetery, Geelong. Mueller himself died just a year after his friend.

During vacations Rupp thoroughly explored the country in and around his father's parishes of Coleraine, Warrnambool, Buninyong and Kingston, making more collections, recording more observations and listing the plants he encountered.[19] Compiled with characteristic clarity and accuracy, these and similar invaluable records remain to enlighten, to delight and sometimes to depress today's ecologists, who are given insights into many landscapes which have disappeared forever.

Rupp's most scientifically rewarding vacation was that of November 1894 to February 1895, when he stayed at Hay, N.S.W. with his sister and brother-in-law, Florence and Arthur Monypenny. Ecologically, he felt very remote from the familiar landscapes of Victoria, perhaps missing more especially the verdant hills of Coleraine, which had appeared so close and inviting. Although there were, he conceded, some 'forests, and great rivers, and rolling fields of green, in Riverina', there also appeared to be overwhelmingly vast areas of 'never-ending level, with scarce even a miserable tree, in some parts, to check the eye', places where 'you can scarcely help feeling over-awed. It is uncanny, and weird, and desolate.' Yet, he concluded, 'Riverina is not a treeless and waterless desert. The greatest wool-growing district of Australia could scarcely be that.'[20]

With customary enthusiasm, Rupp explored widely in the high summer heat—west along the Balranald Road; north to 'The Sandhills'; north-east 'through the Irrigation Area...on to the

At Mungadel.

———

REST in the eucalyptus shade,
Within the flowery-mantled glade,
          Where thrushes love to dwell ;
And many a bird her home has made
In sunny days at Mungadel.

Beneath our feet the waters run,
And catch the gleam of summer sun ;
          And with their music swell
The songs of Nature that have won
Our hearts to peace at Mungadel.

The meadow round is gay and fair
With flowery carpets everywhere ;
          The mallow tall, the bell ;
And myriad buttercups, that wear
Their brightest gold on Mungadel.

The gums with cluster'd blossoms greet
The wild verbena round their feet,
          And, fair to sight and smell,
The yellow lily-stars make sweet
The fragrant air of Mungadel.

No sky so clear, so deep a blue
As that which trends above us, through
          The trees within the dell ;
Whate'er be said, we hold it true—
All things are fair at Mungadel.

Yet soon the sweet November day
Draws to its close, and dies away ;
          So, at its parting knell,
Unwillingly farewell we say
To flowery-breasted Mungadel.

                              M. R.

Rupp's poem, 'At Mungadel', written in November
1894, during a university vacation spent with his sister
and brother-in-law, Florence and Arthur Monypenny,
who then lived at Hay, N.S.W. *(Geelong Grammar School
Quarterly,* December 1894)

Racecourse'; east to the Eli Elwah woolshed; to
South Hay; out to Mungadel (which moved him to
verse); along the Murrumbidgee banks, and around
the township itself to such places as 'the uncultivated
parts' of Hay Park. He apparently recognized and
classified many of the plants encountered with

relative ease, and made collections. However, there
was clearly a case for seeking some assistance from
his esteemed acquaintance, the good Baron in
Melbourne.

Accordingly, Rupp despatched some specimens,
carefully labelled, to Melbourne for authoritative
identification. Quite as avid a letter-writer as Rupp
was to become, Mueller promptly responded, and
indeed part of a longer list of such plant identifications still exists.[21]

Delighted by his botanical discoveries in this new,
if rather daunting, environment, Rupp described
them in a lengthy article for the local newspaper.
Deliberately informative, he proceeded through his
plants, family by family, using Mueller's system of
classification and making a generous bestowal of
botanical names. As at school, he subscribed himself
simply 'M.R.', and had the pleasure of seeing his
first botanical article in a public journal appear in
The Riverine Grazier on 31 December 1894.

Of course the young student writer had problems.
His script, in both roman and cursive form, was
always commendably neat and legible, but his
printed 'g's' had curious idiosyncratic tails, which
led the unbotanical compositor to give some botanical
names an even more forbidding appearance by
rendering the letter 'g' as 'q'. This brought an
immediate letter of correction from the author, who
took the opportunity to mention that contrary to his
earlier statement, 'the poisonous herb Euphorbia
eremophila' was now appearing 'in great quantities
on the plains'.[22]

Thus encouraged, Rupp submitted another
article, 'Some Problems and Theories in Science',
which was published on 11 January 1895. This
touched upon areas which could still be sensitive in
the late nineteenth century—Darwinism, evolution,
geological formation, and the geographical
distribution of certain animals. Rupp's fondly
respected mentor, Professor Walter Baldwin
Spencer, would have been proud of him, but a
correspondent identified only as 'Maorilander' took
him to task over the likely time of extinction of the
New Zealand moa. Rupp promptly defended
himself, but conceded that he should have stated
'comparatively recent times' instead of 'our own
time'.

Undeterred, Rupp penned an interesting and
geographically perceptive description of his beloved
'Western Victoria', replete with scenes of his
relatively recent childhood, and this appeared in the
issue of 1 February 1895.

By the end of this vacation, Rupp had enjoyed

considerable correspondence with Baron von Mueller. Those letters, which Rupp carefully preserved, are characteristic of Mueller's quaintly delightful style, with the salutation well enclosed within the text—for example:

Your observations on the movement of the leaves of Hibiscus trionum is of considerable interest, dear Mr Rupp, and I will publish them under your name. For the additional sending of seeds I am much obliged so also for the revised list of the plants occurring near Hay.
    Regardfully yours
    ferd. von Mueller.

Then followed a postscript longer than the letter itself, seeking information about the fodder value of Lignum, *Muehlenbeckia cunninghamii*, and suggesting that 'among *floating* & submerged plants and *minute* plants you would find much novelty yet for N.S.W.' This, and other letters also contained requests for '*well matured fruits* with perfect seeds', perhaps 'collected . . . by small pay to children'; notes on various plants, their authors and their distribution, and requests for certain species for the Melbourne Herbarium and for exchange.[23] Mueller even enlisted Rupp's aid in the wider cause:

Could you inspire young people, particularly the ladies, to collect all sorts of plants in their vicinity, especially also floating and submerged weeds & minute annuals. The young Australians ought to *love* the native *vegetation* of their natal places!
We have yet much to learn of the geographic distribution of many of the species in the whole inland regions.[24]

Mueller was especially pleased that Rupp had managed to enthuse a Mrs Linton and her children, and accordingly sent her a copy of his *Second Systematic Census of Australian Plants* (Melbourne 1889). Further, 'If Mrs Linton's family make a methodic collection there, a list of the "plants of Hay" could be published as a contribution to geographic phytology'.[25]

Rupp's list had provided 'some new localities . . . which shall be recorded under your honoured name in my works'. Mueller recalled his own brief visit to the lower Lachlan country in 1878, when his investigations of the Riverina flora had been curtailed because of a recent 'severe illness', and he apparently hoped that Rupp's endeavours might compensate for the consequent deficiency in his own knowledge and collections.[26]

Rupp dutifully publicized the Baron's wishes in a final letter to the editor, advising that he had 'sent to Baron von Mueller a list of some 150 or more plants obtained in this district', and suggesting that those, 'particularly the younger', who 'occasionally have a little time on their hands' might make some collections 'when out for walks near the river, or driving on the plains'. The Baron would certainly appreciate and acknowledge any material sent to him, for he was 'much interested in the botany of the Murrumbidgee and Lachlan Rivers'. Seeds as well as flowers and foliage would be especially welcome, and Rupp supplied hints on their collection and despatch.[27]

Having offered Hay Council 'a parcel of seeds, suitable for the park', the young student returned to Melbourne, after an active and botanically stimulating vacation, marred only, perhaps, by the fact that not a single orchid had been found.

At Trinity College the history-loving theologian, who could also play, act and row, was warmly accepted also as an accomplished naturalist—'in science, who can equal Darwin, Tyndall, Spencer, Huxley and Rupp (successor to Charles Eaton)?' asked Andrew Edward Peacock with respectful mock-seriousness in *The Crab-Catcher*.[28] In 1895, the notable educationist and historian, George William Rusden,[29] donated £50 for the establishment of a college museum, and thereby a new opportunity was created for one with special interests and skills. A former Science lecture room upstairs in the Bishops' building was fitted with display cases which were promptly stocked with 'a large number of interesting antiquities, curiosities, and geological and botanical specimens . . . presented by Students and friends of the College'.[30]

G. W. Rusden opened the museum, duly named in his honour, on 1 October 1896. Mr. H. M. R. Rupp, 'one of the resident Students', whose 'especially valuable gift . . . of mounted typical specimens of Victorian plants' received particular mention, was appointed the first curator in the following April.[31]

The Rusden Museum was dismantled in 1919, when the room was converted into a dormitory to help accommodate students returning from World War I. The exhibits, including Rupp's plant collection, 'were walled up under the stair in the Clarke Building, where they lay undisturbed for fifty years'.[32] The Clarke Building was renovated early in 1969 and the old treasures were rediscovered. The plants were found to be 'faded, but generally in reasonable condition', and the warden, Dr Robin L. Sharwood, offered this 'collection of 146 named specimens' to the National Herbarium of Victoria. Dr James H. Willis, a personal friend of Rupp and acting director of the Melbourne Royal Botanic Gardens at the time, was especially delighted to accept the offer and to arrange for his old friend's specimens to be ,added to the state collection. On receipt of Dr Willis's enthusiastic reply, the relieved warden made a marginal note to a colleague: 'So that *was* the right thing to do!'[33] It certainly was, and so after more than seventy years, the 'valuable herbarium' assembled by 'that indefatigable collector, Mr H. M. Rupp'[34] passed from his college to augment the mass of material collected by the illustrious Baron von Mueller and a host of other enthusiasts, both amateur and professional. The Rev. H. M. R. Rupp, former student at Trinity College, would have been both pleased and honoured to know that his work had been deemed so worthy, although he had lived to see some of his other collections gratefully received into herbaria at Sydney, Auckland, Kew, Geneva, Vienna and Cambridge (Massachusetts),[35] and into the private herbaria of fellow enthusiasts whom he had tutored and encouraged.

7 Cures of Souls, Habitats of Plants

Gresford is a very good parish; and you know what those valleys will mean to an orchid man!

—H. M. R. Rupp to A. H. Chisholm, 27 April 1933

Montague Rupp must have decided before leaving Geelong Grammar that he would follow his father's profession, and by the end of his first year at Trinity had, in fact, begun training. In November 1892 the Rev. Charles Rupp, then vicar of Coleraine, and about to celebrate the thirtieth anniversary of his own admission to the diaconate, proudly recorded that he had

been lately utilising, with the Bishop's approval, the services of his son in conducting divine service at Holy Trinity Church, during the vacation of his university course, where he is studying as a theological student, with the view of seeking Holy Orders in this Diocese. It has been exceedingly gratifying...to hear so much kind sympathetic expression of approval, on the part of the congregation, of father and son officiating together in the church.[1]

Service records at Holy Trinity, Buninyong, indicate that Montague also preached in that fine old stone church in vacations during his father's incumbency, August 1895 to February 1897.

After graduation, Montague Rupp applied, on the nomination of the Rev. John Kirkland, for a

(*Opposite*) St Augustine's Church, Beeac, which was planned and commenced during Rupp's ministration, 1898. He returned for the Jubilee Celebrations in 1948. (Photograph: L.A.G., 11 September 1988)

Reader's Licence, and 'having passed our Chaplain's Examination' and 'subscribed a Declaration' of 'assent and consent to the Thirty-nine Articles of Religion of the Church of England', he was duly licensed by Bishop Samuel Thornton of Ballarat as 'Reader at Beeac, in the Parochial District of Colac' on 1 April 1898.[2] Rupp himself recorded that he became a 'stipendiary reader', which meant that he was afforded some sustenance while looking after the northern part of the large parish, and preparing for further examinations.[3] Clearly the parishioners rallied around their reader, for by the spring of 1898 they had constructed the modest weatherboard Church of St Augustine which still stands on the main road, not far from the eastern shore of Lake Beeac.

On Sunday, 21 May 1899, the mandatory *Si Quis* document was read in the new church, formally advising the congregation of the reader's intention of joining the ministry, and inviting anyone to declare to the bishop 'any just cause or impediment' why he 'ought not to be admitted into holy orders'.[4]

Two days later, Letters Testimonial, 'of his good life and conversation' and of his having lived 'piously, soberly and honestly', were prepared for the candidate on declaring 'his intention of offering himself...for the sacred office of a Deacon'. The signatories were the Rev. C. L. H. Rupp of All

St John's Church, Colac, to which Rupp was appointed as curate in 1899 after his admission to the diaconate. (Photograph: L.A.G., 11 September 1988)

Rupp therefore travelled to Melbourne, and on the eve of his ordination declared his 'assent to the Thirty-nine Articles of Religion and to the Book of Common Prayer', his 'Allegiance to Her Majesty Queen Victoria, Her Heirs and Successors' and his 'Canonical Obedience of the Bishop of Ballarat . . . in all Things lawful and honest'.[8] The next day, he was duly presented to the bishop of Melbourne for ordination, as his father had been thirty-seven years before, by the imperishable examining chaplain, Canon George Goodman.

The candidates, eight priests and five deacons, and the assembled congregation of relatives, friends and regular parishioners, were treated to a sermon based on St John ii. 5—'His mother saith unto the servants, Whatsoever he saith unto you, do it'.[9] The

Recently ordained and appointed, the curate of Colac-with-Beeac poses shyly at the tower doorsteps of St John's, Colac, about 1900. (Photograph by courtesy of Mrs R. Cox)

Saints, Learmonth, the Rev. John Kirkland of St John's, Port Fairy, and the Reverend J. C. Carmichael of St John's, Ballarat.[5]

It was further attested that the Right Rev. Henry Edward Cooper, formerly archdeacon of Hamilton, now coadjutor bishop of Ballarat, 'hath enquired of the said Herman Montague Rucker Rupp and also examined him, and hath certified that he is a person apt and meet, for his learning and Godly conversation, to exercise his ministry duly, in the honour of God and the Edifying of His Church'. Nor had he 'written or taught anything contrary to the doctrine or discipline of the Church of England'.[6]

As Bishop Thornton was visiting Inglewood, and as it was deemed 'expedient' that Mr Rupp 'should be without further delay admitted to the Office of Deacon', the candidate was commended to the Rt Rev. Field Flowers Goe, Bishop of Melbourne, for admission to the diaconate at the Ordination Service in St Paul's Cathedral, scheduled for Trinity Sunday, 28 May 1899.[7]

former stipendiary reader of Colac-with-Beeac then returned to his parish as curate. These events did not go unnoticed in his old university, for the Rev. H. M. R. Rupp was shortly listed in *Alma Mater* as its 'Literary Representative' at Beeac.[10]

Further study followed in preparation for priest's orders. After appropriate examinations and observance of the required procedures, Rupp was ordained priest by Bishop Arthur Vincent Green in St Peter's Church, Ballarat, on Sunday, 2 June 1901. The large parish of Colac-with-Beeac was divided, and the newly ordained priest became the first incumbent of Beeac, remaining there until 1903.

By the early spring of 1902, the health of Rupp's beloved stepmother Rachel, then aged sixty-five, had deteriorated alarmingly. On 30 September she died of cancer in the old parsonage which still stands behind the elegant little stone Church of All Saints at Learmonth.[11] It was reported that 'Mrs Rupp was very highly esteemed by the congregation of the local Church of England, and was, with the Rev. Mr Rupp (who is the hon. secretary) one of the most enthusiastic supporters of the horticultural society, her many exhibits being a feature of the annual shows'.[12] Once the 'very lengthy cortege' from the church had reassembled at the cemetery, Rachel Rupp was buried in the highest corner overlooking Lake Learmonth and the beautiful countryside.[13]

Much of H. M. R. Rupp's clerical career is well described in his memoirs, but a summary of professional appointments may be useful to indicate the geographical scope of his pastoral work, and thereby, of some of his botanical investigations. Suggestions of a certain restlessness of spirit—and the indulgence of an agreeable wife and family—are also evident:

1898   stipendiary lay reader at Beeac, in parochial district of Colac.
1899–1901 curate of Colac-with-Beeac, diocese of Ballarat.
1901–1903 incumbent (first) of Beeac, diocese of Ballarat.
1903–1904 curate of Tamworth, diocese of Grafton and Armidale.
1904–1906 vicar of Warialda, diocese of Grafton and Armidale.
1906–1908 rector of Yea, diocese of Wangaratta.
1909–1911 vicar of Copmanhurst, diocese of Grafton and Armidale.
1911–1914 vicar of Barraba, diocese of Grafton and Armidale.

1914–1920 assistant secretary, then secretary, to the Australian Board of Missions, with headquarters in Sydney, and travels in N.S.W., Victoria, Tasmania and southern Queensland.
1920   *locum tenens*, Holy Trinity, Hobart, diocese of Tasmania.
1921–1922 rector (first) of St Aidan's, East Launceston, diocese of Tasmania.
1922–1923 *locum tenens*, Branxton, diocese of Newcastle (December–February).
1923–1924 rector of Bulahdelah, diocese of Newcastle.
1924–1929 rector of Paterson, diocese of Newcastle.
1930–1932 rector of Weston, diocese of Newcastle.
1932–1933 relief duty at Pilliga, diocese of Armidale, and at East Maitland, diocese of Newcastle.
1933–1936 priest-in-charge, Woy Woy, diocese of Newcastle.
1936–1939 rector of Raymond Terrace, diocese of Newcastle.[14]

As a country parson, often with oversight of a vast parish, Rupp was, according to his family, 'an unconscionably early riser', often working in the garden until breakfast began the daily routine. He was a tireless traveller, whether seeking out needy souls or desirable plants—activities which were easily combined. Prior to Christmas, 1930,[15] when his family presented him with his 'first (and only) motor-car—a Morris Cowley' (so that he could 'give up the bike' at fifty-eight), extensive parish journeys were principally undertaken on horseback or by sulky. 'I used to do rides of 350 miles 'way back in Warialda days,' he once recalled.[16] He also recalled some heavy falls, and he came to dislike prolonged sulky journeys, often preferring to use a second-hand pushbike he later procured.

Energetic and compassionate, Rupp developed enduring respect and sympathy for those engaged in rural industries. One of his schoolteacher correspondents was made acutely aware of this fact after rashly airing some criticism of 'the man on the land', for whom, Rupp declared, 'I have fought for half a century and more, and for whom I will do battle as long as I live'. To make his attitude perfectly clear he continued:

I know him in all his aspects. My father was a country parson all his life. I have lived among the old-time squatters of western Victoria, and saw many of them give place to agriculturists and dairy farmers. In New

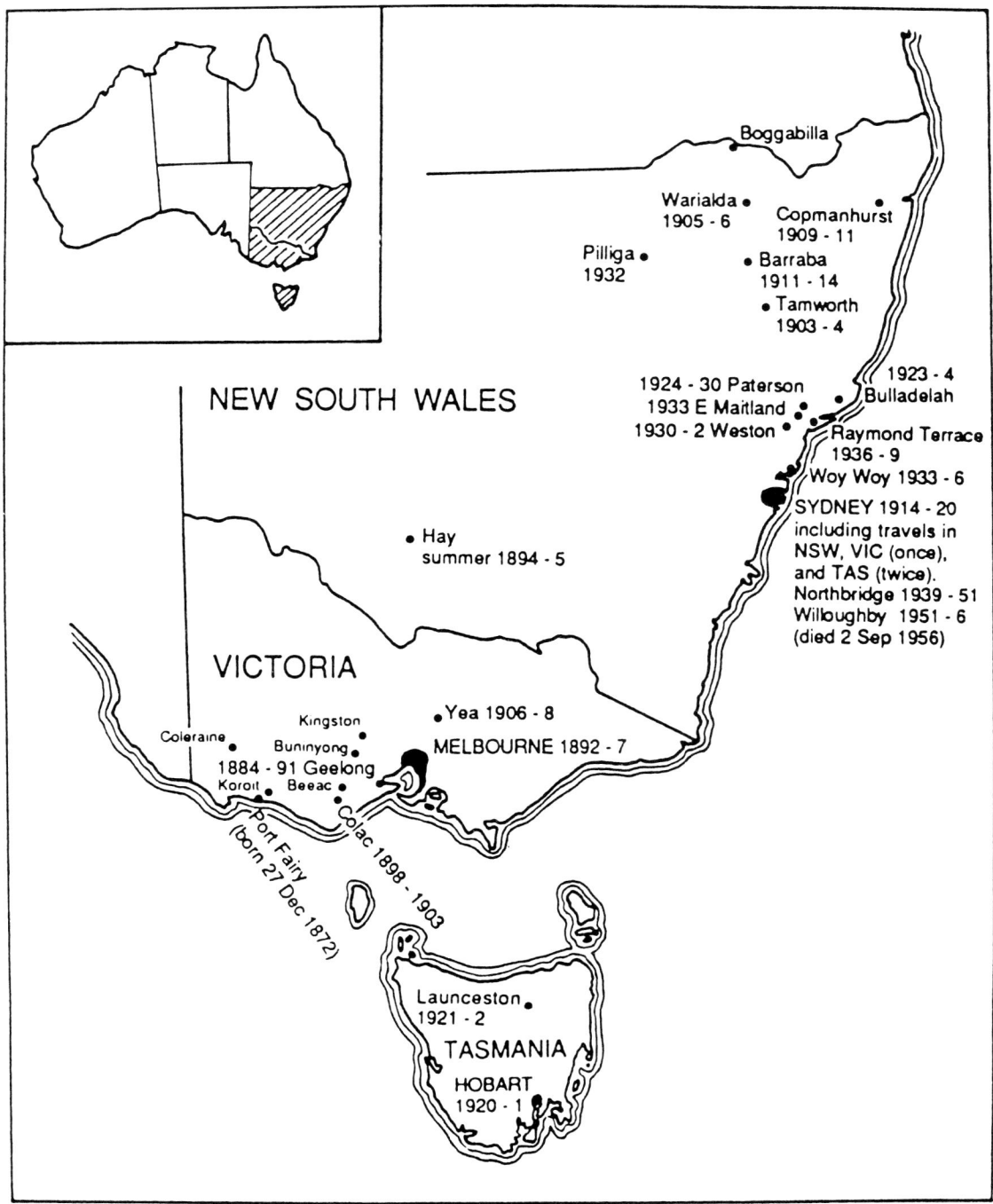

NEW SOUTH WALES

Boggabilla

Wariakla
1905 - 6

Copmanhurst
1909 - 11

Pilliga
1932

Barraba
1911 - 14

Tamworth
1903 - 4

1924 - 30 Paterson
1933 E Maitland
1930 - 2 Weston

1923 - 4
Bulladelah

Raymond Terrace
1936 - 9

Woy Woy 1933 - 6

SYDNEY 1914 - 20
including travels in
NSW, VIC (once),
and TAS (twice).
Northbridge 1939 - 51
Willoughby 1951 - 6
(died 2 Sep 1956)

Hay
summer 1894 - 5

VICTORIA

Yea 1906 - 8

Kingston

MELBOURNE 1892 - 7

Coleraine

Buninyong

1884 - 91 Geelong

Koroit   Beeac

Colac 1898 - 1903

Port Fairy
(born 27 Dec 1872)

Launceston
1921 - 2

TASMANIA

HOBART
1920 - 1

Rupp's life and work centred chiefly around a few cities and a succession of country parishes in south-eastern Australia. (Reproduced by kind permission of Mr Ian Clarke and the Australian Systematic Botany Society)

South Wales I have worked among the big graziers—sheep men and cattle men—among 'selectors' opening up new country on the far western slopes and in the coastal scrubs; I know the corn fields of the Clarence and the dairy farms of the Hunter Valley. Both my girls married men on the land...

There are black sheep and misfits and incompetents in every calling, my own included. But that doesn't warrant me in a wholesale indictment of the class or group which they misrepresent... For some strange reason we are all apt, unless we watch our steps, to judge a class or group of people by its failures... So it seems to me you make the same mistake over the man on the land: you've known some who are deliberately improvident, and some who greedily overstock, and some who can't see a tree without wanting an axe. You let these fools dry up your sympathy for the great majority who don't tempt Providence in these ways, and yet have a hard struggle to make good, and are often overwhelmed by disasters which have beaten all their efforts.. Even milking machines don't make dairying a pleasant or refining occupation; yet it's got to be done. But take him all in all, in my opinion the man on the land deserves every penny he makes, and deserves, too, the heartfelt sympathy of town dwellers who do not know what it is to see their prospects wrecked by drought or flood. Here endeth my homily![17]

No further criticism of 'the man on the land' seems to have been made by this correspondent.

Although he feared that coalminers were sometimes exploited in the course of wider conflicts, Rupp generally supported them also and enjoyed their company. 'Apart from the small percentage of Communists,' he wrote during the Depression, 'I find the miners very likable, friendly and loyal.'[18]

Rupp was reckoned to be a good preacher, with firm views, yet generally tolerant of the views of others. He confessed that although he had 'never been a real "party man"', he was 'a High Churchman by conviction and long experience'.[19] He disliked 'extremes of any sort' whether they were on the one hand the excesses of some Anglo-Catholics, or on the other, the fulminations of 'cruel Sabbatarians...who strain at gnats and swallow camels'.[20] Whenever he sensed that some dispute was attributable to 'narrow-minded extremists', or to their bigotry or intolerance, he defended the victim. 'My sympathies are entirely with the Bishop,' he declared during the widely publicized, if regrettable, litigation over the use of the 'Red Book' which Bishop Arnold Lomas Wylde had authorized within his Diocese of Bathurst.[21]

During his sixteen years in the Diocese of

Newcastle, Rupp comfortably wore the eucharistic vestments according to

the English Use as ordered by the Ornaments Rubric... In matters ritual and ceremonial... I prefer simplicity to complexity, provided that our worship is made as beautiful and reverent as is possible within a fairly liberal interpretation of the laws of our branch of the Church Catholic.[22]

For episcopal visits, Rupp was apt to request the bishop to wear cope and mitre.

Without malice, he loved to recount the story of the fervent young evangelist who 'bailed up a bullocky in the main street' of Bulahdelah about 1924: '"Brother," sez he, "how's your immortal soul?" Without a moment's hesitation the bullocky replied, "Aw—not too bad sonny, 'ow's yer own?".'[23]

In 1926, Rupp was approached to conduct the wedding of one of his most esteemed botanical penfriends. He was delighted at the prospect, and had begun preliminary administrative formalities when further correspondence indicated some difficult obstacles. He was greatly, and sadly, perturbed to find that his friend's rather unorthodox views and beliefs would lead to an untenable position. Clearly the facts had to be faced: 'Our religious standpoints are far apart: but I haven't the remotest intention of straining the friendship of two lovers of Nature by harping on that fact or striking the note of controversy'. Rupp further clarified his position, taking care not to give offence:

My own convictions have been reached and reasoned out through many difficulties and dark passages, and I would not surrender them for—shall I say all the orchids in the world: but I see lots of people who I am sure are much better folks than I, holding views that I could not entertain for a moment; wherefore I conclude that God does fulfil Himself in many ways. I am I suppose (if I must place myself) in the ranks of what some call the 'advanced' wing of the Anglican Church, and am not likely to stray elsewhere.

Happily the potential impasse was amicably avoided. The correspondent, who claimed to be 'a kind of theosophist' simply stated, 'I'm sorry, but never mind, let's go on talking orchids instead, and pay us a visit all the same'.[24] Rupp responded in kind: 'As you remark—"Let's talk about orchids" and hope that they and the birds will still be found when our eyes are opened in the Big Beyond'.[25] Although a decade was to pass before the two ardent naturalists finally met, they continued to 'talk about orchids' and remained good friends for another thirty years.

Rupp came to consider himself 'a very uncon-ventional parson indeed',[26] but perhaps this was so only in a twentieth-century context. Had he served his long ministry during the nineteenth century, as a contemporary of other clergymen-scientists such as William B. Clarke of North Sydney, William Woolls of Parramatta, James Walker of Liverpool or George E. W. Turner of Ryde, he may well have found himself quite in vogue, especially among the more liberally minded.[27]

Whether motivated by a desire to meet new people and to see new places, or by an innate restlessness, Rupp tended to remain for only two or three years in his various parishes. Some clergymen feel that they can give their best to a place in this way. His two most enduring appointments were to the parish of Paterson and to the Australian Board of Missions, each about five years.

Each posting provided its professional challenges, botanical opportunities and personal memories. There was the early satisfaction of building and opening the first Anglican church at Beeac in 1898; the long journey from Victoria to Tamworth, where as senior curate he married Florence Mabel Dowe,[28] on 29 December 1904 in St John's Church—and the subsequent exciting departure 'by the 2 a.m.

The Rev. C. L. H. Rupp (1838–1917) with his daughter-in-law, Florence, and grandchildren, Rachel and Arthur, at Yea, Victoria, 1908. (From Mrs Rachel Cox)

Brisbane express, amidst further showers of confetti. . .en route for Victoria'.[29] Then followed the move to Warialda, where the Rupps' first daughter, Rachel Mary,[30] was born on 29 November 1905. Concerned about his father's health, Rupp shortly transferred to Yea, Victoria, where a son, Arthur Richard Herman, was born on 25 October 1908.

The machinations and logistics associated with some transfers were quite intriguing. By the time Arthur was born, the Venerable Robert J. Moxon, archdeacon of Grafton, had heard that Rupp was contemplating a further change. Accordingly he wrote a confidential letter suggesting that Rupp might consider nomination to the 'new district of Alstonville' between Lismore and Ballina. The stipend was estimated at £150–£200 a year. There followed an exchange of letters and telegrams while Rupp sought certain assurances. Moxon regretted that these 'reasonable requirements' were not acceptable to the parish, and it seemed as though such a transfer would not occur. At the end of October, however, Moxon offered nomination to a parish which would 'entirely suit'. This was Copmanhurst with Yulgilbar, Ramornie and the rising copper-mining settlement of Cangai. The stipend was £180 a year, but there were hopes of an increase to £200. The vicarage was 'a good roomy house, in good repair' and 'except Cangai, all parish work can be done with horse and sulky'. One letter included a map showing the main centres of settlement and the mileages between them.

To the archdeacon's delight, Rupp accepted the post in mid-November. Further details and suggestions were then forthcoming. Copmanhurst had no doctor; perhaps Rupp and his family should have their annual holiday before making the long journey; furniture could be bought in Grafton, or indeed from the former vicar, whose effects were for sale at £45–£50; and, of course, it would be fitting if Rupp were to represent the diocese at the opening of Wangaratta Cathedral on 16 December 1908, but the sooner he arrived after 1 January the better.

Once they had travelled from Yea to Sydney, the Rupps were advised to proceed to Grafton by coastal steamer: 'The steamer from Sydney generally gets to the Heads at daylight and as it takes a long time to get up to Grafton, the passengers and mails usually go on by a smaller river boat arriving about 10:30'. The steamer itself docked at Grafton at about 3 p.m. If an overnight stay were necessary, then the Crown Hotel near the wharf was recommended.[31]

Many people must have been relieved when on Friday, 5 February 1909, the rural dean, the Rev.

C. F. Seymour, inducted the Rev. H. M. R. Rupp 'to the Cure of Copmanhurst' in the sawn-slab and shingle Church of the Holy Apostles which, like its new vicar, dated from 1872.

This appointment made it possible, even obligatory, for Rupp to make his first excursions into subtropical rainforests as he ministered to people associated with timber-cutting, dairying, grazing and mining. He shortly received an encouraging message from Bishop Cooper: 'The Archdn. told me of your visit to Cangai where there seems to be great need for strenuous work'.[32]

After some three years of such 'strenuous work', Rupp was ready to move, this time to Barraba, where after a couple more years another daughter, Eileen, was born on 16 February 1913.

Rupp joined the Australian Board of Missions (ABM) in the spring of 1914 as assistant secretary to the Rev. John Jones.[33] There was barely time to commence deputation work in the Diocese of Grafton[34] and to establish his family in Sydney when there came an urgent call from Tamworth, for Rupp's father-in-law, Richard A. Dowe, had contracted typhoid fever. He died on 8 January 1915, and almost immediately after the funeral, Rupp had to proceed on duty to northern Tasmania.[35] By the autumn of 1915 he was back in northern New South Wales, visiting Glen Innes where he preached in Holy Trinity Church,[36] and saw his father, who was then staying with the Monypenny family. Following visits to Ballarat and other Victorian parishes in May, he returned to the north in the winter, visiting Glen Innes, Casino and Grafton. The vicar of Casino observed to Bishop Cooper of Armidale, 'Rupp left us today after several days in and out. He certainly has found his niche, and does his work remarkably well and looks and is much more cheerful'.[37] Clearly the opportunity to travel about and to enjoy the open air agreed with him.

Returning to Tasmania at the end of 1919 on behalf of the ABM Thankoffering (for Victory) Appeal, Rupp visited thirteen parishes and addressed a public meeting in the Hobart Town Hall at which the bishop of Tasmania presided.[38] Further journeys were made to Victoria, southern Queensland, and to other parts of New South Wales.

Rupp returned to parish work in 1920 through the offer of a locum tenency at Holy Trinity Church, Hobart. The Board received his resignation 'with sincere regret' for he had 'carried out his duties with thoroughness and with ability'.[39] He was succeeded as secretary by the Rev. Hugh Linton.

The short term at Hobart proved very enjoyable, but not so the next couple of years as first rector of St Aidan's, East Launceston, where Rupp felt rejected by 'bigotry and narrow intolerance' worthy of the fierce reformer, John Knox himself, until 'Bishop Stephen of Newcastle...came to the rescue'.[40] Thereafter, all of Rupp's permanent appointments were in the diocese of Newcastle, where he found in parish and synod a personal and ecclesiastical fellowship which he cherished for the rest of his life.

Returning to the mainland, Rupp became rector of Bulahdelah to find timber-cutters, sawmillers, dairyfarmers, teamsters and fishermen and their families living in 'the most wonderful Paradise of wildflowers I ever knew'.[41] The former rectory of weatherboard still stands, like most of the town, against the huge backdrop of the remarkable Alum Mountain, near the foot of which the even more remarkable subterranean orchid, *Cryptanthemis slateri*, was later discovered, causing much excitement.[42]

In and around Bulahdelah, Rupp 'found...the richest botanical ground I had ever explored, and my herbarium got up to 2,000 species and I suppose 5,000 specimen-sheets, while I was there'. But, more significantly, 'Orchids had always attracted me, but it was only at Bullahdelah [*sic*] that I really began to concentrate on them'.[43] In so doing, he found a soulmate in one of his church officers, Dr Hereward Leighton Kesteven, medical practitioner at Bulahdelah, 1920–36, and who to Rupp's further delight, 'thumped out Gilbert and Sullivan with particular relish' on the piano.[44]

Rupp found that the 'Paradise of wildflowers' was an 'unwieldy parish' with seven centres of worship to be incorporated into a regular timetable of services—Bulahdelah, Bungwahl, Mayer's Flat, Boolambayt, Wootton, Markwell and Rosenthal. On 17 September 1924 the Rupps left Bulahdelah for Paterson, where they were met on the railway station by their new parishioners.

Although Rupp apparently believed that his ministry at Paterson, 1924–9, was not as effective as he had wished, his time there was noteworthy in several ways. He clearly performed some active pastoral work, for the number of regular communicants trebled, while he 'personally undertook the financial organization of the whole parish which was practically bankrupt'. A new church of St James was built at Martin's Creek, where people depended 'largely on metal quarries and farming', and an ambitious programme was devised for the restoration of St Paul's, Paterson, and

The Rector of Paterson, 1927. (From Mrs Rachel Cox)

he had been talking to one of his parishioners at 7.30 a.m., Rupp 'had to go out and break the news to his wife that the Dungog train had run him down'. He was 'a fine fellow', Rupp lamented, and one of the truly 'progressive farmers' of the district.[47] Like other clergy, Rupp experienced similar anguish when called to spend 'nearly all night at a very distressing deathbed'. After such emergencies, botanical investigations must have afforded welcome respite.

The decision about the herbarium was not made before much heartfelt consideration. For Rupp, as for many field botanists, his herbarium was not merely 'a receptacle' for pressed and dried plant specimens, it was also 'a treasury of precious memories' of places, people, exciting discoveries, and of long tramps through bushland long-since transformed or obliterated. In short, a personal herbarium was a personal botanical diary, a record of excursions to 'all sorts of outlandish places, nice & nasty'. Unhappily, Rupp had come to realize that his general 'collection has got beyond my management'.[48]

the publication of a history to mark its eighty-fourth anniversary.[45]

Rupp also carried out some profitable botanical field work, including his unforgettable excursion to Barrington Tops, and perhaps even more significantly, he made a crucial decision about the future of his rapidly growing herbarium, which had to be transported from parish to parish and adequately maintained.

Like others, the parishioners of Paterson soon learned of Rupp's extra-ecclesiastical interests. On one occasion, some had the enlightening experience of witnessing their rector's transfiguration from a clerical to a botanical state:

A couple of Sundays ago I nearly trod on a great colony of remarkable fine P. truncata in the Martin's Creek churchyard. This is a comparatively rare species—a large flower on a very stumpy stalk—I had not seen it since 1914, and had been anxious to get it, and had many enquiries for it. Wherefore the local schoolmaster mistook my joy for a temporary aberration.

The schoolmaster must have been impressed, for shortly after he had the rector taking 'a nature-study lesson on Orchids' in his school.[46]

There were tragic times at Paterson, too, as on Tuesday, 20 July 1926, when less than an hour after

Extract from Rupp's notes to accompany the specimens presented to the Botany School, University of Melbourne, 1926. (Photograph: Ian C. Clarke)

NOTES ON VARIOUS PLANTS
REPRESENTED IN THE HERBARIUM OF THE REV.
H.M.R. RUPP, B.A.

FOREWORD.

As my herbarium is about to be presented to the Botany Department of the University of Melbourne, where I was a resident student of Trinity College from 1892–7, I have thought it as well to copy out some notes which I have jotted down at various times in comment upon some of the specimens. Except for some reserve specimens which I am now preparing, I am not including in the collection to be handed over to the University, any substantial assortment of Australian Orchids, of which I have herbarium specimens of about 220 species. These I am retaining, as I wish to continue the study of this fascinating order of plants. I have not, however, omitted from the Notes on Plants such as relate to Orchids; but have merely placed in brackets those which deal with species not represented in the Melbourne collection, as they may still be of some interest.
H.M.R.R.

Paterson, N.S.W., January 1926.

NOTES.

Acacia linearis, Sims. — A. longifolia, Willd. — In N.S.W. these two species seem to approach each other very closely, and I am doubtful of the herbarium determinations.

Acacia stricta, Willd. — In Tasmania I always found this species with very pale green phyllodia; this is noticeable in comparison with

Having made a decision, Rupp wrote to the warden of Trinity College late in 1925 to ascertain the reaction to an offer of 'over 2,000 species' of non-orchidaceous plants to the College Museum. He advised that he had expressed this intention to Dr Leeper some years before, and that now it seemed 'hardly likely that I shall in future give much time to the general flora: specialising is the order of the day, and what I am doing in leisure hours in connection with Orchids appears to be considered of some value'.

Anticipating that there could be some reluctance on the part of the College, Rupp provided a frank assessment of the condition and contents of the collection he had amassed since about 1890 (when he was seventeen), chiefly collected by his own hand (except for the Western Australian species) and fairly representative 'of the flora of Eastern extra-tropical Australia'. A review of the herbarium made 'about 18 months' before had revealed its range and magnitude:

	Locality	Specimens
1	N.S.W. North Coast Districts from Bulahdelah to the Tweed River	678
2	New England Plateau	144
3	Western Slopes, Tamworth to Warialda	460
4	National Park to Hawkesbury River	455
5	South Coast, Illawarra to Bega	211
6	Southern Highlands	49
7	Western Plains	138
8	Miscellaneous, N.S.W.	about 200
9	Ballarat & Western Victoria	119
10	Port Phillip	130
11	Tasmania	518
12	Western Australia	105

To this collection of 3,207 specimens, additions had been subsequently made, and some duplicate orchid specimens would be forthcoming to balance the collection, should it be accepted. However, if the offer were not acceptable, then 'please do not hesitate to say so—I am not thin-skinned'.[49]

The warden, Dr John C. V. Behan, responded immediately and appreciatively, to reveal the unhappy fate of the museum Rupp had so fondly tended and remembered, but expressing hope that the former exhibits might be released from 'a sort of cubby-hole beneath the stairs of the Clarke Building' once proposed extensions had been constructed. How could the herbarium be appropriately housed 'in a manner calculated to be of real benefit to students?'[50]

Rupp anticipated such questions. 'One does not like to look a gift-horse in the mouth, but there are times when it may be wise.' He did not recommend the public display of specimens, for they deteriorate; in fact, the specimens he 'gave the College museum years ago' had been 'installed under glass in this way' and 'probably they have long since perished'. A reference collection, with specimens 'properly arranged in their various orders and mounted on sheets, in numerous brown-paper folios', could be easily stored in kerosene cases, if necessary. He had done this, and by means of an index used in conjunction with tags on the folios, 'anything may be found in a minute'. There were, of course, other and better ways: 'The late Mr Maiden introduced the system of keeping the mounted specimens in large cardboard boxes, placed in order in cabinets. This has been adopted I think at Sydney University.' It was gently suggested that 'the University botanical folks' in Melbourne would doubtless have practical

Extract from Rupp's notes to accompany the specimens presented to the Botany School, University of Melbourne, 1926. (Photograph: Ian C. Clarke)

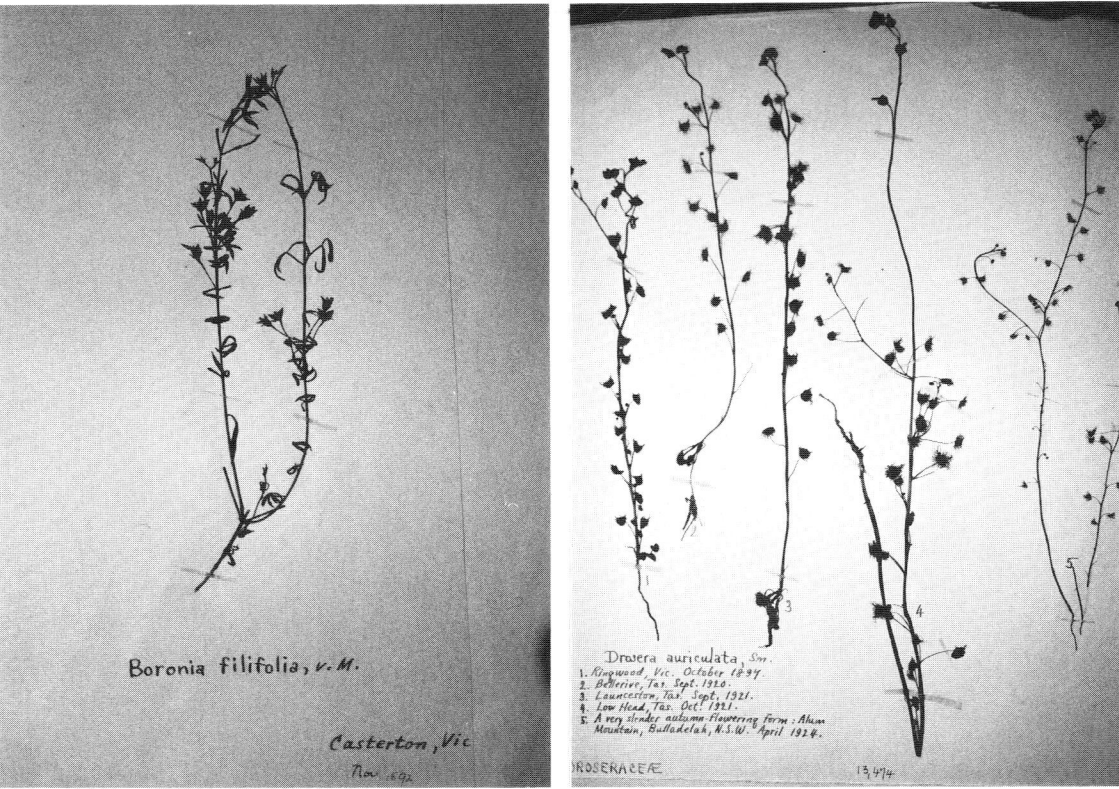

Specimen sheets of *Boronia filifolia* and *Drosera auriculata* from the collection Rupp presented to the University of Melbourne, 1925–6. (Photographs: L.A.G., May 1988)

suggestions to offer the warden concerning the re-mounting and storage of the collection. He continued:

For twenty years I collected widely for the late Mr Maiden, to whose unfailing kindness and encouragement I owe most of the botanical knowledge I may possess. Most of the specimens so collected are in the National Herbarium at Sydney; and I am in regular correspondence with the Curator, Mr Edwin Cheel.[51]

Dr Behan took care to consult the College Council and Professor Alfred J. Ewart,[52] of the Department of Botany, before venturing to reply. Then it was proposed that, notwithstanding Rupp's 'loyal and filial sentiment towards the College', it might be better to offer the collection 'direct to the University', to be housed in an appropriately inscribed cabinet within the Botany Department where it would be readily accessible to those most likely to use it.[53]

Rupp readily agreed: 'Having once (with reluc-tance) made up my mind that if my herbarium could be made useful to other folks, I ought not to retain it when it has obviously outgrown my capacity to give it proper attention, I do not want to delay action'.[54]

By the end of January 1926, Rupp's plant specimens had a new home. As one would expect, they were accompanied by copious documentation, in manuscript, comprising a meticulously prepared index, notes on genera and species, collecting places and ecological features, together with 'General Notes relating to the growth of the Herbarium' which included an interesting biographical summary of the collector himself. In 1988, former Botany Department technician, Mr E. J. Sonenburg, recalled the arrival of two wooden packing cases, each measuring about three feet by four feet, and eighteen inches deep. By August 1926 the specimens were lodged in 'a fine big cupboard . . . of stained oregon', bearing 'a plain brass plate' inscribed in Gothic and Roman scripts:

PRESENTED TO THE
UNIVERSITY OF MELBOURNE
ON THE SUGGESTION OF TRINITY COLLEGE
BY THE
REV. H. M. R. RUPP
A FORMER MEMBER OF THE COLLEGE
DEC. 1925.

The former student's botanical benefaction did not go unnoticed. There were gratifying newspaper notices in both Melbourne and Sydney.[55] Professor Ewart was reported as being 'delighted at the presentation. ''Until now there has been practically no herbarium at the University,'' he said. ''This is the first large donation of the kind that the Botanical Department has received in twenty years''.'[56]

During negotiations, Rupp established correspondence with Professor Ewart, providing additional information about certain specimens, and, towards the end of 1926, sending a further twenty-six species, chiefly orchids and ferns.[57] Rupp was delighted by the resolution of the problem and the recognition of his work. He would take further pleasure in the knowledge that the brass plate is still in position, and

that the Botany Department's Herbarium, to which he provided the foundation collection, now comprises 'some 100,000 specimens representing all major plant groups'.[58]

Confirming his intention to concentrate henceforth upon his favourite plant family, Rupp now began to prepare the manuscript and to assemble photographs for a *Guide to the Orchids of New South Wales* for submission to his friend George Robertson of the famous publishing firm of Angus and Robertson.

The period immediately before Christmas, 1926, brought considerable family anxiety and sadness. Florence Rupp had to undergo major surgery in Sydney, and while she recuperated in one hospital, Rupp's brother-in-law, the amiable Irishman from County Cavan, Arthur Augustus Monypenny, died on 9 December after surgery in another hospital. Monypenny and his wife, Rupp's sister, Florence Emily, had given Rupp that memorable holiday at Hay thirty-three years before, and had provided Charles Rupp with a home at Glen Innes during retirement until his death in 1917. Florence Monypenny later went to live at Artarmon. Rupp

Specimen sheets of Rasp Fern, *Doodia caudata* and Maidenhair Fern, *Adiantum aethiopicum,* from the collection Rupp presented to the University of Melbourne, 1925–6. (Photographs: L.A.G., May 1988)

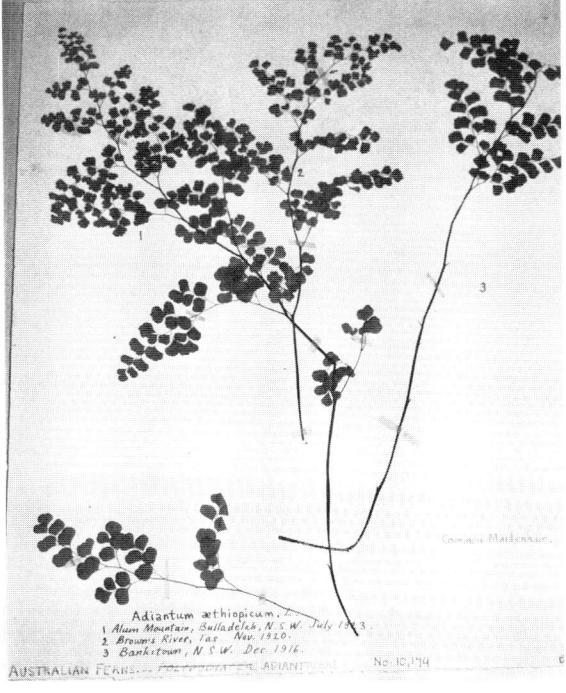

The rector in the greenhouse with his orchids and ferns. Probably at Paterson in the 1920s. (From Mrs Rachel Cox)

was to observe on 'Good Friday evening 1955' that 'my beloved only sister enters her ninetieth year today! And she's livelier than I am.'[59] At her death on 27 August 1961, she had almost attained the remarkable age of ninety-five years and five months, and had survived her brother by five years.[60]

As early as spring 1928, Rupp indicated to Bishop Long of Newcastle that he was ready for a new pasture, for he was feeling the effects of constant travel by pushbike and sulky over indifferent roads, suffering from insomnia and worrying about the effectiveness of his work. However, he decided that an appropriate time to move would be after the anniversary celebrations into which he and many parishioners had put so much effort. These were planned for November 1929, by which time there had been some adverse events. In January, Paterson suffered the effects of fire, drought and some closure of quarries, and in September there occurred the worst flood since 1893. 'The river rose 36½ ft., and was within 20 yds. of our house,' Rupp recorded.[61]

Late in October 1929, Bishop Long advised, 'The only impending vacancy is at Weston. Do you feel inclined to tackle coalfields work?' The response was immediate: 'I have no idea of my own qualifications for coalfields work, my only experience of it having been a three months loc.ten. at Branxton: but if my Bishop and the parish are willing to try me I should not shrink from it'.[62]

In the first week of November 1929, Rupp visited Weston to assess the bishop's offer: 'I liked what I saw of Weston and Abermain, and I should be quite willing to tackle a parish of that kind. But the financial difficulties appear to be very serious.' He felt that he could manage if the stipend were not less than £300.[63] The 'financial difficulties' were of course due to, and indeed part of, a wider problem of national and international concern. At Tyrrell House, Newcastle, on Wednesday 6 November, Bishop Long received the Weston parochial council, 'a deputation of ten men... Fine fellows they were—Very worried about their Parish.' They were assured that the 'Diocese would stick to them until the "lockout" is over', and financial settlement could be made when work recommenced.[64]

On 26 November 1929 the bishop nominated

Rupp to be rector of Weston, and the next day officiated at the anniversary celebrations at Paterson. By mid-January 1930, the bleakness of the economic outlook, especially on the coalfields, moved Rupp to seek an interview with the bishop before he left for the forthcoming Lambeth Conference. Rupp explained:

I am not afraid to tackle the Weston job, but I am facing conditions of which I have had no experience, and my knowledge of those conditions is only derived from the newspapers and irresponsible local talk. I should therefore value your advice very much, and I feel that it would probably save me from at least some pitfalls of ignorance.[65]

Apparently reassured, Rupp moved his household to Weston at the end of January 1930, as the Depression clouds darkened. With the bulk of his herbarium then in Melbourne, he probably found this move a little less arduous than usual. However, there remained not only the orchid herbarium, but also the living plants—the considerable contents of the bush house which had become a feature of the Rupp rectories. In these rustic structures, Rupp battled against the assaults of heatwaves, droughts, storms and frosts, of slaters, millipedes, grasshoppers, slugs, snails, rats, and even the 'sheer fiendish mischief' of spotted bowerbirds, while he coaxed unidentified or otherwise interesting orchids to flower, so that they might be examined, photographed, dissected, described, sketched, pressed, dried, or simply enjoyed.

By May 1926, the Paterson rectory's bush house, with its assorted pots, hanging baskets and sections of tree trunks, had boasted nearly sixty species of native orchids. These included the striking *Phaius australis* and *Calanthe triplicata* as well as other terrestrials such as *Cryptostylis*, *Thelymitra*, *Diuris*, *Caladenia*, *Lyperanthus*, *Eriochilus*, *Acianthus* and a dozen species of *Pterostylis*, and epiphytes such as *Dendrobium* (14 species), *Sarcochilus* (9 species), *Bulbophyllum* (4 species) and *Cymbidium* (4 species). About a third of the species represented had come from the bush around Paterson; some had been brought from the previous parishes of Barraba and Bulahdelah, while others had been sent from elsewhere, more especially from Tamborine Mountain, Queensland.[66] No doubt by the time the removalists arrived, this collection was even more representative.

In February 1930, Bishop Long sent Rupp a farewell message as he prepared to leave for London, taking the opportunity to express good wishes for work in the new parish. Rupp remained optimistic:

'Notwithstanding the prevalent depression, and the grave problems of present and future, I think we are going to be very happy at Weston. The people—men as much as women—are most kind and cordial. First impressions go a long way with me.' Rupp wished his episcopal friend 'God-speed in all that lies before you and a safe return', to draw the final jocular response that 'troubles will end as soon as they get me out of the country'.[67] Within five months however, the bishop was dead. His clergy were devastated.

The industrial situation on the coalfields grew more sensitive and serious. Miners were beginning to lose support from some sections of the wider community and hearts continued to harden. Rupp's parish council was virtually bankrupt, but he enjoyed the miners' company, noting with great satisfaction that contrary to the expectations of some people,

those terrible fellows, the miners, have been falling over each other in helping us ever since we came: one is making my shelving, others got poles and put them up for my son's wireless; three got me a big lorry-load of tea-tree for a bush house and helped put it up: and altogether we find them very good fellows.[68]

The new rector's 'first impressions' of Weston were therefore 'very pleasant', as he rescued his 'poor transported orchids from a miscellaneous heap in a shed, and...established a good many of them in their new home'.[69] The recipient of this news, the naturalist-journalist Alec H. Chisholm, felt that Rupp, having enjoyed considerable botanical success at Paterson, would find the coalfields to be something of a botanical desert, and accordingly had declared his concern in the *Sunday Pictorial*. Having described Rupp as a man of 'retiring disposition' and 'one of the most distinctive of Anglican clerics', Chisholm commended him on the history of the Paterson Church and on the intended book on New South Wales orchids, then lamented:

It is a loss to Australia that Mr Rupp has been transferred to Weston-Abermain, for he will not have nearly the scope for orchid-study there that he had at Bullahdelah [*sic*], and more recently at Paterson. In both these localities he discovered orchids of outstanding interest—so much so that these little towns will always be familiar names to orchid-students the world over.[70]

Rupp himself was quite enthusiastic, and a few days before leaving Paterson, he sent Chisholm a firm assurance: 'But Weston/Abermain is a perfect *paradise* of wild-flowers—infinitely better than Paterson... Given a decent season, I'll eat my hat if I don't get more orchids over there in my first year than in 5½ here.'[71]

Perhaps Rupp should have known that such a categorical statement would bring a gleam to any journalist's eye. There promptly appeared a small headline: 'Eating a Parson's Hat'. This time Chisholm confessed that when he reported that 'the noted botanical parson possessed a retiring disposition' he

did not bargain for the fact that his [the parson's] friends, both clerical and unregenerate, would shake their heads at him and say, 'Well, well, who would have thought that!'...the point on which I did slip was in suggesting that Weston-Abermain...would not be so prolific in orchids as his old locality, Paterson...and he assures me now that he will eat his hat—a large order for a parson—if he doesn't get more orchids in one year at Weston-Abermain than he got in five years at Paterson.

But—dreadful thought!—supposing orchids are declared 'black'.[72]

Thereafter, Rupp gleefully kept Chisholm informed of the progressive 'score' as more species were discovered, until he was able to announce in triumph after nine months in Weston: 'I've saved myself an indigestible meal. My orchid list here is up to 65—two more than the total of five years at Paterson.'[73] Within a couple of months, the number had increased to sixty-eight species, and the 'parsonic hat' was clearly secure.[74]

Meanwhile, Angus and Robertson's Halstead Press had proved to be 'exasperatingly slow' in producing the *Guide to the Orchids of New South Wales*. Due to have been published in August 1929, the book finally appeared in April 1930.[75] Edwin Cheel, curator of the National Herbarium at the Sydney Botanic Gardens, provided the Foreword, noting that 'It has long been felt by naturalists that such a book as the one now offered...is urgently needed'. Priced at seven shillings and sixpence, and printed on glossy paper, the book was well illustrated by eight of Rupp's line drawings and eighty-four photographs, many of which had been used to illustrate his *Sydney Mail* articles of 1926-8.[76] The work was warmly reviewed in the *Mail* as 'an extremely valuable addition to the literature of Australian flora'. In fact, the reviewer had 'come across few books of late more compact of interest and information or more beautifully turned out'.[77]

The *Guide* was also briefly noticed in the *Sydney Morning Herald*, in the *Garden and Home-Maker* magazine, and by Estelle Thomson in the *Brisbane*

Courier. Rupp's old friend and fellow orchid enthusiast, Alexander G. Hamilton, brought the book to the notice of teachers in the New South Wales *Education Gazette*:

Very many teachers in country places are greatly interested in the local orchids, and would like to know something about them. Their peculiar and often bizarre forms, their great beauty, and their intricate structure attract attention from the most casual observer...there has long been a demand for a book on them.

Now...the Rev. H. M. R. Rupp has given us just the book wanted, and one which will be of the greatest assistance to orchid lovers...

The book reflects great credit on the author and publishers, and will be most useful to all lovers of these beautiful flowers.[78]

The Melbourne *Australasian* also was very encouraging. The reviewer alluded to the long-felt need for such a popular guide, and continued:

The author expresses a fear that from his desire to inform the 'popular' mind and at the same time be useful to the botanist, he may fail to win the approval of either. He need have no such fear. He would be a well-informed botanist who would find this little book useless, and he would be an unambitious amateur who would not be grateful for the scientific information here given. Mr Rupp is one of a very few men in Australia who possess an extensive acquaintance with orchids.[79]

When A. H. Chisholm published a similarly enthusiastic review in the *Sunday Pictorial*, Rupp brightly responded:

Well, I haven't got enough spurious modesty to pretend the book is rotten—but I'm feeling how much better it could have been! Had I known it was likely to be held up so long, I should certainly have improved upon some of the pictures. And [in] spite of all the proof-correcting, there are a few blunders...

The 'get-up' of the book is admirable—I only hope the contents will prove worth the setting they have been given.[80]

By the end of June, 240 copies had been sold, and 'considering the times and the nature of the book', the author was happy enough.[81] He was even happier by October of the following year when George Robertson was 'hinting at a possible second edition'.[82] Unfortunately, with Robertson's death in August 1933 that prospect faded.

Joy over the success of a function to help reduce parish debts, followed by the exciting arrival of the Morris Cowley for Christmas 1930 and the gratifying discovery of 'a good chap opposite' who was a born mechanic, were offset by the sadness of travelling to Tamworth for the funeral of Florence Rupp's mother, Mary Dowe, who died on 23 December.

Prospects at Weston seemed a little brighter for the new year, but by the spring of 1931 Rupp described his position as 'extremely precarious and full of worry', despite the fact that his parishioners were 'most loyal' and battled to keep the parish solvent. Personal anxiety was greatly increased when a disaffected and litigious resident threatened court action over some advice Rupp had allegedly given to another party, albeit in confidence and with the best of intentions. Rupp declared that he 'had nothing to conceal or to be ashamed of, and would have gone to gaol if necessary'. Legal advice was obtained by both sides, and the vestrymen stood by their rector 'to a man' to the extent of asking the bishop to relieve him of Sunday duty when the plaintiff, after reducing a demand for damages from £200 to £10, 'dropped the case like a hot brick'.[83] Thus, on the first Sunday in September 1931, the archdeacon of Newcastle visited Weston while the rector of Weston 'was bundled off for the weekend to Lake Macquarie', where he temporarily forgot his worries 'in the joy of unlimited orchids'.[84] Throughout this regrettable affair, Rupp's chief concern was 'that the Church would be dragged in however the case went'[85] and 'mud would have been slung at the Church'; understandably, 'it was a terrible worry'.[86] On the brighter side, Rupp was able to advise Alec Chisholm that in terms of species, 'the coalfields orchids now stand at 73'.[87]

In October, Rupp notified Bishop Batty that, without wishing 'to appear a pessimist or alarmist', his position would 'shortly become quite untenable'. The reasons were clear enough. The Abermain mines were 'in a bad way'; Hebburn No. 1 pit was 'working badly' and 530 men had been dismissed from No. 2; the lessees of the parish hall had 'given up' because their clubs were no longer functioning, yet the people themselves remained 'very loyal and warm-hearted'. The bishop invited Rupp to call, and later urged him not to become 'unduly depressed'.[88]

On Boxing Day, 1931, Rupp took his family to Port Macquarie for a couple of weeks ('perfect weather, lovely scenery, A1 surfing and fishing, and plenty of orchids') and returned to Weston feeling 'a little fitter to face the music'.[89] Unhappily, the music proved discordant, and a further crisis came in February 1932, when the parish council 'reluctantly decided' to notify the bishop that it could not continue to function. Rupp agreed with the decision, declaring that 'it's just bad luck for me and

not the people's fault'[90]; in fact, he did not 'know how they have carried on so long', for there were 'half a dozen parishes on the brink of similar collapse'.[91]

In March there was an enquiry into the affairs of the parish of Weston. It was then claimed that two-thirds of the population were on the dole and one-third were not earning half-time wages. Hebburn No. 1 pit was working only three days a fortnight and No. 2 was closed altogether with 1,100 men out of work. Weston and Abermain each had an estimated population of about 4,600, although Rupp noted, 'many of our people are camping down at the Lake',[92] where coal could be found and fish could be caught.

Rupp now had physical problems as well as mental anxiety, for in the autumn of 1932 his 'ancient enemy', bronchitis, returned with asthmatic symptoms, and he fell victim to an attack of sciatica. Despite 'electrical massage at hospital' he was obliged to remain seated while conducting the Good Friday services.[93] Cranking the car in its more reluctant moments was out of the question.

In his sadly reduced state, Rupp sought Alec Chisholm's advice on submitting illustrated articles for publication in order 'to supplement, even if only by a very little, a very slender income which is growing more and more precarious'.[94] Rupp had already written some of his botanical reminiscences in a work entitled 'Gardens of Nature' which he had sent to George Robertson in 1928, without any obvious result.[95] Now that conditions had so dramatically worsened, expert opinion was sought:

Now I don't imagine myself a literary star, but I have...had a good many things accepted for publication, and people capable of judging have told me that I can write. (By the way, those 'Gardens' have never seen daylight yet.) I have some hundreds of photographs, many of them of scenes well off beaten tracks...and a good proportion of them, I know, though I says it as shouldn't, are quite worth reproduction. I could write something of interest, I think, about them all. But I don't know how to go about getting my wares inspected...I know nothing of the illustrated papers and magazines that Australia now produces... Can you advise me?

Rupp added that if the advice 'should be "let the cobbler stick to his last", the cobbler will take it smiling'.[96]

Chisholm promptly suggested that approaches be made to the Melbourne *Australasian* and the Sydney *Australian Woman's Mirror* (published by *The Bulletin*, this threepenny journal sold 'over 167,000 copies' weekly). The appended Bibliography reflects the extent to which the advice was earnestly followed. By mid-May 1932, just prior to leaving Weston, Rupp had submitted material to both journals, and while the *Australasian* had initially demonstrated a disappointing reticence, the *Mirror* had published and paid for the first of what proved to be a series of twenty articles which were to appear over the next four and a half years. Rupp was very grateful to his literary mentor:

I must thank you again for putting me on to the 'Mirror'—they pay a good deal more liberally than the 'Mail' did. I haven't had time to explore any of the other avenues you suggested yet—except that I sent an article to the 'Australasian' about three months ago...and that's the last I've heard of it.[97]

The painful period of strikes, lockouts, mine closures, unemployment and financial hardship finally took its toll in April 1932, when the unfortunate rector and his parish virtually collapsed together. Rupp's doctor addressed one problem, the diocesan authorities addressed the other, and the bishop endeavoured to attend to both.

Rupp recorded that he 'had a bit of a nervous breakdown',[98] and that he had been advised to take three months' leave of absence. Following an interview with the bishop, he requested leave from the end of May to the end of August 1932, with the understanding that household effects could be stored in part of the rectory.

Heartened by the archbishop's permission to officiate in the diocese of Sydney and by the knowledge that the bishops of Newcastle and Grafton were seeking a position for him, Rupp prepared for his departure. There was a last act of kindness for which he was particularly grateful. On 13 May 1932, a 'good friend...F. J. Fieldsend, came out...with his lorry and took all my orchids to his own beautiful fernery, where he is going to look after them, as he says, "until you tell me you're ready for them again". It was entirely his own suggestion.'[99] A fortnight later the Weston rectory was vacant.

The problem of temporary accommodation was resolved when Florence Rupp's sister kindly offered a cottage, Willinga, in Collaroy Street, Collaroy. During June, July and August 1932, Rupp enjoyed freedom from parish worries and plenty of work on orchids, so that he was rapidly refreshed and restored. Late in July he told Bishop Batty that 'the change and rest have made a new man of me', and that he was considering four months of temporary work offered by Bishop J. S. Moyes of Armidale. He

continued: 'I have managed to scrape in pocket-money by free-lance journalism, and we are not on the dole yet; but another month will see us at the end of our resources'.

Bishop Batty still had no viable vacancy, and suggested that Rupp leave his furniture at Weston and accept Bishop Moyes's offer. Weston would be worked from nearby parishes. As a gesture of goodwill, the bishop sent £25 as a grant-in-aid which was very welcome.[100]

On 3 August 1932 Rupp submitted his formal resignation from the parish of Weston and made preparations for the move to Pilliga, between Narrabri and Walgett. He had planned to travel by train and prepare the way, while his son Arthur conveyed Florence Rupp and some essential household effects in the Morris Cowley, but apparently they all travelled in the car.

They arrived at an unpropitious time, for the country had suffered a hard season. 'More desolate, drought-stricken country I have rarely seen,' Rupp noted,[101] just before a welcome fall of rain. 'I won't guarantee to eat any hats here over comparing the orchids with Paterson!' he told Alec Chisholm. In any case, 'in the Pilliga Scrub...the rabbits are in such myriads that I doubt if orchids would have a chance'.[102]

Ecologically, Rupp found the country 'extraordinarily interesting', apart from its remarkable propensity to bog man, beast and vehicle. He found only one orchid, *Cymbidium canaliculatum*, which because of its 'amazing range' fascinated him. The bird life provided further delight while in spare hours he worked on his 'Recollections of an Amateur Botanist' or 'My Australian Hobby', and even favoured Bishop Batty with some natural-history observations.[103]

Logistically, Rupp found Pilliga 'an absurd parish', with Gwabegar nineteen miles away and Baradine forty-three miles, both to the south on the same road, so that the large parish seemed to have no centre from which one could work efficiently. Boundaries have since been rearranged.

More worrying uncertainty followed the departure from Pilliga at the end of December 1932. On the first day of the new year, Rupp advised Bishop Batty that he had returned to the Hunter and had 'taken a cottage' at 71 George Street, East Maitland. There was some temporary work at East Maitland, but Rupp's spirits were understandably low: 'my prospects, frankly, are nil. I don't quite know what will happen to me—if this goes on much longer it will mean that I shall be *incapable* of returning to the work in which I have served my best for 33 years.'[104]

There was 'an unofficial ''sounding'' as to whether I would go back to Copmanhurst...where I was vicar 1908–11,' Rupp told the bishop, but 'I could not entertain the idea.' There were further 'soundings' concerning temporary work at Lambton with Jesmond or at Bellbird. Aberdeen was also suggested, but Rupp was not prepared to ask his wife to 'camp' at Aberdeen as she had done at Pilliga. He would, however, attend to Sunday duty and visit the schools, if required.[105] Acceptance of temporary work at Gresford seemed to provide the best solution,[106] but then came the offer, as from 1 July, of the parish of Woy Woy. On 12 May 1933 the

The rector of Woy Woy and his wife — an informal snapshot of about 1935. (From the late Mrs Eileen Cox)

Rupps, accompanied by their son, drove to Woy Woy, talked to the people, admired the place and accepted. Rupp was at his new post by the end of June, determined 'to have a bush house' as soon as possible.[107] He found it 'a lovely spot: the Rectory...right on the banks of Brisbane Water at Blackwall, about a mile from Woy Woy proper'.[108] By October, a 'tiny bush house' had been built—a sure indication of a return to normal life.[109]

Although, as was often the case, the parish needed 'building up', the Rupps enjoyed their time at Woy Woy and apparently intended to remain there until

retirement. Despite depressed conditions, Rupp found that the parishioners 'rallied wonderfully as soon as they felt they could trust me', and he became 'quite proud' of them.[110] However, recovery was slow. Even in 1935 Rupp declared, 'When I got to know the place, I was astounded at the poverty here. Things have improved a little, but quite half our folks are merely on relief work.'

In fact, he found 'more poverty and distress here than I ever saw on the coalfields'. Of about 250 church families, there were perhaps 100 which 'could hardly contribute regularly' to the church, while perhaps 150 might manage sixpence per week. Some might even contribute one shilling per week, but the parish was hardly affluent.[111] Rupp declined to state any stipend 'which I considered they ought to pay me'. He told the members of the parish council that he knew of their situation, and would leave it to them to raise what they could towards support of the church. Rupp was convinced that this 'was the right attitude under the circumstances' and so it proved.[112]

Rupp's earnest ministrations did not go unappreciated. The people at Ocean Beach rallied to build a church, restoration work was undertaken on the church at Blackwall, and Rupp stirred interest in a small neglected cemetery at Booker Bay where victims of the wrecked paddle-steamer *Maitland* were buried in May 1898. Following a serious burning accident, one parishioner sent a heartfelt tribute directly to the diocesan registrar: 'it was the good old Doctor and the carefull nursing that pulled me through, and the carefull attendence of the revern Mr Rupp. he is a very hard working man. he is the best Clergy that we have had here. we would not like to lose him.'[113] There were lighter moments, too, as when one admirer gave the rector 'a most appalling second-hand pith helmet' found in an 'old clothes shop'. It was believed that 'I would look very nice in it in summer', lamented Rupp, who mischievously contemplated means by which the headpiece might be run over 'accidentally' while he was on his rounds!

Rupp's restorative influence in this parish over three years doubtless influenced the decision announced in Bishop Batty's letter of 17 February 1936: 'at a meeting of the Presentation Board held this morning, you were unanimously elected to the Rectory of Raymond Terrace'. The bishop urged acceptance with the assurance that the initial stipend would be £345 a year and that two-thirds of removal expenses would be met. While there was 'no prospect' of a parish car, the former rector had received 'help in the way of horse feed, etc. to a value [of]...about £25 a year'.[114]

Rupp replied the following day:

Your letter...came as an absolute shock...I had not the remotest idea of leaving Woy Woy—we had made up our minds to stay here until I retire...in about two years; & my idea was to do a lot of scientific work that I can't do while I am in charge of a parish, & make myself generally useful to the Church relieving men who want a spell, and that sort of thing...

Having discussed the matter with his wife and family, Rupp accepted four days later, observing that 'Mrs Rupp belongs to neither the interfering nor the critical type of parson's wife!'[115]

Of course there was a catch. Rupp found that while St John's Church at Raymond Terrace was 'a gem', the rectory was 'in a shocking state' although 'if the proposed renovation and remodelling are carried out it will be delightful'. He confessed to one of his botanical friends that although he and his wife were 'very sorry' to leave Woy Woy, the new position would mean 'freedom from financial worry for the rest of my time'. Now in his sixty-fourth year, he felt that this final opportunity should be taken even though it would mean camping alone for months in 'the old detached kitchen' while the restoration work was carried out. 'It will be a great home when it is done—fancy, stone walls 2 ft. thick!' he declared with admiration.[116] Over half a century later, the building remains a splendid example of Georgian architecture.

Having enjoyed a warm farewell function at Woy Woy (and having suggested that his successor should work from a motor boat rather than from a motor car), Rupp, with the carriers' help, was established in Raymond Terrace rectory 'by candle and lamp-light' between the hours of seven and eleven on Wednesday night, 18 March 1936. His wife took the occasion to have a well-earned holiday, and it was well that she did, for the old building had been empty for about three months, and was 'swarming' with 'white ants and sundry other vermin'. Rupp was in constant terror that his beloved books and precious orchid specimens would be consumed. The bishop's recent testimony must have seemed most pertinent: 'One thing that emerges very clearly from my musings on the subject is the gratitude which I and the diocese generally owe to you for the faithful service you have given under such very inadequate conditions'.[117]

Bishop Batty inducted Rupp to this, his last parish, on St George's Day, 23 April 1936. Problems were already evident. Rupp had quickly ascertained that many parishioners were 'relief workers' families

or on the dole'; the Morris Cowley, 'pretty nearly done for' by the Woy Woy appointment, had to be given some much-needed repairs, and travel was thereby curtailed for a month; work on the rectory, begun undesirably late, proceeded so slowly that it seemed that the requirements of the six foolscap pages of typed specifications would never be fulfilled. By July, only two rooms had been fully renovated, and Rupp told the bishop that, had he known what was in store, he would have probably remained at Woy Woy.[118]

In October, shortly after Florence Rupp had been able to set up a normal household in the now-restored residence, there occurred a crisis of a far different kind. A returned soldier, depressed by hardship, committed suicide, and the Rupps took the devastated wife and her three children into the rectory. It was 'a frightfully sad case altogether', Rupp commented.[119] On being informed, the bishop immediately sent £5 to help in 'this emergent case', and Florence Rupp further demonstrated her compassion by taking the widow for a change to the Collaroy cottage while church people looked after the children.[120] Rupp took up the cause in earnest, supporting a claim for a pension for the family of 'an M.C. man...badly knocked about' during the war, and condemning the fact that 'apparently Govt. departments apply literally the principle of giving to him that hath'. He averred that there were those 'in receipt of substantial salaries (in one case £10 week, in another £8) who are receiving full War pensions', yet his parishioner had been refused.[121] He was still writing to the bishop and interviewing officials over two years after the tragedy, achieving some success. These acts of spontaneous kindness were not forgotten. The lady was among the most affected mourners at Rupp's funeral twenty years later.

In August 1937 Rupp wrote to the bishop seeking leave from a clergy retreat (not for the first time!) and broaching the matter of retirement. He would soon be sixty-five, and his doctor suggested that he should retire. The bishop (then in his fifty-ninth year) urged Rupp to stay, and expressed the hope that he would write the monthly missionary page for the *Newcastle Diocesan Churchman*. Rupp shrank from any additional writing because he already had 'been drawn into an extensive correspondence...with botanical societies and workers' throughout Australia and beyond, and because his practice was to write out his sermons 'in full' before delivering them. However, he would undertake the task for three months if no other offer were immediately forthcoming.[122]

During the weekend of 25–6 September 1937, 'the 75th anniversary of the present St John's Church' was marked by an episcopal visit, special services and a parish picnic. After Christmas, Rupp 'boarded the R.M.S. *Orford* for Burnie, Tasmania'. Friends at Wynyard took him 'all over the lovely N.W.' and he returned, greatly refreshed, by the *Ormonde* on 11 January 1938 to an 'atrocious summer'.[123]

In October 1938, Rupp advised the bishop of his intention to retire by the end of April 1939. The response was heart-warming: 'I really have appreciated the solid and excellent work you have done, and the example of priestly zeal and efficiency you have set'.[124]

Shortly after, the bishop assured Rupp's old adversaries, the administrators of the Clergy Provident Fund, that 'they are dealing with one who has been all through his ministry a most faithful and devoted servant of the Church'.[125] He was a man who 'leaves the Diocese in completely good standing', and who, presumably, was without doubt worthy of his pension.

Bishop Batty also extolled Rupp's worth publicly. In a monthly diocesan letter he declared:

I cannot forego the opportunity of saying once more how highly I estimate Mr Rupp's ministry in all the parishes he has held in this Diocese. No man has a higher ideal of the Priesthood than he: and no man devotes himself more wholeheartedly to putting this ideal into effect.[126]

The bishop reiterated this view when addressing Synod in May 1939: 'Mr Rupp...has given fifteen years of very excellent service in this diocese, and I know of no one who has a higher conception of the pastoral office than he, or a greater desire to give effect to that conception in his ministry'.[127]

On Easter Saturday in April 1939, Rupp delighted in celebrating the marriage in St John's of his younger daughter, Eileen, to Ashley Clarendon Cox, then manager of Belmont near Scone, and brother of Leslie Clarendon Cox, who had married

the elder daughter, Rachel, two years before.[128] In the following month, Rupp retired, and with the aid of a small legacy which had come to his wife, they moved to Sydney. Once again he was pleased to be succeeded by his old friend, the Reverend Hugh Linton.

Montague Rupp had served the Church for over forty years, often in 'hard' parishes, within five dioceses in three states. Ten years after his retirement he recalled that the highest income he had ever received was £400 a year when New South Wales Secretary of the Australian Board of Missions and while rector of Raymond Terrace. Something less than £300 had been his average annual stipend,[129] with relieving work at East Maitland yielding about half this rate during the Depression. There had been many lean times, when only his wife's 'little private income' enabled them to withstand the 'horror of debt'.[130] In December 1931, when the basic wage was £3 10s. 8d. a week, even the 'very loyal and warm-hearted' parishioners at Weston could barely find between them £1 a week for the offertory.[131] Happily, Bishop Batty was also warm-hearted, and though he was hard pressed at that time to meet requests and demands from throughout the diocese, he sent £10 'as a personal contribution towards the Weston Stipend Fund'.[132] For some years, annual holidays were possible largely through the bishop's bestowal of £5 grants from the 'Clergy Holiday Fund', but even this had to be reduced to £4 in 1937. The situation at Woy Woy had been so serious that in the spring of 1934, when the basic wage was £3 8s. 0d. a week, Rupp declared, 'I am receiving a lower stipend from this parish than when I started my ministry in 1899'.[133] Even then, he stuck to his principles: 'I am not going to accept as stipend money raised from "stunts" and dances, etc. I have had too much of that; and I think it is absolutely demoralising to both parish and priest.'[134]

So it was that the rewards of the Rev. H. M. R. Rupp's spiritual and scientific endeavours lay chiefly in the endeavours themselves.

8 Man of Letters, Man of Science

One of life's pleasures for me is corresponding with orchidy folks.
—H. M. R. Rupp to Fred Fordham, 26 June 1933

As a Man of Letters, Montague Rupp had few equals in his field—he was the consummate correspondent. His letters, like his memoirs, were characterized by a delightful literary style—direct, lucid, informative and enquiring; comfortably conversational, free of unnecessary embellishment and imbued with a delicious sense of humour. Whether in the characteristic clear, bold hand, or in meticulously accurate typescript—in black or red depending on the state of the ribbon—Rupp's letters (almost invariably signed 'H. M. R. Rupp', underscored by a single elongated zig-zag line) entertained, enlightened and challenged a rather unusual circle of professional and amateur naturalists and botanists for sixty years and more, throughout Australia and beyond.

The volume of correspondence Rupp generated, especially during the last thirty-five years of his life, was truly astonishing. Sometimes he wrote to the same person on consecutive days, or even twice on the one day, and many letters were enriched by botanical sketches, detailed lists of plants or lengthy postscripts, prompted by 'second thoughts', by the discovery or arrival of additional information, or by the long-awaited blooming of a plant in the rectory bush house. Some letters were composed over two or three days, and there is clear evidence that carbon copies of at least some outgoing letters were kept. Perhaps it was the bulk of this correspondence that ultimately led to some regrettable destruction.

Many of Rupp's correspondents recognized his letters as potential works of reference, and consequently preserved them. Most letters were replete with observations on the geographical distribution of orchid species, flowering seasons, flower colour and perfume, and morphological features, both 'typical' and variant. There were comments, too, on Rupp's first encounter with certain species and on subsequent efforts to cultivate them; there were suggestions on species to seek in likely localities at given times; there were offers of specimens and requests for specimens. Rupp was keenly interested in taxonomy and taxonomists, noting the appropriateness of certain names and descriptions, the judgements of other authors and the material available to them. He developed a special interest in the Tasmanian botanist, Ronald Campbell Gunn (1808–81), and gathered material for an intended biography.

The letters indicate that Rupp's own judgements were based upon a convincing combination of judicious reading, wide experience in the field and a painstaking scrutiny of material. He also possessed a remarkable power of discernment (a quality he saw in others as 'the true orchid eye') and a sharp, retentive memory for places, seasons, habitats and details of structure—a faculty he was distressed to find, after his eightieth birthday, had 'grown very treacherous' and had become 'as crooked as a snake'.

Rupp and his friends corresponded not only by means of letters, sketches and photographs, but also by hosts of little packets and boxes, light metal containers (perhaps fondly intended for slices of wedding cake) and quite large parcels, containing an incredible variety of dried, pressed, living, dying, dead and even pickled specimens—leaves, stems, pseudobulbs, rhizomes and tubers, buds, flowers and seed-pods, or whole plants. The benign rector astonished postmasters, postmistresses and postmen wherever he lived. Generally he had cause to appreciate and praise their indulgence. 'Our local P.O. people are very good,' he once noted, 'the postman brought the box right to the door at 8 a.m.' As a result, in this instance, a 'wonderful Diuris collection...turned up...in first-rate condition',[1] and botanical science took another short step forward. Shortly after he observed:

the Postal Department is pretty good with specimens, and I never have cause to complain. In fact I have cause for gratitude in that they invariably deliver the parcels I get, however bulky. If the parcel is labelled BOTANICAL SPECIMENS, WITH CARE (or PERISHABLE, if specimens are living) they don't delay.[2]

Notwithstanding these praiseworthy efforts at a time when costs were relatively low and attitudes to service were commendably high, it was inevitable that some specimens should fail to endure. Some withered or decayed *en route*, others were squashed beyond recognition. Rupp tended to attribute many of such disasters to inadequate care over packing on the part of the sender. He found that even some acknowledged experts, like Dr Rogers of Adelaide, were notorious in this regard. The advice was simple—in the absence of such desirable material as sphagnum moss, 'damp blotting-paper' provided good packing in 'any small container', securely wrapped, and labelled as suggested. 'In the past fortnight,' Rupp once wrote, 'I have received over 100 living specimens, and 95% arrived in good condition.'[3]

Soon after moving to Tamworth in 1903, Rupp, the erstwhile correspondent of Baron von Mueller, contacted the then director of the Sydney Botanic Gardens, Joseph Henry Maiden ('who was a *good friend* to me always'[4]), to initiate a correspondence and a traffic in specimens which lasted for twenty years. When, in 1924, Maiden, without notice, had Rupp's illustrated 'Notes on the Habits of Certain Orchids' published in the *Australian Naturalist*, the unknowing author was brought to the attention of a significant band of enthusiasts, many having a similar interest. The appearance of three further articles on orchids in the same journal in 1925-6, of two others in the *Victorian Naturalist* and yet three more in the *Proceedings of the Linnean Society of New South Wales* during the same period, attracted even wider notice.

Rupp was formally accepted for membership of the Naturalists' Society of New South Wales in June 1924, of the Linnean Society of New South Wales (on the recommendation of Edwin Cheel) in July 1927, of the Field Naturalists' Club of Victoria in March 1934,[5] and on 7 August 1934 he was elected one of two foundation vice-presidents at the first general meeting of the Orchid Society of New South Wales.[6] Additional opportunities were thereby provided for publishing observations and discoveries, and for becoming acquainted with other labourers in the field, especially when it was possible to attend meetings.

When Rupp finally met J. H. Maiden on that memorable afternoon at Turramurra, 'just a few weeks before his death' in November 1925, the nucleus of a group of pen friends and personal acquaintances, eminent in botany and in other branches of natural history, had already formed. Rupp had established contact in 1923 with W. H. Nicholls,[7] and in 1924 with A.H. Chisholm,[8] who apparently introduced him late in 1924 or early in 1925 to Hilda Geissmann (later Mrs Curtis, or 'Hilligei' to her close friends[9]) of Mount Tamborine, southern Queensland. Then the publication of Rupp's 'Cult of the Orchid' in the *Sydney Morning Herald* on 1 December 1925 attracted the notice of Mrs Edith Coleman,[10] who, like fellow Melbourne naturalist, Charles Barrett,[11] was another valued friend by 1926.

In April 1926, Rupp was delighted to acknowledge having 'several keen correspondents in Victoria and Tasmania, one in Queensland, and Dr R. S. Rogers in S.A.; but in our own State I have hitherto been playing rather a lone hand'.[12]

It should be noted that four of the earliest Victorian correspondents—Barrett, Chisholm, Coleman and Nicholls—became prominent in the Field Naturalists' Club of Victoria, of which Rupp also became a member. The most esteemed of the Tasmanian contacts was probably Mrs Florence Perrin of Launceston.[13]

Rupp's memoirs clearly indicate how highly he valued his friends and rejoiced in his ever-widening coterie of 'orchidy' correspondents. He shared enthusiastically in their exciting discoveries and successful cultivations, followed their travels with avid interest, encouraged and advised them, extolled the worth of their endeavours, and referred one to another. Early in 1925, Rupp assured Hilda Geissmann: 'your letters are really appreciated—

and they're always as welcome as the flowers of the bush we both love'. He further advised:

W. H. Nicholls is a fine chap, worth helping. He got into touch with me nearly 12 months ago, but Mr Barrett told me all about him recently...he's doing wonderful work on orchids. He is, bit by bit, preparing what promises to be quite a classic on Victorian Orchids. He does exquisite water-colour figures of the details and his photographs are splendid. I'll post you some that I have mounted... I had told him of the many plants and pictures you sent me, and he asked for your address some time ago.[14]

In July 1926 Rupp declared that W. H. Nicholls 'is a wonder. I think he's the soundest orchid man we've got',[15] and later, 'If his work is ever published he will rank as the 20th century Fitzgerald'.[16] Although he long deferred to the opinions and judgements of 'the dear Doctor' (Rogers), 'our leading orchidologist',[17] Rupp still valued, twenty years later, Nicholls's 'opinion on all orchidological matters very highly'.[18]

Such deep respect did not of course ensure that the two experts maintained a state of blissful botanical agreement. As a mutual friend, Mrs Pearl Messmer, remarked to Joyce Scrivener in a letter of 6 May 1942:

If you could share them as I do, you would have a good laugh at the arguments that go on between friends Rupp and Nicholls, to say nothing of the recriminations, with asperity. I, being friendly with both, get the opinions of each about the other's ideas. At present they are having a 'box on' about Caladenia Patersonii var. hastata and C. hastata, whichever the poor thing ends up by becoming...no doubt the sparks will fly... My Father and his friends were exactly the same, and used to scrap like dogs, over the fishes, while we lesser fry can just sit by, and enjoy our quiet chuckles.

With due deference, Rupp won this round. The Spider Orchid in question is still known as *Caladenia hastata* (Nicholls) Rupp.

Rupp found Hilda Geissmann to be 'a great enthusiast', 'a fascinating personality', 'a remarkable—and very fine—character', who not only maintained a generous flow of letters and plants, but also sent 'a wonderful assortment of splendid photographic studies' which Rupp displayed at a Naturalists' Society exhibition in Sydney.[19] Others were similarly impressed by this outstanding lady, who became renowned for the wide knowledge she accumulated while exploring the bush with her great Thornton Pickard half-plate camera. Charles Barrett, Bernard O'Reilly, Francis Ratcliffe and Frederick Colliver, as well as Rupp and A. H. Chisholm, all paid their tributes. Ratcliffe recorded: 'The least aggressive person I ever met, one thing alone could stir her to anger—cruelty. She fought cruelty with a quiet resolution which spared no-one.'[20]

Hilda Geissmann cast her anti-cruelty net widely, to ensnare sundry governors who indicated an over-fondness for duck-shooting, zoologists who sought to shoot specimens of flying-foxes, and even botanists who slipped specimens into plant presses. Rupp endeavoured to assure her on the botanical question: 'I don't think it's really cruel to press orchids...we pick them and lots of them wither and are thrown away before their time. I get much pleasure from preserving many of my treasures.' He continued:

If you would care to have any more pressed specimens, it will be a pleasure to send you some. I have been very fortunate in getting in touch with Dr Rogers—Mr Maiden switched me on to him over some Bullahdelah orchids—he has sent me a lovely lot of South and Western Australian specimens, chiefly Caladenias, Prasophyllums, and Pterostylis.[21]

Late in May 1925, Rupp sent Hilda Geissmann more of his 'reserve' specimens, 'in the hope that they will be of interest', adding, 'I love to write about these things to anyone who loves them as I do, and I'm sure from your letter that you are like-minded on the subject. ''Mad on bush weeds'' is the way some people describe me.'[22]

Rupp, delighted to receive 'such nice letters all aglow with the love of Nature', readily appreciated that 'the Postal Department will be glad that you and I have got into correspondence'.[23] When, however, 'the Postal Department' failed to deliver one letter and specimen, a sensitive situation developed, and correspondence lapsed for months. On discovering the reason for the misunderstanding, Rupp hastened to give reassurance: 'among the letters of my bush-lore friends none have been more valued than yours', but even then the Man of Letters felt constrained to admit that at one stage, 'my orchid correspondence began to grow at such an alarming rate that I had to put the brake on'.[24] However, the brake was but lightly applied, and their correspondence—warm, uninhibited, yet essentially scientific—spanned twenty years or more, with the unconventional parson even signing some of the more chatty of 105 letters still extant as 'Ruppie'.

Edith Coleman, who by mid-1926 was sending letters and specimens 'pretty nearly every week', Rupp immediately recognized as another 'great enthusiast and good observer'.[25] Shortly afterwards he noted, 'I send Mrs Coleman lots of things, because she I think, sends me more specimens, living and pressed, than any other correspondent I have— and that is saying a good deal'.[26]

'Have you met Mrs Edith Coleman?' Rupp asked Alec Chisholm. 'If not, you *must*—I am sure you will like her—she's just A1, and a splendid naturalist.'[27] Rupp's admiration never waned. Nearly twenty years later, when paying his final tribute, he stated that the correspondence begun in December 1925 (or early in 1926) had 'continued unbroken till her death' in 1951:

She wrote so much for publication that I often wondered how she found time for her private correspondence at all, but her letters were always delightful. . .

Every subject on which Mrs Coleman wrote she illuminated; for she was no merely superficial observer. . .her work has a place in that great fabric of scientific truth. . .and it shall not perish.[28]

Part of that work was a study of pollination in orchids.

During the 1920s, Rupp made many new botanical friends, while maintaining contact with earlier ones, such as Dr Frederick Arthur Rodway,[29] then of Nowra but formerly of Barraba. While at Bulahdelah, 1923–4, Rupp sent orchid plants, including the Lawyer Orchid, *Sarcochilus olivaceus*, to the German-born naturalist and artist, Adam Forster. Within a couple of years Rupp had despatched 'hundreds of things' to this 'very nice old man'. He was 'a real artist', whose work was 'very beautiful & orchids are among his happiest efforts.'[30] By October 1924, Rupp was the proud owner of 'a beautiful water-colour reproduction' by Forster, of the Cradle Orchid, *Corybas aconitiflorus*.[31] More correspondence followed, especially concerning a species of Double-tail Orchid, *Diuris venosa*, the first plant Rupp actually named and described in 1926.[32]

Letters, specimens and sketches continued to pass between them until the artist's sudden death in April 1928. Ten years later, some of Forster's paintings were used to illustrate the work of another of Rupp's friends, Thistle Y. Harris,[33] whose remarkably successful *Wildflowers of Australia* is still in print after half a century.

Other friends of the 1920s included Erwin Nubling,[34] Ferdinand August Weinthal,[35] and Alexander Greenlaw Hamilton.[36] Dr H. L. Kesteven of Bulahdelah has already been mentioned, and while at Paterson, 1924–9, Rupp found congenial 'orchidian neighbours' in John Tucker and his

wife.[37] By this time, Rupp also counted Obed David Evans,[38] then at Sydney University, and Edwin Cheel[39] of the Sydney Botanic Gardens, among his 'very good friends', but both were necessarily rather hard-pressed by official duties to devote much time specifically to orchid research.

Having now decided to concentrate upon the family Orchidaceae, and accordingly disposed of the larger part of his herbarium, Rupp was 'very anxious to get into touch with somebody else in New South Wales who is interested in the investigation of these fascinating plants'.[40] In short, he sought a fellow-specialist who was relatively uncommitted to other botanical pursuits and endowed with some leisure time and enthusiasm, for there was 'such a big field for investigation'. Evans suggested George Vance Scammell as a likely candidate[41]—and so he proved,

George Vance Scammell, BSc, ARACI, FCS (1903–86), director, F. H. Faulding & Co. Ltd., 1928–64, commissioner, Australian Red Cross, 1942–5. He began corresponding with Rupp in April 1926, and later provided most of the line drawings for Rupp's *The Orchids of New South Wales,* 1943. (Photograph from the late Mrs Lorna Scammell)

for on 19 September 1924 he had read, as an undergraduate of twenty-one, a paper to the Sydney University Botanical Society beginning 'What is an orchid?'

Somewhat tentatively, Rupp wrote to Scammell on 5 April 1926. Rapport was immediately established, and within a fortnight the amiable parson had invited the nascent company director to come and stay in 'a spare corner' of the Paterson rectory. Further, he had confessed

I really *do* other things besides writing letters—but it often happens that just after I get a letter I have time to answer it & I have found that if I don't take opportunity by the forelock, my correspondence, which is now considerable, is apt to slip into arrears.

Then came the cheerfully frank admission, 'and once I get hold of a botanical correspondent I like to keep him!'[42] There was no likelihood that silent neglect would give offence. Within four and a half years, Rupp sent this new correspondent nearly seventy letters, somewhat formally at first, then, after Scammell's visit to Paterson in the late winter of 1926, addressed 'Dear George'. In the following year, Rupp declared, 'I have a reputation as a lightning conductor in answering letters' and reiterated, 'but I find that if I don't answer at once in cases where I can, I get badly in arrears'.[43]

With energy, courtesy and purpose, Rupp extended his 'mailing list' and cultivated his penfriends as carefully as the orchids in his bush house. The geographical distribution of correspondents was as significant as that of the plants they so diligently sought and the location of a new and potentially responsive orchid enthusiast was as joyful an event as the discovery of a new locality record or even a new species. Thus on receipt of a large parcel of orchids from W. W. Mason at remote Cape Tribulation in North Queensland, he exclaimed, 'This chap's going to be a good find!'—and he was.[44] Such collectors compensated for Rupp's lack of opportunity to travel far from his parishes. His strong desire to visit North Queensland and Western Australia was never fulfilled.

Rupp intended to explore the Barrington Tops, between Scone and Gloucester, in January 1927, but owing to the 'drought and desolation' of the season, he decided instead to spend his customary post-Christmas holiday in Melbourne, where he could meet some of his penfriends in person: 'I've got lots of relatives there, and I particularly want to meet Nicholls and Mrs Coleman, who corresponded so

frequently with me. Am going to stay with Chas. Barrett for a start.'[45]

This proved to be 'a great time', for between about 5 and 20 January 1927 Rupp stayed a couple of days with the Colemans at their summer cottage at Healesville, and most of the time at Elsternwick, with Charles and Florence Barrett 'who were kindness itself'. The Barretts invited 'nearly all the known Melbourne orchid enthusiasts...in two instalments' to spend an evening with their delighted guest, who eagerly 'compared notes with Pescott,[46] Nicholls, Tadgell,[47] Green,[48] and many others'.[49] Another evening was 'spent...with Nicholls at his house' at Footscray, and 'the whole of one day at the National Herbarium' at South Yarra, where P. F. Morris assisted him to examine the orchid collection, including 'a lot of...Caladenias'.[50]

About this time, Rupp contacted Kew Gardens and the British Museum, in the first instance about Robert Brown's *Caladenia testacea*, which proved rather puzzling in Victorian and Tasmanian collections. Early in 1930, correspondence with Kew was revived after the then director, Dr A. W. Hill,[51] had visited Sydney. This time, Rupp did not need to take the initiative, for Hill made an outright request: 'As I know that you are a student of the Orchids of Australia, it has occurred to me that you might from time to time have duplicate or superfluous specimens of Australian orchids or other plants which you would be willing to present to Kew'. The cost of sending parcels of specimens would be defrayed.[52]

By now, Rupp had experienced something of an embarrassment of riches as far as the mailbox was concerned. 'I have several times thought of doing what you ask,' he replied from Weston, 'but my correspondence on orchids has grown to such dimensions that I have become rather canny of deliberately adding to it.' However, he concluded, in characteristic fashion, that although reserve specimens were depleted, he could send some newly described species (for example, *Diuris venosa* Rupp; *Prasophyllum acuminatum* Rogers; *P. ruppii* Rogers and *P. rogersii* Rupp) as well as Brown's *Caladenia testacea* and a *Caladenia* hybrid. These would be despatched 'when I make up my first packet for you...possibly today'.[53]

Ten species of orchids were appreciatively acknowledged, with 'the putative hybrid' being considered 'particularly interesting'.[54] This prompted the despatch of a 'putative' *Pterostylis* hybrid, together with other material, and the news that the *Guide* had been published 'about Easter'.[55]

Dr Hill was delighted with these accessions: 'Specimens from you are especially appreciated on account of your special knowledge'.[56]

Closer to home, the Rev. E. N. McKie, Presbyterian minister at Guyra, N.S.W., was sending 'interesting things' from the New England district by the spring of 1930, and the two clergymen corresponded regularly until McKie's death eighteen years later.[57] Other highly valued contacts made or maintained during the 1930s included Mrs Olive Rodway (née Barnard) of Hobart, Tasmania; Kenneth Macpherson of Proserpine, Queensland; Mr and Mrs D. J. Barr of Thornfield, Bellingen, N.S.W. (who, having 'contracted orchid fever', took Rupp on excursions through the Dorrigo rainforests in November 1934[58]) and Cyril T. White, the government botanist of Queensland.[59] By September 1939, Rupp had also been in touch with Frederick J. Rae, director of the Melbourne Botanic Gardens, over the identity of some Hyacinth Orchids, *Dipodium* spp.[60]

Some welcome relief from current personal, economic and industrial worries came at the end of November 1931, when the botanical world was startled, in Rupp's words, by 'what . . . may fairly be termed a sensational discovery'. Mr Ernest Slater of Bulahdelah,[61] while 'scraping away fallen leaves and debris from about the roots of the orchid *Dipodium punctatum*' discovered 'a curious little object' which he suspected 'might be of orchidaceous character'. As previously arranged, Mr Slater sent the *Dipodium* plants to Mr and Mrs F. J. Fieldsend of East Maitland,[62] taking care to include the new curiosity as well. The Fieldsends sent this puzzling specimen to their friend Rupp, who confirmed that it had the small withered flowers of an orchid, the like of which had first been discovered only about three years before in Western Australia—'an orchid which apparently germinates, grows, and comes into flower beneath the surface of the soil'.[63]

Rupp immediately communicated the exciting implications of this discovery and 'within a week', Ernest Slater, Frederick Fieldsend and Dr H. L. Kesteven sent four more specimens. Rupp thereupon named and described the plant as *Cryptanthemis slateri* in a paper submitted to the Linnean Society of New South Wales.[64] The excitement was slow to abate. Eighteen months afterwards, Rupp observed with concern: 'So much interest has been aroused by Cryptanthemis that I tremble for the welfare of Dipodium, which is a beautiful thing'.[65]

With the aid of a grant from the Australian and New Zealand Association for the Advancement of

Fred Fordham (1890–1978), schoolmaster and amateur orchidologist, who corresponded with Rupp for twenty-five years. (From a snapshot taken at Martin's Gully School, Armidale, about 1950, by courtesy of Mrs Jean Trotter)

Science, in mid-June 1933 Rupp travelled from East Maitland to Bulahdelah to seek more information and material with Dr Kesteven's assistance, and he returned in October.[66] Rupp sent some of the rather precious material to Dr T. G. B. Osborn, Professor of Botany at Sydney University, who undertook to refer part of it to Kew, and this occasioned more correspondence with Dr Hill in 1934.[67] This fascinating saprophytic subterranean orchid continues to attract scientific interest.[68]

Some of Rupp's most admiring and enduring correspondents were state schoolteachers. In October 1931, while at Weston, Rupp began corresponding with Fred Fordham, then in charge of the public school at Brunswick Heads, N.S.W.[69] Letters began with some discussion on the possibility of a symbiotic association between the Sun Orchid, *Thelymitra aristata* Lindley, and the Pink Rock 'Lily', *Dendrobium kingianum* Bidwell. Rupp responded with typical enthusiasm: 'I have been anxious for several years to get into touch with someone "orchid-ly" inclined on the North Coast'.[70]

It proved a remarkably profitable association. Over 220 letters were exchanged during the next twenty-five years, with Rupp's final brief greeting,

Participants in the celebrated Brunswick Heads camp of August–September 1936. *Standing (l. to r.):* James Edgar Young, William Henry Nicholls, Rev. H. M. R. Rupp, Miss Florence Irby, Miss Thistle Y. Harris (who later married naturalist David George Stead), Cyril Tenison White, Mrs Hilda Gladys Curtis (née Geissmann), Mrs Pearl Rae Messmer (née Finckh), Miss Mae Smales, Dr Colin Prentice Ledward. *Sitting:* Miss Winifred Irby, Miss Doris Goy (who later married botanist Lindsay Smith). Other participants included Fred Fordham (the chief host, then headmaster of Brunswick Heads school), Ellis Le Geyt Troughton (curator of mammals at the Australian Museum) and Anthony Musgrave (entomologist at the Australian Museum). (Photograph by Fred Fordham, by courtesy of H. S. Curtis, Brisbane)

perhaps written within a few days of his death, bringing the fruitful and enjoyable correspondence to a dignified conclusion. For the most part the letters dealt with orchidological matter, but there was a liberal sprinkling of nice personal touches, for example: 'My name is pronounced ROOP rather than as spelt. Everyone is bothered with it at first, and I am called all sorts of things. Strictly there should be ü over the u, but my father never used it, nor do I.'[71] After nearly nine years of correspondence, Rupp suggested 'haven't we reached the stage yet when we can drop "Mister"? I'm on if you are! Thanks for Liparis (?) specimens.'[72]

Thereafter, it was 'My dear Fordham' and 'My dear Rupp', but never 'Fred' or 'Monty'. In fact, Rupp was rather sensitive about forms of salutation. He despaired of 'some correspondents who persist

in addressing me "Dear Rev. Rupp", thereby reducing me to impotent fury'.[73] Would one write to the prime minister, 'Dear Sir Menzies'? he once asked, and on another occasion (when not writing to Fordham), he explained, 'Sorry I've got 3 initials, but at least use one! Maybe we're fussy, but all clergy hate being written up as Rev. Smith, Rev. Jones, &c.!'[74]

One important result of the Fordham association was the organization of a 'naturalists' Camp' in the spring of 1936, when there was an 'invasion' of Brunswick Heads 'by an orchid-hunting regiment'.[75] The original invitation, issued in 1934, was for Rupp to spend a holiday with the Fordhams at an appropriate time so that orchid studies could be pursued, but the idea appealed to other enthusiasts. One of the instigators and participants was another

W. H. Nicholls and H. M. R. Rupp with Fred Fordham's daughter Helen, in Fordham's bush house at Brunswick Heads, N.S.W., 5 September 1936. (From the Fordham Album by courtesy of Mrs J. Trotter, née Fordham)

of Rupp's botanical friends, Mrs Pearl Messmer of Lindfield. While she and Rupp promoted the idea, Fordham investigated the availability of accommodation and the most likely localities for profitable excursions.

Rupp set out in the Morris Cowley from Raymond Terrace on 21 August 1936. Other participants travelled from widely separated places—W. H. Nicholls from Melbourne; C. T. White, J. E. Young and Miss Doris Goy from Brisbane; Dr Colin P. Ledward and Miss Mae Smales from Burleigh Heads; Mrs Hilda Curtis from Tamborine Mountain; Misses Florence and Winifred Irby, from Deep Creek, near Casino (bringing their pet sugar squirrel, Mirram); Anthony Musgrave and Ellis Le Geyt Troughton from the Australian Museum, Sydney, and Miss Thistle Y. Harris, who was then teaching at St George Girls' High School, Kogarah. Rupp stayed with the Fordhams, betraying a weakness for milk coffee and toasted salmon sandwiches, and endearing himself to the daughters of the household, Jean, Margaret and Helen, and to their dog, 'the inimitable Ginger'.

After some days of intrepid excursions into the Brunswick River rainforests and surrounding country, the naturalists, laden with well-filled notebooks, plant presses and collecting jars, went their own ways. Anthony Musgrave, entomologist at the Australian Museum, promptly published an illustrated account of his time 'with orchid hunters in the Mullumbimby district' which is still a pleasure to read.[76]

Rupp drove Mrs Messmer, Miss Harris and W. H. Nicholls back to Raymond Terrace, via Grafton and Armidale, where he saw his son Arthur, then teaching at The Armidale School, and daughter Rachel, then teaching at the New England Girls' School. After staying overnight at Armidale's Imperial Hotel, the party proceeded south to Muswellbrook where Rupp had the misfortune to miss the Sydney road and, like many before and since, ultimately found himself in Denman. It would seem that the previously developed euphoria over the botanical delights of Brunswick Heads then began to wane. Despite many stretches of steep and corrugated roads, the Morris 'behaved angelically the whole way', a total of '1,029 miles' since leaving Raymond Terrace, with virtually no mechanical problems, except one, which 'Dick' the excellent NRMA mechanic at Mullumbimby promptly resolved.

Having gained Raymond Terrace, Rupp and his brave little Morris still had troubles, for in driving the passengers to the railway station, there was 'a most aggravating delay at Hexham punt', so that they missed the train 'by 2 minutes' to test the virtue of patience even further.[77] Nevertheless, Rupp retained happy memories of this trip, with its good fellowship and rich harvest of plants and knowledge. He regretted that most of Nicholls's best material, including a prized Lily-of-the-Valley Orchid, *Dendrobium monophyllum*, was stolen on the way to Melbourne, and he took steps to make good the loss.[78]

The Sun-orchid, *Thelymitra aristata* Lindl., growing in symbiotic relationship with *Dendrobium kingianum* Bidw. Photograph by Rev. H. M. R. Rupp at Brunswick Heads, N.S.W., September 1936. (From the Fordham Album by courtesy of Mrs J. Trotter)

H. M. R. Rupp and Mrs Pearl Messmer pose amid the pneumatophores of a veteran mangrove during an excursion from Brunswick Heads in 1936. Note Rupp's large vasculum or metal collecting box for conveying plant specimens back to base. (From a photograph by Mrs Hilda Curtis)

For six years or so thereafter, letters to Fred Fordham were apt to include little quips and what were termed 'doggy-rel' verses for the children and their dog. When one daughter proved a little tardy in acknowledging some of these homespun compositions, there came a reminder:

Helen Fordham, if you linger
     Longer ere to me you write,
I'll just pen a note for Ginger
     Asking him your toes to bite.[79]

The children's responses brought great delight to one who had such a sense of fun. Enclosing 'another crazy production' of his own, Rupp once declared that it was in response to a letter which was 'quite a masterpiece'[80]—a warm tribute to the girls' literary skills.

Late in 1939, Rupp sent Fordham a letter of introduction on behalf of another friend, 'a Kogarah youth', Alick W. Dockrill, an assiduous 'pursuer of the hapless orchid', a 'very nice lad' whose discovery of 'a new *Bulbophyllum* on the McPherson Range'

clearly demonstrated possession of 'the true "orchid eye" '.[81] A. W. Dockrill, who now lives at Atherton, subsequently proved the efficacy of his 'orchid eye' in numerous publications, including papers on *Bulbophyllum* and *Cymbidium* and on taxa of the subtribe Sarcanthinae. His important *Australian Indigenous Orchids* was published in 1969. In 1953, Rupp named the Bird Orchid, *Chiloglottis dockrillii*, in his honour, commenting 'he has earned it. He has collected many admirable orchid specimens for the Herbarium here, and was the discoverer of the big *Prasophyllum uroglossum*.'[82] Like so many other correspondents, Alick Dockrill was to visit Rupp at home to learn a little more about the Orchid Man at first hand.

Another greatly esteemed teacher–correspondent was Trevor Edgar Hunt, whom Rupp dubbed the 'philosopher of Ipswich'.[83] He was apparently introduced in turn by both C. T. White and A. H. Chisholm. In March 1941, White sent to Rupp for identification, 'two small orchid plants', accompanied by flower sketches received from Hunt

Pearl P. Messmer.

Hanna M. Irby

Winifred E Irby

Mae Smales

C.P. Ledward.

W.H. Nicholls

Hilda Curtis

C.T. White

THE . ORCHID FAIRY.

The Orchid Fairy is a minx —
  She's always up to tricks :
She sets you riddles like the Sphinx,
  And puts you in a fix.

She hides her flowers underground
  Or turns them upside down,
Or "inside out" they may be found,
  And then you are "done brown."

But still, she is a Fairy Queen ;
  And so, of course, may do
Surprising things that might have been
  Amiss in me — or you.

H.M.R.Rupp.

A page from young Jean Fordham's autograph book, with a poem by H. M. R. Rupp, and the signatures of some of his colleagues at the orchid camp held at Brunswick Heads, August–September 1936. (By courtesy of Mrs Jean Trotter)

at Ipswich. Rupp tentatively identified them as *Sarcochilus dilatatus* and *S. hillii*. He also remarked that Hunt's district was 'a new one to me'; that he was 'always most grateful at any time to receive specimens of orchids from anywhere', and that he would 'be only too pleased' to assist with determinations. Rupp then had in Queensland at least three active contacts—Mrs Hilda Curtis of Mount Tamborine, and two doctors, Colin Prentice Ledward,[84] the quiet, reserved physician of Burleigh Heads, 'a regular correspondent', and Hugo Flecker,[85] radiologist of Cairns and one of the founders of the North Queensland Naturalists' Club.

Ink quickly flowed. 'Please don't apologise for your letters,' Rupp assured his newly acquired young friend, 'nothing gives me more pleasure than my

Trevor Edgar Hunt (1913–1970), Rupp's 'philosopher of Ipswich' with whom he exchanged nearly 400 letters between March 1941 and September 1955. (Photograph by courtesy of Mrs Betty Mathieson)

orchid correspondence.'[86] Then eighteen months later, 'Do write when you can—I love my orchid letters'.[87] Trevor Hunt proved to be astonishingly responsive. By September 1955, when Rupp sadly announced that he had submitted his 'swan song' to the *Victorian Naturalist* in the form of 'a few Diuris descriptions'[88] and that he was unlikely 'to do any more orchid work', or indeed recover from his current severe bout of cardiac asthma, they had exchanged nearly 400 letters.

Rupp had gratefully informed Alec Chisholm after corresponding with Hunt for nearly five and a half years: 'Many thanks for introducing Trevor Hunt. He and I have never met yet, but we have become close pals, and our correspondence paper and envelopes would set up a young stationer in business.'[89] They finally met early in 1951, nearly four years after their joint paper reviewing the genus *Dendrobium* in Australia had been published by the Linnean Society of New South Wales, and two years after their review of the genus *Bulbophyllum* had been published by the Royal Society of Queensland.

Other teachers with whom Rupp maintained a longer or shorter botanical correspondence included 'the pedagogue from the Genoa River near Cape Howe', Norman Arthur Wakefield,[90] whose early interests embraced ferns and orchids. Rupp noted in February 1942 that Wakefield was in the army, and in December 1943 that he was 'somewhere in New Guinea', where, Rupp added ruefully, 'he had seen plenty of orchids but is not allowed to send any away'.[91] Another was Warren W. Abell, a Queensland teacher who during the 1940s and 1950s was stationed at such outposts as Gadgarra via Yungabarra, Durong via Tingoora and Kybong before being transferred to Ingham. Rupp unerringly predicted, 'I think he'll be a gold mine...when he gets going properly'.[92] The celebrated Thomas Stephen Hart was another pedagogical correspondent,[93] with whom Rupp exchanged information on the genera *Caladenia* and *Pterostylis*. C. Keith Ingram (now of Mt Tomah, N.S.W.), a keen teacher–naturalist who became an inspector, was also a correspondent who, like Rupp, numbered the Althofers of Dripstone among his friends. Miss Kathleen McIlrath, art teacher at Murwillumbah High School (and now living at Banora Point) was another. She in turn introduced 'a youthful enthusiast from Murwillumbah', John Young, who made bold to call upon Rupp at Northbridge in August 1947, to present a most intriguing plant, 'an indisputable hybrid *Glossodia major* × *Caladenia caerulea*'. Young's immediate reward was to be treated to a botany excursion 'behind Roseville golf-links' where 'the chief excitement centred in our discovery in an easily accessible cliff crevice, of a colony of *Rimacola*, with a number of budding racemes'.[94]

John Young subsequently introduced another teacher–botanist, John Leaver, who contacted Rupp in 1952 from his school on the Upper Macleay River. Yet another teacher–correspondent was Lionel Gilbert, while he was stationed at Nabiac, near Taree, N.S.W., conveniently close to rainforest, eucalyptus forest and coastal heathland.[95] At the time of Rupp's death, Leaver and Gilbert were both teaching at Wauchope.

Rupp was in contact with several more correspondents during the 1940s. Five were sharp-eyed ladies whose botanical and artistic skills were very quickly appreciated and commended. Returning from 'an interesting excursion' to Pulbah Island, Lake Macquarie, in October 1941, Rupp was pleased and surprised to hear again from an earlier correspondent, Joyce Scrivener. Early in 1926, when about sixteen and already a keen student of botany, she was advised by her mentor Edward Jesse Gregson of Mt Wilson, to contact Rupp about some of her discoveries, including the Horned Orchid, *Orthoceras strictum*, and Tongue Orchid, *Cryptostylis subulata*. Now, over fifteen years later, she issued an invitation for the Rupps to visit the family home, the guesthouse Taihoa at Mt Irvine in the Blue Mountains. Rupp gladly responded, and spent 'a wonderful week' with the family in November 1941.[96] He found the Scriveners 'such kind and jolly people',[97] with Joyce and her younger sister Gwen remarkably well acquainted with local bush tracks and orchid haunts. A spirited and mutually informative and enjoyable correspondence naturally ensued.[98]

Both sisters were ardent bushwalkers and naturalists, but while Joyce was especially attracted to the botanical characteristics of orchids, Gwen, a competent artist, found the painting of orchids to be more rewarding then pressing them. Accordingly they made a very effective team. When they did decide to form a herbarium, Rupp promptly sent specimens of fifty-nine species from his 'reserves'.[99] Quite a lot of correspondence was devoted to the genus *Prasophyllum*, the Leek Orchids, more especially to the 'little chaps of the section...Genoplesium' which Rupp considered 'veritable *imps*, delighting in discomfiting the silly humans who presume to try and put them in their places'.[100] For a while it seemed that one form might be published as 'Prasophyllum scrivenerae Rupp', and there was a flurry of exchanges of letters, specimens and Gwen Scrivener's drawings, which Rupp described as 'splendid', admitting that he could not 'get within coo-ee of their excellence'.[101]

The Scriveners further demonstrated their generosity by sending at times of wartime shortages, welcome hampers of fruit, vegetables, walnuts, chestnuts and even (at Rupp's request) some bark from the Sassafras tree, Doryphora sassafras, so that he could experiment with its alleged tea-making qualities. There were reciprocal visits, and in the autumns of 1943 and 1944 Rupp returned to Mt Irvine, the second time accompanied by his wife and son, Arthur, then convalescing after war service. In spring 1944 Joyce Scrivener married Dr Archibald C. Telfer, another orchid enthusiast who was also one of Rupp's correspondents.

Through the interest of Dr Colin Ledward of Burleigh Heads,[102] Rupp was pleased to record in October 1945 that he had made the aquaintance of two more sisters, Dorothy and Jean Gemmell,[103] of Glen Aplin, which he noted with geological satisfaction was situated 'on the granite belt above Stanthorpe'.[104] Curiously, again one sister, Jean, was principally a collector and classifier, and the other an artist. With some knowledge of the Glen Innes area, Rupp was very interested to learn about the orchids to be found in the tableland country to the north, more especially species of Diuris and Pterostylis. Several of the Gemmell sisters' discoveries proved to be new records for Queensland, and for the next six years or more the despatch of specimens became quite brisk. Rupp noted with warm approval that 'the Gemmell specimens usually reach me perfect, packed in a peculiar spongy lichen of the granite'.[105]

Both sisters visited the Rupps at Northbridge,[106] where there was much animated conversation about their orchid discoveries and their mutual botanical friends—Hilda Curtis, George and Peter Althofer, Dr Colin Ledward and Miss Mae Smales. Jean Gemmell maintained her records and concluded that at least 104 orchid species grew on the Stanthorpe granite belt. Rupp was delighted by such investigations and considered Dorothy Gemmell's 'coloured drawings of flowers' were 'almost too good to be true!'[107] He would have been further delighted to know that some were ultimately published, with her sister's text, in 1988.

Publication of Rupp's 'Orchid Flora of the Blue Mountains' in the Australian Orchid Review in June 1947 elicited an immediate response from Miss Isobel Bowden of Woodford,[108] and an enthusiastic

correspondence ensued. Miss Bowden noted in her diary:

The most interesting thing at present is a list of orchids of the Blue Mts in the Orchid Review prepared by Rev. Rupp. I am surprised I have so many kinds. Tomorrow I will have another hunt . . .

I think I will keep a separate orchid book for the Blue Mts and see how many of Mr Rupp's list I can find.[109]

It was soon apparent that this lady with 'the real orchid eye' shared Rupp's love for 'the pygmy Prassies', those small-flowered Prasophyllums which had so intrigued him at Mt Irvine. At the outset, Rupp declared: 'I am only too pleased to do anything I can to help folks who are keen on studying our Orchids. I have rather a large circle of correspondents, and that is really why I usually answer letters very promptly; otherwise they are apt to be overlooked.'[110]

By then, Rupp had already discussed Miss Bowden's initial letter with Mrs Pearl Messmer, and it was decided that a spring excursion to Woodford should be held as soon as possible. In the event, the postwar petrol shortage thwarted the first attempt. Miss Bowden called on Rupp at Northbridge, and at the Herbarium in the Sydney Botanic Gardens, where she is still remembered as an intrepid investigator.

Rupp later revived a topic of mutual interest: 'I hope the impish little "Prassies" will turn up in abundance sooner or later. They seem to be temperamental little folks, and it is difficult to foresee when they will appear; one just has to watch.'[111] Miss Bowden did watch, diligently and effectively. Late in January 1948, Rupp declared, after receipt of Prasophyllum specimens, 'You are certainly beating us all with these little imps this season!'[112] A few days later, he wrote, 'you are putting us all to rout in the matter of the pygmy Prasophylls!' and concluded, 'Please go ahead; we shall have P. Bowdenae yet! Congratulations.'[113] There was not long to wait. Within three weeks, Rupp had telephoned Miss Bowden to advise that one of her discoveries would bear her name.[114] As Rupp had predicted, Miss Bowden had proved 'a perfect goldmine'.[115] More Prasophyllums continued to arrive at Northbridge, neatly packed in damp Maiden-hair Fern inside matchboxes or paint-boxes, and Rupp sent more 'cordial congratulations'.

It was soon established that Miss Bowden had discovered five new species of Prasophyllum. Rupp considered her 'a marvel'[116]; she was 'the blessed Miss Bowden', 'Diana the Huntress, alias Miss

Bowden',[117] and 'the indefatigable Miss Bowden' who had sent specimens of no fewer than twenty species of *Prasophyllum*, including the five hitherto undescribed. How 'orchids must tremble when she is on the warpath', Rupp chuckled, as meticulously packed parcels of rarities arrived in the mail.[118]

In October 1952, shortly before Rupp's eightieth birthday, his friends Norman and Marjory Loader took him for 'a lovely time' in the lower Blue Mountains, where Miss Bowden showed them 'over her orchid preserves from Springwood to Woodford'.[119] Miss Bowden, who also numbered Mrs Messmer and Alick Dockrill among her friends, followed some of Rupp's old orchid trails to Paterson where she visited his former neighbour, Mrs Hunter, during a holiday trip in September 1953.

Isobel Bowden had her own views on Rupp's predilection for writing letters and publishing papers: 'the fine old man...was anxious to put into print all the ideas as well as knowledge which he had, for as he said, others could then work on it—even if sometimes he proved to be wrong'. With regard to the Prasophyllums:

Mr Rupp was not interested in 'splitting' species but, as he said, he had observed a lot and if he did not try to communicate what he had seen, the experiences would die with him and he could not be of use to future botanists. He did believe however, that there were more undescribed species in the pigmy group.[120]

Accordingly, with the warm encouragement of the then director, Mr Knowles Mair, Miss Bowden offered her correspondence from Rupp and from W. H. Nicholls to the Royal Botanic Gardens, Sydney, in 1968,[121] and students of orchidology will long be thankful that she did so.

Rupp's most important professional contact in the 1940s was made with James H. Willis,[122] who joined the National Herbarium of Victoria in 1939 as a botanist. Rupp had contacted him by early August 1942, and for almost fourteen years relished corresponding with such a highly competent professional botanist who was not only a similarly accomplished letter-writer with wide interests, but also a superb calligrapher.

They exchanged nearly 200 letters, relating chiefly to orchidological matters and to the publication of papers, especially during Dr Willis's term as editor of the *Victorian Naturalist*. News of the activities and opinions of botanical associates and other correspondents was also included. Among those mentioned were Mary Atkinson and Neil Burrows of Tasmania; Mervinia Masterman and Harold

Goldsack of South Australia; Rica Erickson of Western Australia; Trevor Hunt of Queensland; Edith Coleman, Norman Wakefield, and of course, W. H. Nicholls, of Victoria. Even the author of this script unknowingly caused a sigh: 'Oh dear, here's a letter from Lionel Gilbert, with a long MS on Ferns and Orchids in the Rain Forests, which (the Orchid part) I'm pleased to peruse and comment on'.[123]

Six years after their correspondence began, Rupp and Willis met in Melbourne in the spring of 1948. Rupp noted, 'at last I met Willis, a prince of good fellows'.[124] Thereafter he tended to write even more enthusiastically, confessing on one occasion: 'As for me, I'm rather like Tennyson's brook, except that I write instead of babbling',[125] and later, 'I write because I am a perfectly incorrigible letter-writer; but anyhow I'm glad to know that my screeds are welcomed'.[126] Fortunately for the biographer, the 'screeds' were not merely welcomed; they were also treasured and preserved.

Other wartime and immediately postwar contacts included the surveyor–botanist William Hunter of East Gippsland[127]; Dr H. D. Gordon and Dr Winifred Curtis of the Botany Department at the University of Tasmania[128]; Ernest Todd[129] (now of Toronto, N.S.W.), who shared an interest in the orchids of the coalfields, and his friend Melville Walter Nichols, a Kurri Kurri shop assistant with whom Rupp exchanged specimens and at least ten letters between 1940 and 1956.[130] Yet another was 'a Brisbane lad' with the RAAF in New Guinea and later at Morotai, Geoffrey Piper, who at one stage sent material 'pretty well every week'.[131] This was sometimes accompanied by 'colour sketches' executed by Piper's 'mate, who is extraordinarily gifted with brush and colour... His name is Winchester'.[132] There was Donald Savage, an eighteen-year-old friend of T. E. Hunt, described by Rupp as 'a youngster both keen and intelligent, using every opportunity to explore and hunt for orchids',[133] and E. J. Smith of Kalbar, via Ipswich, also sent material.[134] F. W. Schmidt, a Sydney wool-buyer, contacted Rupp in July 1948 concerning a *Caladenia* he found near Kurnell, and correspondence continued at least until March 1951.[135] Important additional contacts were made in north Queensland about this time, including A. J. Parker of Proserpine; W. F. Tierney of Cairns; Keith Kennedy of Paluma, near Townsville[136]; W. W. Mason of Cape Tribulation, as already noticed; and Arnold Johnson of Hambledon, via Cairns, who sent 'a box of orchids' containing two species Rupp had wanted 'for years'.[137]

On 27 July 1945 Rupp was elated by the arrival of 'a very interesting letter' from George William Althofer,[138] proprietor of the Nindethana Australian Plant Nursery at Dripstone, near Wellington, N.S.W. It was a model letter. Unsolicited and unexpected, it proposed a 'theory of flowering cycles' and offered orchid specimens from the local area. While Rupp conceded that a 'theory of flowering cycles may quite possibly prove correct', he was more inclined to believe that 'seasonal conditions probably have a good deal of influence in the matter', but it seemed 'certain...that some orchids "rest" over several seasons and then reappear'. He doubted that this occurred 'regularly' according to any cyclic pattern. As far as orchid specimens were concerned, he would be delighted to receive any from the Wellington district, for it was 'one of the comparatively few areas of the State into which I have never been'.[139] Thus began ten years' correspondence, with an exchange of at least eighty letters, accompanied by a spate of specimens.[140] Rupp joyfully informed Trevor Hunt that Mr Althofer

wants to know if I would like him to send specimens of the orchids of his district. Of course I have replied No, keep your old orchids. (Sez you!) For years I have wished I had a correspondent in one of our western areas where there are mountain ranges; and Wellington should be ideal. He says he knows at least 20 different species.[141]

Wellington *was* ideal. There immediately developed much earnest discussion over the genus *Diuris*, the Double-tail Orchids. Althofer's first shipment of specimens arrived at a busy time, and uncharacteristically Rupp's acknowledgement was delayed. 'I like being bombarded with orchids', he assured his new penfriend, 'but the fire got too hot for me to cope with, averaging 5 packets per diem!'[142] George Althofer kept up the fire, and on 19 September 1945 there came a 'wonderful Diuris collection, which turned up...in first-rate condition'. Rupp immediately 'sent an S.O.S. to Mrs Pearl Messmer' of Lindfield to help him examine this 'most fascinating lot', which provided 'such a feast of good things', including at least nine new species. 'There is much to be learnt about the genus Diuris yet,' Rupp admitted, for some species seemed to hybridize freely in certain areas, whereas elsewhere, hybrids were 'very rare, and the species are clear-cut'.[143]

In October 1945 (when the Gemmell sisters were assiduously collecting yet more *Diuris* specimens near

Rupp's sketches of dissections of *Diuris* flowers from the Central Western Slopes of N.S.W. From the original, in the Marjory Loader Collection, prepared for a paper read to the Linnean Society of N.S.W., 30 June 1948. 1. *D. althoferi* Rupp (= *D. abbreviata* Benth.), 2. *D. cucullata* Rupp (= *D. abbreviata* Benth.), 3. *D. cuneilabris* Rupp (= *D. platichila* FitzG.), 4. *D. paiachila* Rogers (now considered hybrid between *D. pardina* Lindl. and *D. behrii* Schldl.), 5. *D. flavopurpurea* Messm. (= *D. platichila* FitzG.), 6. *D. platichila* FitzG., 7. undetermined, but allied to *D. semilunulata* Messm., 8. 'another undetermined form', 9. *D. aurea* Sm., 10 *D. brevissima* FitzG. ex Nicholls, 11. 'undetermined form from Kerr's Creek', 12. *D. lineata* Messm. (= *D. platichila* FitzG.).

Stanthorpe), George and Peter Althofer visited the Rupps at Northbridge. Thereafter correspondence was vigorously maintained on *Pterostylis, Caladenia, Prasophyllum* and *Diuris*, including *D. althoferi* which Rupp formally described and named.[144] Rupp confided to Dorothy Gemmell:

The genus Diuris is driving me asylum-wards this year!...most of the yellow species seem to vary so much that they run into one another and make identification extremely difficult. I have been receiving a large number from our Central Western Slopes, and have fairly tangled up in them.[145]

This plethora of difficult material was rendered more overwhelming—and to some extent frustrating—when some of it arrived in a crushed or sodden condition, shrivelled or desiccated. Rupp even transitorily felt 'a bit orchid-weary, though not for all the world would I have folks stop sending them'.[146]

The flow of specimens from the Wellington district maintained its exciting, if daunting, momentum. George Althofer began to feel a little diffident, but he was immediately reassured: 'Don't apologize for sending me orchids! Nothing can please me more than the specimens I get from you and other orchidy people. Only I am afraid I cannot always give satisfactory replies to the enquiries that accompany them.'[147] Thereupon, five parcels arrived in seven days, 'with 74 specimens!'[148] In mid-November 1947, Rupp confessed to George Althofer that he was

still rather bewildered among the multiplicity of Diuris forms sent down by you and your brother this season. I have been busy dissecting, pressing, and mounting individual flowers of as many of them as possible... They say one can't have too much of a good thing; but I'm beginning to wonder!

A week later Rupp advised that he had 'between 30 and 40 Diuris flowers dissected and mounted at the National Herbarium, from your collections. Many of them present great difficulties.'[149]

It was hoped to resolve some of these difficulties with the assistance of 'Mrs Messmer...a very acute observer with perhaps as much knowledge of N.S.W. orchids as anyone in the State'. Finally in mid-1955, Rupp, confident that his friend would 'make a job of it', presented his notes on some of the troublesome species, including *Diuris punctata* Smith, to Alick W. Dockrill who competently addressed many of the problems.[150]

In September 1950 Rupp visited the Althofers at Dripstone. Travelling to Wellington by train, he spent a day or two at the brothers' nursery, before his friends Norman and Marjory Loader (then of Castlecrag, now of Dural) drove him back to Northbridge in a very well-laden car. Rupp declared, 'My visit to Nindethana will always be one of the very pleasant memories of a fairly long life'.[151]

Rupp's correspondents were not of course confined to Australia. In addition to contacting

Edwin Daniel Hatch, FLS, with whom Rupp shared a keen interest in the relationship between Australian and New Zealand orchids. Without ever meeting, they produced a major paper on their mutual interest for the Linnean Society of New South Wales in 1945. (Photograph, 1984, from Mr Hatch)

botanists at the Kew Gardens, the British Museum (Natural History), the New York Botanical Gardens and members of other overseas institutions, he enjoyed correspondence with the celebrated Oakes Ames of Harvard University, and with some New Zealand botanists, first H. B. Matthews in 1928, then Miss Lucy Cranwell, and later E. D. Hatch of Auckland. In 1942 Dr H. H. Allan suggested that Hatch refer his specimens of *Pterostylis* and *Prasophyllum* to Rupp for examination and possible identification. Hatch and Rupp shortly produced a joint paper on the relationship between the orchid floras of Australia and New Zealand, and they remained in contact for another twelve years or more.[152]

Another valued New Zealand correspondent was Cedric Smith, who approached Rupp in July 1947 suggesting an exchange of specimens for some Stewart Island material, and a month later, a 'wonderful collection' arrived to invigorate Rupp's long-standing trans-Tasman interests.[153] Rupp continued to 'enrol' additional correspondents even when he was approaching eighty. One was Stanley Goessling St Cloud who contacted him in September 1951 concerning an undescribed species of *Dendrobium*. Rupp considered that St Cloud 'evidently has the "orchid eye" ', and by May 1953 described him as 'my Cairns huntsman'. St Cloud called at the Northbridge home soon after.[154]

Apart from the sheer pleasure provided by correspondence with these, and many other, 'orchidy folks', Rupp was afforded thereby the means of extensive consultation. He had little time for those fiercely independent souls who purported to be scientific without considering any opinions but their own. This attitude tended to lead to the production of a work which 'simply crawls with mistakes that could have been avoided' had the author 'condescended to ask other botanists for help'.[155] Even the revered Baron had offended in this way: 'Mueller was a great botanist, but he was not a good man at collaborating—with Australian botanists at any rate'. Furthermore, 'The dear old Baron...was a fair terror in the matter of nomenclature. If he did not like a name he had no scruples about substituting another in those distant days long before the adoption of an *International Code of Botanical Nomenclature*.'[156]

Thus our effusive Man of Letters was also the cautious Man of Science. As he frequently warned others, pressed material could prove unreliable in the first instance, although essential for subsequent checking and for permanent reference: 'I mistrust determinations worked out from dried specimens only—that is why I am not keen on Bentham.'[157]

There was no satisfactory substitute for fresh material, but 'the difficulty is to get contemporary living specimens from widely-parted areas'.[158] There were other pitfalls—'colour cannot be relied upon',[159] and one had to watch for albino forms and freakish variants, as well as for whole ranges of intermediate forms between two apparently quite distinct species: 'I have learnt...to be very wary about describing a new species from a solitary specimen unless the collector can guarantee that at least several plants were found'.[160]

Edaphic and climatic factors had also to be considered, as well as physical features such as altitude; there were 'good seasons' and 'bad seasons' and the possibilities of hybridization. As for making a specific judgement:

'Lumping' is just as objectionable as 'splitting'; i.e. while it is true that one can cause confusion by making mere variants from a type into unnecessary new species, one can cause equal or worse confusion by dumping into one species a number of forms that differ very conspicuously in important characters. One must try and avoid both errors.[161]

Rupp addressed this fundamental biological problem in a paper contributed to the *Victorian Naturalist*: 'What Constitutes a New Botanical Species?', with, of course, special reference to the family Orchidaceae.[162]

Rupp's advice to Trevor Hunt was clear:

If you are wise you will not aim at the unattainable goal of perfection—it has the nature of a mirage. Bentham's *Flora Australiensis* was hailed as the last word on its vast subject; and look at it now—still a classic and a monumental work, but as full of holes as a cullender. And yet no less indispensable.[163]

As for modifying one's published opinions, 'you'll find that all botanists are liable to publish descriptions which require altering or adding to; it is almost inevitable where material is comparatively scanty'.[164] If one made a blunder, then let it be acknowledged: 'My own worst bloomer was to describe as a new species (*N.Q.Nat.*, Dec. 1935) *Cleisostoma cornutum* which turned out later to be nothing more exciting than *Sarcanthus tridentatus*! The flowers (2) on the first plant received were mutilated, and misled me.'[165]

As suggested, Rupp's letters provide interesting glimpses of his methods of working. When complimented by George Scammell on the quality of the pressed specimens he sent to his friends, Rupp made light of the fact:

As to my pressing, you are the fourth correspondent who has commended me about it, & I am moved to think that the less one bothers about methods of pressing the better! for I have never followed the careful advice given by various authors. I blush to say that except in the case of exceptionally succulent victims, I never even use blotting-paper! Plenty of ordinary newspaper between boards, & a big heavy stone on top, is the only recipe I have. I used to strap the boards sometimes, but I prefer the weight. Except that now and then a poor thing has been squashed a bit too much, I nearly always get satisfaction from this cheap and trouble-free method. The epiphytes of course are difficult subjects, and never look *really* well (I mean those with hard stems and leaves), but I have never seen any that did.[166]

It will be noted that there was no mention of the care taken to *arrange* specimens for pressing in such a way as to show their habit and structure to the best advantage.

Rupp fixed his dried specimens to herbarium sheets with 'very fine strips' of gummed paper so that they might be removed for close examination if necessary, and then replaced. By early May 1926, he had 'just gone in for boxes' for storing his collection. Hitherto he had used 'home-made folding covers' but 'nice strong boxes made at Firth's, Redfern' were procurable for '3/8 per dozen'.[167] Once stored, the plants were protected 'against the inroads of mites' by means of 'naphthalene flakes or crushed moth balls... Some folks object to the odour—personally I don't mind it.'[168]

The detailed examination of withered or withering flowers, or those which had already been pressed and dried, called for special skills. In 1932 Rupp recorded the receipt of Helmet Orchids, *Corybas unguiculatus*, which appeared as a leaf and 'a shrivelled stem with a little blob at the end of it'. Undaunted, he 'removed it and boiled it, and at once it returned to normal shape and form: then a dip into absolute alcohol, and it stiffened out and became manageable, and quite good for pressing'. This little operation immediately extended the known range of this species northward 'by hundreds of miles'.[169]

Dried flowers could be rendered sufficiently soft and pliable for examination or dissection by soaking them 'for at least an hour' in a 'softening solution' comprising 65 per cent water, 20 per cent methylated spirit and 15 per cent glycerine. This solution also preserved flowers 'just as well as formalin or spirits'.[170] Methylated spirit alone 'made delicate flowers so brittle', but a 10 per cent formalin solution was 'the best medium for preserving and sending away flowers'[171] (although not by post). The strength of the formalin and the time of immersion had to be watched carefully, for unless appropriately diluted it 'takes the colour out very quickly'.[172]

Rupp once added a postscript to a letter for George Scammell:

Did I tell you of my 'discovery' re formalin + meth. spirit? You may find it useful. Mrs Coleman sent me our new Corysanthes species (dilatata) from Healesville in July, and being busy I put it in formalin (weak), which preserved the colours. After 48 hours, being still unable to deal with the plants, I decided to sacrifice the colours as the specimens soon go 'soggy' in formalin and can't be handled. So I put them in meth. spir. which hardens but bleaches. No bleaching has occurred except to the leaves—the flowers are still perfect. Have treated a number of Dendrobs the same way. Green seems to bleach, but no other colour. Why?[173]

It would be interesting to read Scammell's chemical explanation, but weak formalin solution can act as a 'colour-fixing' agent at least in some cases, and Rupp was also aware that whereas methylated spirit removed colour, rectified spirit did not have the same effect.[174]

In the case of the more difficult groups, such as the pygmy Prasophyllums, Rupp considered that fresh material was essential: 'I just love the pygmy Prassies—when I get 'em alive and not past maturity',[175] but, as he remarked to W. H. Nicholls, 'I distrust any exposition of the pygmies, my own certainly included, made from boiled-up or softened dry specimens. I... have always found that the softened dry specimens have undergone certain changes which can only *approximately* be rectified by the softening process.'[176]

Once examined, macroscopically and by dissection, plants could be matched with existing descriptions, or if this could not be achieved, a new description might appear to be necessary. However accurate and vivid, whether in technical English or botanical Latin, descriptions of the general habit and intricate morphology of plants were, and still are, greatly enriched by some complementary graphic representation. For orchid enthusiasts, a seemingly unattainable standard had long since been set by R. D. FitzGerald in his monumental folio work, *Australian Orchids*, produced with exquisite lithographic plates (subsequently hand-coloured) by the New South Wales Government Printing Office between 1875 and 1894. Although incomplete, this was the basic illustrated reference work, despite shortcomings which experts like Rogers and Rupp felt moved to indicate by their annotations on many loose plates which still exist.

It was not until 1923 that W. H. Nicholls, the bookbinder, began his self-appointed task of emulating R. D. FitzGerald the surveyor, by setting out to depict every known species of Australian orchid—a number now known to exceed 600. In April 1940 'Nicholls... sent over in 6 huge MS volumes Part I of his magnum opus' for Rupp's perusal and comment. Rupp readily appreciated that the work 'represents years of toil, often under great difficulties, and the coloured plates are delightful to look at', but was apprehensive that 'the yelping pack' of critics might attack the magnificent work on details of colouring and literary style. Some careful editing was both required and merited.[177]

Rupp had tremendous admiration for skilfully executed botanical art—for example, the work of FitzGerald and of his lithographer, Arthur J. Stopps; of W. H. Nicholls ('a very excellent artist' whose project was 'a magnificent effort'[178]); and of Erwin Nubling, Gwen Scrivener, Dorothy Gemmell and George Scammell. Though he made quite competent line drawings, Rupp recognized his limitations. He once told J. H. Willis: 'It is well understood by now (or ought to be) that I am no artist—I never had any training whatever.'[179]

One way out of this difficulty was to use a camera. In April 1926 Rupp had 'a splendid lot of photos' taken and sent by Hilda Geissmann, W. H. Nicholls and G. V. Scammell.[180] By then he was himself an experienced amateur photographer. Certainly he was taking photographs on quarter-plate glass negatives in 1905. When, in July 1926, George Scammell raised the matter of plant photography, Rupp replied:

It was out of the question for me to go in for orchid-photography properly. What I do is merely to take ordinary 1/4 pl. pictures now and then for my own use—though they are not much use. I have tried enlarging them with an old magic lantern. . .I should never dream of suggesting my pictures as making any serious contribution to the study of orchids—but they have a certain amount of usefulness in giving general impressions where I am powerless to give more.[181]

Rupp was accordingly overjoyed, and somewhat overcome, by Scammell's offer of a spare camera:

It is very kind of you indeed to offer me that camera: indeed I hardly know what to say about it. Although I used to do a lot of photography, and am fairly good at landscapes and trees (there are a few of mine in Maiden's *Forest Flora*), I really don't know much about it. I started with a Victor No. 1, and had the luck to strike a remarkably good lens. Then about 1911 a friend gave me a Victor No. 6 fitted with an Ensign lens and bulb attachment, and I have used that ever since, and have never used a dark slide camera. But I confess I should like to have a go at some of the orchids my 'orchidery' is promising me this season.[182]

In three days 'the camera and accessories arrived safely' and Rupp, with '*no* experience with other but box cameras' proceeded to master the new technology.[183] Some additional lenses arrived, and Rupp delightedly prepared a vase of Greenhood Orchids, *Pterostylis curta* and *P. nutans* for his first attempts at close-up photography.[184] The efficacy of this new acquisition and Rupp's rapidly acquired skills were amply demonstrated three months later when a profusely illustrated article appeared in the *Sydney Mail*. Rupp used the occasion to express his gratitude once again: 'Did you see my article and pictures in the Sydney Mail? The pictures of course

I owe to you—my old camera would never have done them.'[185]

Fifteen years later, Rupp considered that this camera with its 'Rectimat symmetrical lens' was still doing 'fairly good work and will last me out'. At a foot from an object, and with a supplementary lens fitted, the camera took quarter-plate photographs 'about life-size', many of which illustrated the *Guide to the Orchids of New South Wales* (1930) and contributions to journals and newspapers in the late 1920s and 1930s. Regrettably, wartime shortages of materials virtually terminated this enjoyable pastime.[186] After discharge from the army, Arthur Rupp made three large albums and mounted 650 of his father's photographs taken between 1905 and 1945. During this exercise, Rupp discovered 'such a lot of good negatives of which I can't find any prints' and he yearned for supplies of photographic paper and chemicals once more.[187]

Many years before, George Scammell had compiled an album of photographs and presented it as a gift, not long after sending the camera. Rupp had responded immediately:

I am moved to wonder whether there is some occult power in my 'make-up' which mesmerizes my 'orchidy' friends and impels them to shower good things upon me. . .your delightful album is the latest arrival. It is really too good of you to go to all that trouble. I am most grateful.[188]

Yet another album was used for the preservation of two dozen 'Photographic Studies of New Zealand Orchids by H. B. Matthews, Remuera, Auckland'. Taken 1921–4, these photographs with captions in Rupp's hand chiefly depicted species of *Pterostylis, Corybas, Thelymitra* and *Calochilus*.[189]

Dr R. S. Rogers of Adelaide also sent gifts— Charles Darwin's *Fertilization of Orchids*, George Bentham's *Flora Australiensis*, Vol. VI, 1873 (containing the Orchidaceae) and Maurice Maeterlinck's *Life of the White Ant*.[190] These were followed by Ridley's *Flora of the Malay Peninsula*, and many loose plates from R. D. FitzGerald's *Australian Orchids*. Fred Fordham also sent some parts of FitzGerald's work, as did Dr F. A. Rodway of Nowra (who 'collaborated with me in collecting native plants for over 40 years'). Yet more odd plates came from W. J. Enright of Maitland. Rupp fearlessly revised the nomenclature on many of these, and annotated the descriptive texts. Ultimately, he had quite a number of duplicate plates which he in turn passed on to those who had helped with his work on orchids during retirement—the Scrivener sisters and Mrs Marjory Loader.[191]

Another treasured gift was received from Dr Bruce Sinclair, the Rupps' physician at Paterson: this was Robert Brown's foundation work *Prodromus Florae Novae Hollandiae et Insulae Van Diemen* (London 1810). This rarity may well have had an interesting provenance, for Dr Sinclair was Baron von Mueller's great-nephew.

Rupp was truly moved by such

kindnesses that come along from all sorts of unexpected places and friends—hard cash can't provide these and the spirit that prompts them.

I was almost staggered last year [1925] when a Maitland man, whom I had only 5 minutes conversation with, sent me out his Fitzgerald (2 parts missing) with a message that I could 'look after it' for him as long as I was here![192]

Within his means Rupp, too, was both reciprocally and spontaneously generous, with his time, knowledge and advice, photographs and sketches, dried specimens, living plants, scientific papers and hospitality. His more affluent friends readily responded, knowing that any gift or other assistance would be greatly appreciated and fully utilized.

Although botanical matters necessarily determined the purpose and content of most of his correspondence, it was inevitable that the Man of Letters and of Science would reveal more about himself than his methods of working and botanical judgements. There were also revelations of personal views and beliefs and of certain character traits, all depending of course upon his relationship with the correspondent. Some correspondence was maintained at the strictly botanical level, but sometimes, as in periods of personal anxiety, or of national, industrial, social or political crisis, there was a virtual outpouring of the soul, and such letters assume a personal and historical, as well as scientific, significance.

Normally jovial and enthusiastic, Rupp experienced many times of stress and depression, which in the spring of 1937, even affected his letter-writing:

I am metamorphosing from an A1 correspondent to a Zx one. I used to love letter-writing—now I have to force myself to it, and am always in arrears. Signs of the approach of old age I suppose...I don't seem to be able to get into the bush. I have lost all my old liking for solitary rambling.[193]

And a little later: 'I used to have the highest of

reputations as a correspondent—but I am afraid I've lost it'.[194] It was then that the fortnight in Tasmania proved so restorative, and the customary vitality returned.

Rupp's long experience of living in the country and his special interest in the bush made him very concerned about the seasons and weather conditions. 'There is *nothing* in the bush here,' he wrote from Woy Woy in May 1935, 'we have had no rain (beyond 2 showers) since February and everything is parched up.'[195] Three months later he continued the theme: 'The weather here continues the policy of frightfulness without abatement. I have never experienced such a season in my 30 years of N.S.W.'[196]

Rupp was particularly distressed by extremes of temperature, by winds (hot or cold) which he declared were 'very bad for arthritis',[197] and by drought conditions. In October 1936 he remarked that he had not been in good health: 'I never am in droughts—they have an extraordinary effect on my nerves'.[198] Five years later he described a further drought as 'truly most deplorable and desperate; it depresses me very much. One just has to watch treasures dying.'[199] At the end of December 1944, he recorded: 'It has been somewhat a gloomy Christmas over here; the heat, dust, and smoke have been almost unendurable, and the talons of Drought are digging deeper into the very life of N.S.W. every day. There are no indications of any relief.'[200] Then in August 1946, he was again ill at ease: 'All through my adult life, *drought* has had a most depressing and enervating effect on me... I can't get my mind off the dying stock and the destruction of fair bush- and grass-lands. I was never a townie, you know.'[201] In the summer of 1952, it was the same story: 'All my life, drought—real drought—has had physical effects on me, and these react psychologically'.[202] The literary parson even found that, during droughts, the Muse deserted him,[203] and during the war he proclaimed, 'I am far more scared of the drought than I am of the Japs'.[204]

If in lieu of droughts, or worse, in addition to them, there were strikes or political blunders, then inhibitions were really thrown to the winds, more especially as Rupp became older and subject to increasing infirmity. His sympathies for the South Maitland coalfields miners and for those engaged in rural industries have already been noticed. Although these sympathies persisted, Rupp, like many others during and after the two World Wars, modified social and political views which had hitherto been relatively uncomplicated. When there was the promise of 'another big shipping strike' in 1925, he spoke for many by declaring:

We seem to be bent on self-destruction with our incessant class-warfare, & it is so difficult to get at the rights & wrongs of it—there is so much folly & selfishness on both sides. One of these days we shall have the Asiatics swooping down & settling our madness for us whether we like it or not.[205]

And shortly after:

It seems to me our whole Commonwealth is going stark staring mad. I used to be a Labour supporter once—but the whole Labour Movement appears to me to have turned traitor to its own ideal. That ideal originally was, 'Equal right for every man and woman to make the very best of his or her life'. Whereas now 'Labour' shrieks aloud to the rest of the world 'What's yours is mine & what's mine is my own'. Labour has sold its soul to satisfy its greed—& it isn't satisfied & never will be. . . But there—let's talk orchids![206]

As the Depression loomed, 'People growl at "the ways of Providence"'—but I guess Providence growls about our wastefulness. If we spent some of the millions we throw away on useless M.P.'s and wildcat State undertakings, on elementary water conservation, we could smile at droughts.'[207] During the Depression, the rector of Weston considered:

We Australians seem to have a genius for misgovernment. It is appalling that a young country like ours should be in the condition it is in today, & though general extravagance is partly the cause, to my mind misgovernment is mainly responsible—& we seem determined to go from bad to worse. N.S.W. is in a parlous state, & we are almost on the brink of civil war.[208]

Then in April 1941, Rupp declared: 'I have not one atom of sympathy for the gasworkers (and I'm not usually that way when strikes are on). . . . It's a damned sick world (I make no apology), and we're one of the sickest parts of it.'

He could find neither solace nor solution in the personalities and policies of the political parties of the time.[209] Ten years later he was of the same view, for clearly party was 'so much more important than country'.[210] In fact it was all nearly too much: 'What between drought and strikes and the ugly world-lookout, this old man begins to tire of life sometimes.'[211] Just over a year later he advised: 'No—I'm not under the sod yet—though occasionally I begin to think it wouldn't be a bad place, what with the drought and strikes and disgusting political jobbery'.[212] The passing of seven more years brought little reassurance: 'What with drought and fires and locusts and the foulness of political and civic life, I guess N.S.W. is a good place to be out of these times'.[213]

Thus Rupp found that, after considerable wartime anxiety, peace brought only temporary joy and relief. At the beginning of 1946 he considered that whatever was abandoned, the vegetables in the garden had to be 'kept going' for 'only Labour Councillors and Communists and other bloated aristocrats can afford to buy them at the villainous prices prevailing'.[214] Towards the end of the year, it seemed that yet again:

Sydney is in for a very unhappy Christmas I'm afraid; indeed it looks as if we are in for something like civil war. The bitterness of feeling engendered by the lawless and cruel strike of the gasworkers is growing dangerous, and the spineless acquiescence of both Federal and State Governments is feeding the flames. It seems to me that we really have no government at all—mob rule and anarchy are the order of the day. I can't see any rift in the clouds.[215]

Then, in January 1947:

Milk rationing starts tomorrow. Life is becoming rather burdensome for ancients like myself; gas strikes, meat strikes, transport strikes, any darn sort of strikes the Commos choose to work up; a puppet Government that dares not lift a hand against their lawlessness. . . and besides all these ills, drought, drought, drought.[216]

A month later, a general coal strike was 'promised in a few days' and '30,000 tons of spuds from Tasmania' were rotting on the wharves because no one would move them, while millions were starving. It 'gets my goat' fumed the 'very unconventional parson',[217] as he continued to wonder about benefits derived from the sacrifices made during the war by his son and the sons of others.

Rupp thereby came to bemoan the fact that although he had 'never been a conservative', he was loath to support any political party. On the one hand there was 'the tyranny of the Labour Unions'[218] and his disillusionment 'with the Chifley crowd', and on the other, there were the Liberal and Country Parties, forever 'cutting each other's throats' so that he had 'not much time' for them either.[219] Thus, declared Rupp, he felt 'like crying with Tennyson':

Ah God, for a man with a head, heart, hand
Like some of the simple great ones gone
For ever and ever by!
One still strong man in a blatant land,
Whatever they call him, what care I?

Aristocrat, democrat, autocrat, one
Who can rule, and dare not lie![220]

Quite early in his correspondence with George Althofer, Rupp exploded:

'Dark days', literally and metaphorically, seem to lie ahead of this State. I have been a 'barracker' for Labour for years, but I'm fed up. No capitalist ever showed more indifference for the public welfare than the gang running this country at present. By which remark you may perceive that I'm not given to concealing my views![221]

A few months later, more 'views' were forthrightly expounded:

strikes, here, there and everywhere are making Australian life pretty miserable for all folks who have the brains to see beyond tin hares and beer. We are reducing democracy to a Gilbertian farce, which can be at the same time a dire tragedy. Always, all through history, the prostitution of democracy to selfish cliques has ended in a dictatorship. I'm beginning to grow pessimistic.[222]

When George Althofer mentioned how wonderful it would be if some assistance could be obtained to establish a state-sponsored native orchid garden, Rupp again declared pessimism. He felt certain that governments, 'with their materialistic outlook and their contempt for anything except increasing facilities for ''workers'' to do as little work as possible and to have plenty of cash for horses and dogs and beer, would never help such a project'.[223]

It was certainly widely held that politicians and unionists did have a case to answer. In 1949 Rupp, like many other citizens, was still dragging home dead branches from any nearby bush to provide fuel for household warmth and backyard cooking fires. Even these primitive tasks were made arduous during the flood rains of winter that year, while 'the rulers of our country (i.e. the Miners' Federation)' apparently called the tune. 'I'm not a model of patience and cheerfulness,' Rupp conceded.[224] Furthermore, the gas and power restrictions 'have driven me ''stony broke'' ', he noted, but finally in July the situation was somewhat relieved by the arrival of a load of firewood and the donation of a Primus stove, an item then 'unprocurable except at blackmarket prices'.[225]

Concerned about his wife's indifferent health, and often plagued by rheumatoid arthritis (for which he took tar-water, prepared from 'genuine Stockholm tar' obtained from Washington H. Soul Pattison & Co. for sixpence an ounce), Rupp felt frequently

'jumpy and grumpy' thanks to 'the weather and the coal-miners'. He was still alarmed about the state of the country in the early 1950s: 'There isn't much news else, beyond the usual strikes, blackouts, and gas and power restrictions,' he gloomily told Trevor Hunt; in fact, 'between Communism, drought, and infernal heat, what hell the world is today'. Even 'business morality seems stone dead', with promises followed by inactivity, and if this were not enough, he was shortly afterwards afflicted by gout. 'Shades of our great grandfathers,' he sighed.[226] He had been happier with the miners of Weston in the depths of the Depression.

There were other glimpses of periods of despondency. In the autumn of 1946 Rupp exclaimed in mock despair: 'Why was I ever fool enough to encourage folks to study orchids?'[227] Early in summer he pronounced 'my brain is addled with all the orchids stuffed into it'.[228] Before summer was over he felt that 'It's high time you younger blokes took over and put me on the shelf!'[229] Nevertheless he worked on, even then (March 1947) purchasing a second-hand Remington typewriter for £10 to replace the machine his son Arthur had 'repossessed' after lending it on enlistment in 1940.[230]

The Man of Science also made the surprising admission: 'I'm colour-blind in a queer sort of way—no one appreciates colour more, but I can't distinguish between certain colours unless they are side by side, and not always even then'.[231] This meant that often he could not 'distinguish between dark greens, reds and browns easily'.[232] He further confessed to knowing 'nothing of insects' and to having 'an ingrained repulsion for them—can't help it'.[233] Apparently even the assurances of O. Sargent, K. C. McKeown, Mrs Edith Coleman, Mrs Rica Erickson and Fred Fordham that some flies, bees, wasps, gnats and other insects play crucial roles in the pollination of certain orchids were not sufficient to modify this aversion.

Rupp had no discernible prejudices as far as his beloved orchids were concerned. He once entreated Fred Fordham: 'Please *never* suppose that any orchid you find is too commonplace for me!'[234] Naturally there were some declared 'favourites'. The pygmy Prasophyllums, Rupp's 'imps', retained an almost endearing appeal as a group, while some species won special admiration, for the Man of Science never lost his aesthetic sense:

Sarcochilus fitzgeraldii F. Muell., the Ravine Orchid of northern New South Wales and southern Queensland. 'I think it is my favourite of all Australian orchids I know.'[235]

Dendrobium tofftii Bail. of North Queensland, 'is a magnificent thing, almost the finest of all our Dendrobes'.[236]

Dendrobium moorei F. Muell. of Lord Howe Island, has 'absolutely the whitest flower I know'.[237]

Dipodium punctatum (Sm.) R.Br., the Hyacinth Orchid, Rupp noted, was especially remarkable for its 'amazing range' from southern Tasmania to the hinterland of Cairns.[238]

In 1946 Rupp considered that *Disa uniflora* Bergius, of South Africa, was 'the loveliest orchid in the world',[239] but seven years later, with the discovery in New Guinea by Captain Neptune Blood of *Dendrobium ostrinoglossum* Rupp, he modified this view. 'I do not think there can be a lovelier orchid in the world,' he wrote, adding that it had been 'naughtily suggested I should name it after the discoverer!'[240]

Rupp had little scientific or horticultural interest in the ostentatious and often exquisitely beautiful blooms exhibited at orchid shows. Largely derived from the painstaking hybridization of striking exotic species, these orchids attracted Rupp's attention as plants of great beauty, but no more. This prejudice did not necessarily extend to those who grew these plants. Rupp had an especially high regard for George Hermon Slade, 'an enthusiastic Sydney orchid grower', and for Murray Moodie, a long-standing friend who gave 'a square deal' at his nursery. He accompanied Rupp on profitable botanical excursions to such places as Brookvale and Jannali, where once they spent a whole day in 'a dampish patch—perhaps 50 acres in area' to discover twenty-three species of orchids.[241] In the spring of 1944, however, Rupp felt that the interests and affairs of the Orchid Society were becoming unduly dominated by commercial growers, and he submitted his resignation 'on the ground that I am a botanist and not a cultivator of exotic orchids'. To its credit, the council of the society responded by electing him an Honorary Life Member—'Coals of fire on my head' sighed the remorseful recipient of the honour.[242]

The Man of Letters once delighted in describing 'rich treasures of orchids' where one could pick great bunches and yet leave the masses of blooms apparently untouched. But the Man of Science later despaired, for 'sheep and orchids don't combine—at least not in a way favourable to the orchids'. He lived to see such precious 'treasuries' destroyed 'on land which has now been orchid-less for probably 50 years'.[243] Consequently he became wary: 'I am not keen on guiding people to exact localities—orchids are disappearing fast enough without that'.[244]

In publishing records of distributions, regional terms such as 'Blue Mountains' or 'Atherton Tableland' were sufficient. He came to deplore 'development' projects which in their consuming sprawl obliterated the habitats of prized species, and to regard the preservation of the indigenous flora as a patriotic duty. Accordingly he even remonstrated with (at least) one man whom he found chopping out 'a nice clump of the beautiful Dwarf Apple down at the bottom of his garden...he "doesn't want any of that bush rubbish"'. A good Australian—I don't think.'[245]

As indicated in the last chapter, the Man of Letters and Man of Science was foremost the Man of God, and this role made its own epistolary demands. For forty years, there were 'godly folks' as well as 'orchidy folks' to be kept informed, and 'administrative folks' to be kept satisfied, not necessarily on strictly godly matters, nor indeed in strictly godly terms. It seems clear that Rupp enjoyed a pleasant, friendly, if deferential, relationship with most of the bishops under whose jurisdiction he worked—Samuel Thornton of Ballarat; Henry Edward Cooper, first of Ballarat, then of Grafton and Armidale, and his coadjutor, Cecil Druitt; and successive bishops of Newcastle, Reginald Stephen (a Geelong Grammar and Trinity College alumnus, and former bishop of Tasmania), George Merrick Long (another 'Trinity man' whose sudden death affected Rupp greatly) and Francis de Witt Batty.

Rupp apparently corresponded with his Fathers in God and their officers frequently and fluently, sometimes with an even breezy shrewdness, as in December 1930 when he wrote to the newly elected Bishop Batty:

My dear Bishop,
                    I hope the appearance of another letter will not incline you to think 'this man may well become a nuisance'.[246]

As a young clergyman, Rupp had been quickly introduced to the mysteries of routine forms and letters associated with parish returns, accounts and assessments, the nomination and appointment of church officers such as parish councillors, lay readers, and trustees of church lands, buildings and cemeteries. There were surrogate's declarations to be made in accordance with the procedures for celebrating marriages, various undertakings relating to synod decisions, diocesan ordinances and to canonical obedience. In Rupp's case, there was a decade of intermittent but frank correspondence over clergy insurance and superannuation funds after he was 'taken aback' by the advice in 1905 that he had been 'assessed' at £6 10*s*. 0*d*. a year for each fund, payable half-yearly from his meagre stipend.

There were other professional calls on his facile pen. Matters concerning land acquisition, use or disposal, and proposals to construct new church buildings demanded particular attention while Rupp was serving at Beeac, Copmanhurst, Barraba, Paterson, Woy Woy and Raymond Terrace. Faculties from the bishop had to be obtained before installing items of use or of ornament in existing churches or in new buildings, as at Beeac (near Colac), Bereen (near Barraba), Martin's Creek (near Paterson) and Ocean Beach (near Woy Woy). The bishop had also to be contacted regarding special visits for dedications, anniversaries, confirmations and parish missions, and he expected to be advised if parish problems or personal difficulties arose, and when holidays were planned. It is interesting to note that several of the prompt replies Rupp received from Bishop Batty, for example, were composed, typed and signed by the bishop himself.

When Rupp resented some policy or action which he considered was unjust, unwise or ill-advised, he promptly said so, as in the tussles over contributions to provident funds, and when his effort to help a needy family by selling some land was thwarted by a decision he found 'beyond. . .comprehension'. As indicated, he suffered great distress during the Depression. Nearing crisis point, he wrote very directly to Archdeacon Henry A. Woodd, vicar-general, declaring that he had served the Church

for 33 years, and whatever my faults and mistakes, there is nothing in my record that I have cause to be ashamed of, except in the private and personal sense in which every honest man is ashamed of his shortcomings. Yet after all these years, and after a slight breakdown which I couldn't help, the Church says I must go and make shift for myself until 'something turns up'. . . Anyone would think I was under a ban for something dark and dreadful: one or two laymen have said so jocularly, but it isn't humorous to me. I can't go on like this indefinitely: my faith in God doesn't waver, but my faith in the Church's sense of a fair thing is another matter. . . I felt I ought to let you know how I feel.

Nevertheless, Rupp had 'no hard feelings about the *diocese of Newcastle*' for, 'I know how things are, and

I quite believe that the Bishop is doing his best to try and fit me in. But try and put yourself in my place.'[247]

Between 1924 and 1939, Rupp despatched over 250 letters in his inimitable script to the Newcastle diocesan authorities, advising, responding, requesting, suggesting, enquiring and chiding.[248] There is also evidence that he penned at least 100 letters to the Grafton and Armidale diocesan officials between 1904 and 1915.[249] However the ecclesiastical administrators may have regarded the Rev. H. M. R. Rupp, they could never have supposed that he was reluctant to keep in touch.

Although erudite scholars of Trinity once 'lamented the fact' that Rupp 'lacked the classical mind', his later achievements could well have reminded them of some words of the poet Juvenal: '*insanibile scribendi cacoethes*', for Rupp did indeed demonstrate 'an incurable itch for writing'. His writings, carefully compiled according to the then accepted rules of grammar, often featuring pithy Biblical phrases, poetical quotations and charming archaisms such as 'shew', and punctuated by frequent colons and semi-colons, often bore a superficial resemblance to Miles Coverdale's matchless version of the Psalms as preserved in the Book of Common Prayer.

So far, some 1,700 of Rupp's letters to individuals and institutions have been traced, many of them through the commendable efforts of Dr J. H. Willis and Mr Don Blaxell. Still more are believed to exist. Collectively, this correspondence preserves a wealth of personal, ecclesiastical and botanical information of tremendous significance. Its very richness causes profound regret that the letters to such notable naturalists as Dr R. S. Rogers, Dr Hugo Flecker and Mrs Edith Coleman have not been found, for they would be of the utmost importance.

The Man of Letters once declared: 'I love to get letters, and to write when I can. They help to take one's mind off the frailty of the body.'[250] Naturally, he took pleasure in the knowledge that his literary offerings were appreciated: 'I am glad that my letters are so acceptable. I write a lot; I cannot bear to be idle. . .and when I get tired of doing other things I can always turn to my correspondence.'[251]

We can only be grateful that he did.

9 Retirement and Recognition

It came as a great surprise; I had no idea I was even in the running.

—H. M. R. Rupp to G. W. Althofer, 7 June 1955

When Rupp retired from the full-time ministry in May 1939, he was midway through his sixty-seventh year, but seventeen very productive years still lay ahead. The faithful Morris Cowley was sold, and on 5 July the Rupps at last became metropolitan dwellers at 24 Kameruka Road in the pleasant leafy Sydney suburb of Northbridge. Still physically fit and active, Rupp declared the place to be 'a "bonzer" spot—overlooking a bush reserve and a bit of Middle Harbour—yet only 5d. per bus to Wynyard'.[1] Here, in a brick house built on sandstone foundations on a precipitous sandstone site, sixty feet by two hundred, (with '36 rock hewn steps down to the front gate' and 'space for an orchid or two'[2]) Rupp lived very contentedly with his wife, his books, his herbarium and his well-stocked bush house, for over twelve years. There was time to enjoy the nearby bushland at the head of Sailor's Bay, to make excursions to outlying hunting grounds, to continue taxonomic work, and to maintain the astonishing output of correspondence. There were opportunities to attend meetings of the Linnean, Naturalists' and Orchid Societies, sometimes giving the address, and to receive family members and botanical friends.

Rupp continued to attend meetings of the Newcastle Synod, and in fact returned to St Peter's, East Maitland, 'one of the finest churches in Australia',[3] for relief duty early in 1940.[4] In June and at Christmas he returned to the diocese to assist the rector of Terrigal.[5] Although ecclesiastically uncomfortable in the Diocese of Sydney, Rupp established a warm friendship with the Rev. (later

Canon) Ernest Cameron, and frequently assisted at St Luke's, Mosman. He found that this was 'one of the few Sydney churches where I feel at home. I never was an extremist; but from a Sydney Diocese standpoint, I'm one of those wicked High Churchmen'.[6] Rupp also 'felt at home' at St Thomas's, North Sydney, which he regularly attended, and he sometimes assisted (as at Christmas, 1941) at All Saints' Church, Cammeray.

Rupp's removal to Sydney in mid-1939 must have seemed a godsend to another parson's son, Robert Henry Anderson, chief botanist of the Sydney Botanic Gardens. He immediately offered Rupp 'a job at the National Herbarium, overhauling and revising the Orchidaceae'. By December he was hard at work on this project, believing it would take at least two years, for 'no one' had touched the specimens 'for years'. But, he added, 'it's right into my hands; I love it'.[7] Rupp continued this task, if sometimes intermittently, virtually until his death.

In the best bureaucratic tradition, while staff workloads were in effect full, the coffers were demonstrably empty. However, after due consideration, the Department of Agriculture offered Rupp 'the magnificent "honorarium" of £26 per annum' for his services[8]—a slender, if welcome, supplement to the modest (and fixed) Church pension of about £4 a week. Rupp found another helpful sideline in supplying, from his garden, plants of the popular Crucifix Orchid, *Epidendrum obrienianum* (a hybrid between *E. radicans* Pav. and *E. evectum* Hook.f.), for Woolworth's store. In

February 1946 one delivery of '14 dozen plants' obligingly 'sold out immediately',[9] and three years later another such late summer harvest yielded about £8. The chief question posed by the store was, 'When will the next lot be coming?' When the reverend supplier received threepence per plant, the secular retailer received one shilling; when he later received sixpence, the retail price rose to one shilling and sixpence.[10] This was the closest Rupp ever came to being numbered with 'the commercial growers'! Over forty years later, this hardy epiphyte, which Rupp loved, for it 'never stops flowering',[11] is still to be seen at his former home.

When the Second World War began in Europe on 3 September 1939, Rupp was busily constructing a bush house behind his new home. After the news broke, he wrote to Fred Fordham:

And now the mad egomaniac of Germany has dragged us into the maelstrom of war again. Well—it is going to be a horrible business, but I hope we shall not look back until the devil's organization of Nazism is destroyed root and branch. There can never be any real peace until it is. One can't say much just now—but that's my firm conviction.[12]

The war years brought more crises than the memoirs indicate. In April 1940 the Rupps' son-in-law, Ashley ('Don') Cox, husband of Eileen, was seriously injured when thrown from his horse out in the bush, and in the following month, Rupp sadly observed, 'Our little home-world is tumbling about our heads...my son and one son-in-law are enlisting'.[13] Arthur Rupp and Rachel's husband, Leslie Cox, joined the AIF. This close personal association with the war made Rupp even more outspoken about his father's ancestral country: 'We have an appalling task ahead, for Germany must be dealt with so drastically that never again shall she be allowed to produce inhuman monsters of the Hitler–Goering–Ribbentrop breed'.[14] He pleaded for justice to 'all the unfortunates who fled to us from Nazism' and decried the witch-hunts for spies:

If Nazi agents crept in among them here and there, as doubtless they did, it was the laxity of our own authorities that let them in, and why should we make the refugees whom we invited pay for our stupidity? It would be the grossest kind of Hitlerian injustice to do so; and in my opinion doing evil that good may come of it never pays in the long run.[15]

The war effort had to be total, with the 'complete conscription of wealth, industry, and man-power', but even as he wrote, he heard the dark news from France, and added a postscript: 'So the French military clique prefers Nazi slavery for its people: and betrays its ally. My God!'

During July 1940 Rupp was confined to bed with bronchial trouble, but by late August was back at the Herbarium, where he was soon immersed 'deep in African and American orchids'. Towards the end of the year there was further anxiety when Florence Rupp successfully underwent major surgery, and Rupp, 'reputedly a very good cook' with a 'batch of apricot jam' to his recent credit, looked after himself while Arthur was preparing to leave with the 8th Division. Rupp's horror of drought was revived during 1941 and 1942, but worse, Japan entered the war on 7 December 1941 with its devastating attack on Pearl Harbor.

It was immediately appreciated that the Herbarium, so close to the Garden Island naval dockyard, was potentially in grave danger. R. H. Anderson took instant and decisive action, formally advising the Department of Agriculture that irreplaceable type specimens and about 500 rare books should be packed forthwith and sent to the country for safety. Through a supreme effort by the Herbarium staff, the first consignment of '90 herbarium boxes containing approximately 2,600 specimens' left Sydney by rail on Friday, 19 December 1941, bound for the Glen Innes Experiment Farm, where a Herbarium botanist, Douglas Oakley Cross (later a noted Macquarie Street allergist) supervised the storage arrangements.[16]

Shortly after, Rupp recorded:

We are sending away as many type specimens from the Herbarium as possible, to Glen Innes... If a raid comes, the Herbarium will probably suffer, being close to Woolloomooloo... As I had listed all orchid types (about 64) I soon got them all out, and now I am helping to pick out some of the others, which are not listed and so we have to hunt through the boxes... I am not doing anything about my own herbarium, as we are moderately safe here unless the Japs are ever able to secure land bases.[17]

The orchidological parson was a realist, and this essential work continued for a couple of months at least. Early in March 1942, Rupp noted:

We are very busy at the Herbarium picking and packaging representatives of as many species as possible to be sent up country. I have packed 14 boxes of Indomalayan orchids this week; the Australian species have all been despatched. We got all the types away first. I am very glad of this work, for idleness under the present anxiety would be fatal. And I am adjudged too old for the duties of a warden in this hilly suburb. So I have to realize that I *am* in my 70th year.[18]

With the fall of Singapore in February 1942, the Rupps' anxieties centred around the fate of their son. They were, 'like many thousands of others . . . in an agony of apprehension'. Rupp was appalled by the 'inconceivable mismanagement of the Malayan affair', but as far as Australian affairs were concerned, he maintained that the prime minister, John Curtin, was 'courageous and right' and a 'better leader' than others would have likely been. However, Rupp became 'shocked and angry at the evidence of broken morale' as Japanese victories mounted. This made him 'lash out in fury', for

If this is the measure of our stamina, then indeed our brave men . . . are wasting their lives, for we're not worth saving . . . I am disgusted with the Churches, too. They are giving no lead. Talking a lot of platitudinous twaddle, and bidding us to humiliate ourselves and change our hearts with a view to coaxing the Almighty to intervene on our behalf and hunt the Japs away. Oily dope I call it. My idea of tomorrow's Day of Prayer is that we should pray to see ourselves as God sees us, realize our follies and stupidities, and ask for courage to face their consequences, and grace and strength to overcome the powers of evil at whatever cost to ourselves. But I'm a bit of a rebel, and always hated conventional complacency.[19]

Two weeks later, Arthur Rupp arrived home, unannounced, after a remarkable series of escapades. The family was still thankfully rejoicing over this happy reunion when Sydney itself revealed its vulnerability. On the last night of May 1942, Japanese midget submarines penetrated the harbour defences, and Rupp like many Sydney residents, heard the sounds of war at first-hand. Even in recounting this emergency, he could not resist a jocular touch:

The noise of the explosions and guns woke me up; at first I thought it must be a.a.c. [anti-aircraft] fire. I got up and had a look round, but the street lights were still on. Further explosions rattled our windows, and I guessed that something more than practice must be doing; but I did not think of submarines. Mrs Rupp, who is rather deaf, heard nothing; but the storms last night woke her, and she said to me this morning, 'I heard the guns this time; I wonder how many more submarines they got'![20]

Rupp considered that 'Sydney had taken its submarine raid very calmly; though I suspect that for the time being many Manly–Narrabeen folks will take the long bus route in preference to the ferry'.[21]

Having acknowledged this 'little excitement', Rupp declared, 'What a relief it is to turn one's thoughts on to something other than war!'[22] Accordingly there were solitary trips in the bus to seek orchids in the bush at Castle Cove and at Castlecrag, where lamentably, 'swanky stone houses and their grounds' had, within two years, wiped out colonies of three species of *Prasophyllum*; there was an enjoyable outing to East Hills with the Naturalists' Society, and there were even more enjoyable visits to the Scriveners at Mt Irvine, and profitable plant-hunting excursions to such places as Pulbah Island (Lake Macquarie), Oxford Falls, Brookvale, Janalli and the Lane Cove River with fellow enthusiasts— Mr and Mrs E. Bryce, Mrs Pearl Messmer and her son Bruce, Murray Moodie, P. A. Gilbert, A. M. Olsen and Erwin Nubling. There was also a trip to Normanhurst to spend a 'whole day examining literally *volumes* of the most careful and exhaustive drawings and methodical notes' that Rupp had ever seen. These were Erwin Nubling's studies of *Prasophyllum*.[23] In the spring of 1943 came the pleasure of exploring the bush with Murray Moodie's friend, Dr Archibald C. Telfer, who married Joyce Scrivener the following year.

Despite such welcome diversions, the war was never far from Rupp's thoughts: 'What an inexpressibly hateful business it all is! to think of all human energy and wealth all over the world being mobilized for the purpose of slaughtering one another, at the will of a few insanely ambitious homicidal maniacs!'[24] As reports of atrocities in Europe increased, the queasiness Rupp felt about his paternal ancestry became more evident:

I can find no words to express *my* feelings about those inhuman devils of Nazis. There are thousands of good Britishers who, like myself, are partly or even wholly of German origin; and this monstrous brood of fiends out of hell have made it an everlasting reproach to everyone that they should have any blood-connection with the land that produced such unspeakable wretches.[25]

Arthur Rupp was commissioned and posted to New Guinea with the 2/7 Battalion early in 1943. By April he was gravely ill with scrub typhus. After some months in hospital he was invalided back to Australia, and although discharged in autumn 1944, his condition long gave cause for concern. There was also concern over Rupp's own health, for between August and December 1942 he was seriously affected by acute neuritis, a complaint which had been diagnosed some eighteen months earlier.[26] By November 1942 he had lost over thirteen kilograms (to weigh '10 st. 7½—bit of a comedown from 12 st. 9!') and had been admitted to the Royal North Shore Hospital. Here he was 'overwhelmed with kindness' from his family and 'the hospital staff'. Friends also rallied. Mrs Messmer called 'every 2nd or 3rd day, with flowers, fruits, sweets, or books' and the Scrivener sisters sent flannel flowers and boronia, to fill the ward with 'the fragrance of the Mt Irvine bushland'. Together with 'injections and massage' but 'not much medicine', these kind attentions proved therapeutic. By December it was possible to write letters again, to make botanical judgements, and even to return to the Herbarium 'for an hour or two' after an absence of nearly three months, but there had been times when he had felt 'so close to The Other Side' that he 'didn't want to come back'.[27]

Customarily healthy and vigorous, Rupp was appalled to find how 'humiliatingly weak' he had become. It was an effort to hold a pen, or to use the typewriter, and the dissection of specimens was out of the question. There was an embarrassing incident *en route* to the hospital for further injections, when a tram started rather too promptly for his fatigued condition, and he 'fell on to an unfortunate woman's lap'.[28] This weakness was all the more disturbing because of a special project on which he had been engaged, and now for the first time he was feeling rather old and infirm. He was, however, also extremely resilient.

R. H. Anderson had determined upon a new Flora of New South Wales to replace Charles Moore and Ernst Betche's 1893 *Handbook*, which contained R. D. FitzGerald's 'arrangement and description' of the Orchidaceae. In June 1941 Rupp readily accepted the responsibility of producing a successor to his *Guide*—a new work, *The Orchids of New South Wales*, to be the first instalment of the proposed Flora. By September 1941 Rupp was busily engaged on this exciting assignment, and by early August 1943 had produced a script for 'final revision' in which R. H. Anderson and Miss (later Dr) Joyce Vickery participated: 'We are very busy on my MSS now— Mr Anderson and Joyce Vickery criticise and I respond. It's rather good fun—we constantly have little digs at each other all in the best of friendly spirits.'[29] Each had a copy of the script so that the botanical keys and the descriptions of some 240 species could be subjected to scrutiny from different viewpoints. Rupp found his colleagues extremely

critical, true devil's advocates, but also helpful, good-humoured, and 'great people to work with', although 'Anderson is a wee bit too Linneany for my liking now and then'.[30]

At the end of August, Rupp reported: 'Mr Anderson and Joyce Vickery and I are going like scalded cats on my MSS now—the printers are pushing us for more copy. We got the proofs of the first 20 pp. or so yesterday.'[31] Early in October 1943, at the anxious time of Arthur's admission to the 113th Army General Hospital at Concord, Rupp was 'still grinding at the book; but we've reached the glossary and index stage now'.[32] He typed the whole script himself before it went to the Herbarium secretary, Miss Phyllis Goddard, for final preparation.

Yet, however clear the script or sound the botany, wartime shortages and restrictions combined with official indifference hardly provided the most congenial conditions for such a project. Anderson had hoped that the book would appear as a publication of the National Herbarium under the full auspices of the government, but Rupp felt that neither Anderson nor Miss Vickery saw much hope of this, predicting that 'red tape and meddlesomeness' would obstruct production altogether. In any case, he considered that the government 'wouldn't waste money on piffling botanical works'. Accordingly, 'Miss Vickery, who was equally keen about it, guaranteed the cost of production minus the illustrations, from her private means, hoping to be repaid by sales'.[33]

Apart from three of Rupp's own sketches, George Vance Scammell provided all of the 'exquisite line drawings', which, in Rupp's view, 'could not have been bettered by anyone'. At one stage it was suggested that illustrations might be omitted altogether, to which Rupp replied, 'no illustrations, no manuscript'. Friends and fellow enthusiasts rallied, and their generosity was put on record in the book.[34] One supporter, Dr Archibald Telfer, sought information on how the publication would be financed, when the sum of £25 had still to be raised to produce the line-drawing blocks for the modest number of '23 plates selected'. Rupp explained the position and Telfer was promptly thanked for 'a generous contribution' of £10.[35] George Scammell donated the same amount to help ensure that his own drawings would be used.

There were irritating delays, and increases in costs at an economic level which seems strangely remote from our own experience. The princely sum of £58 for the illustration blocks and binding virtually put the project at risk. The Orchid Society was especially helpful, and the problem of institutional association with the book was nicely resolved by the wording of the imprint: 'Issued from the National Herbarium, Sydney, as a part of the Flora of New South Wales'.

By early November 1943, Rupp understood that the book was 'imminent'; a month later, he expected it in about a fortnight. Although the titlepage was dated 'December 1943', Rupp still had not seen the book at the end of January 1944. It was, however, available by 27 February, and to the author's delight, 'well received', with '700 copies on order already'.[36] By 20 March, 710 copies had been sold—in fact, sales became 'embarrassing... we never anticipated such a demand.'[37]

Rupp considered that 'whatever may be the verdict on the contents, the book is admirably produced, and reflects great credit... upon the Australasian Medical Publishing Co.' of Glebe.[38] It was produced in two forms—in paper wrappers at six shillings and hard-bound at nine shillings, all bearing Rupp's own cover design depicting *Dendrobium falcorostrum*, *Pterostylis furcillata*, *Diuris venosa*, *Bulbophyllum weinthalii* and *Corybas fordhamii*. This information was inadvertently omitted from the book itself.[39]

With his full understanding and appreciation of the work involved, R. H. Anderson did not believe that the author should have to wait unduly for some reward. Accordingly, at Christmas time 1943, when the book's appearance was expected daily, Rupp received

a pleasant surprise at the Herbarium... in the form of a cheque 'on a/c' in connection with the book. They thought I should have *something* for the actual time and labour bestowed on it—and I can't say that I disagree with them! In addition I am to have a royalty of 6d on every copy sold. So if I live to see 2,000 copies sold, I'll be rolling in wealth to the tune of about £50. It won't pay as well as 'The Headless Horseman of the Burning Mountain' etc., but still—I didn't do it for what I could make out of it.[40]

In May 1944, Rupp joyfully reported: 'The book is selling like hot cakes. I've had royalties on 900 copies, and we can't fill orders. The printers hope to give us 500 more this month.'[41] There were even rebukes 'for not ordering more than 1,000 copies' in the initial print run.[42]

At the book's first appearance, Rupp frankly declared that 'Much of the credit is really due to Miss Joyce Vickery... indeed I wanted her name to go on the title-page as collaborator, but she wouldn't

consent. I don't hesitate to say it would have been a much less accurate job without her assistance.'[43]

In due course, Miss Vickery was apparently recouped for her generous underwriting of the project, and Rupp received 'about £40 in royalties'.[44] Relatively more lavish were the rewards afforded by several enthusiastic reviews, among which those in the *Australian Orchid Review* (June 1944), the *Journal of the New York Botanical Garden* (March 1945) and in *Nature* (July 1944) written by Professor E. J. Salisbury of Kew, probably gave special satisfaction. Of course, there was criticism as well as praise. J. H. Willis and W. H. Nicholls were uneasy about some aspects, for example, the 'rufa group' of *Pterostylis*, the Rufous Greenhoods. Rupp conceded that George Scammell's plate of *Pterostylis squamata* probably *did* depict 'a form of *P. rufa*, though not the type form', but he remained 'entirely unrepentant' about the determinations and descriptions of other species.[45] However, he later amended his description of *P. boormanii* and sent copies to Melbourne![46]

Immediately before the book's appearance, Rupp was given an important commission. On Tuesday, 2 February 1944, he caught the night mail train to Glen Innes in order 'to inspect the Nat. Herb. specimens stored there, and pack most of the types for return',[47] since Sydney was no longer considered to be under serious threat. By late July, most of the material had been returned to the Herbarium, where on 24 February, Rupp was 'greeted' by '16 boxes of Orchidaceae... i.e. about 800 specimen sheets',[48] which had to be checked and replaced in the general collection. Later in the year, another aspect of his Herbarium duties was to work assiduously through the orchid collection of the engineer–botanist Henry Deane (1847–1924), a friend of R. D. FitzGerald whose *Australian Orchids* he endeavoured to complete. By August, Rupp was busily engaged on 'A Critical Review of R. D. Fitzgerald's "Australian Orchids"'. He wondered whether he was appearing presumptuous in attempting this. 'Maybe—but it's time someone did it' he told Fred Fordham.[49] At this time, Rupp was working at least two days a week at the Herbarium.[50] In the spring he enjoyed a holiday near Blandford with his daughter Eileen and her family, before returning to proceed with such tasks as writing '3,000 words on Australian Orchids' for the *Australian Children's Encyclopaedia* edited by his old friend Charles Barrett.

The peace celebrations of May and August 1945 brought both joyful relief and serious concern. Arthur Rupp's health still caused anxiety, and Rupp commented 'I'm afraid we cannot "forget the night just gone'' while our beloved son is still feeling his way towards the dawn'.[51] Happily the dawn soon broke, and to the family's great joy, Arthur was able to return to teaching at 'Shore' (Sydney Church of England Grammar School) at the beginning of 1946. In December 1948 he married the matron, Lillian Williams, in the school chapel.

In the spring of 1945, there was a minor embarrassment when Rupp found that his 'new species of 1944', *Dendrobium ancorarium*, was identical with Bailey's *D. adae* of 1884.[52] But Rupp accepted this as part of a taxonomist's lot. There were talks to the Naturalists' Society and the Linnean Society on searching for orchids in northern New South Wales, and there was the unusual task of accompanying Mr Douglas Cross of the Herbarium on a trip to the Gosford district to inspect the properties of those who were seeking licences to sell wild flowers under new legislation.[53] Furthermore, Rupp claimed that 'never before have I been snowed under with specimens as during the past eight weeks or so'.[54] While this was pleasantly challenging, it actually led to a little 'orchid weariness', a state which Rupp usually managed to overcome with ease. There was also some discomfort and concern caused by attacks of 'galloping pulse', but the doctor was reassuring and Rupp 'went away for a week or so with friends at Bowral where the mountain air and the lovely surroundings quickly rejuvenated me'. The ascent of Mount Gibraltar, and botanizing on its 3,000-foot summit, were clearly invigorating, and he returned home 'feeling nearer 30 than 70'.[55] No doubt a contributing factor to this renewed vigour was the fact that his host was George Greene, an old schoolmate of Geelong Grammar days, 1889–91.[56] Their reminiscences would have been worth recording.

Towards the end of the year, Rupp was further elated when first a new orchid was referred to him from North Queensland (*Mobilabium hamatum* Rupp) and then two more species were referred by Charles A. Gardner, government botanist of Western Australia. One was a new Sun Orchid, which Rupp named *Thelymitra cucullata*, and 'the other . . . the first discovery in Australia of a *Monadenia*, hitherto supposed to be exclusively African'.[57] Rupp considered this 'the most exciting thing at present in the Australian orchid world'. 'How the dickens did it reach Australia, 5,000 miles across the ocean from its family circle?' he asked.[58] He described the plant as being 'very close' to the South African *M. micrantha* (now *M. bracteata*) 'but different', especially in the column,[59] and described this Albany discovery

as *M. australiensis*. It is now considered to be identical with the South African species, somehow introduced to become successfully naturalized[60]—a rare thing among orchids.

Amid the industrial strife then evident, Rupp wished one of his correspondents 'as happy a Christmas as our miner-ironworker-wharfy bosses permit'. He had long since declared that he was 'fed up with politics'[61]; now he was disturbed by 'the trend of affairs in Australia . . . One longs for the old Labour Party with its high ideals and straight goers. I'm afraid it's dead . . . it wants a bad knock-out to bring it to its senses if it's got any left.'[62] It was much more pleasant to contemplate recent excursions to Caringbah and Kurnell, and to Asquith, where he discovered 'a most beautiful specimen of *Thelymitra chasmogama*'. Days later, he had still not 'got over the excitement'.[63] Thus closed 1945.

During the immediate postwar years, 1946–8, Rupp found ample grounds for disgust with 'Communist outrages on our social and industrial life', apparently unchecked by 'hopeless politicians'. He considered the situation 'deplorable; but characteristic of the aftermath of War'.[64] To his cousin May Garrard (née Wilson) he confided: 'I have always . . . been sympathetic towards the so-called Labour Movement; but to-day I feel disposed to paraphrase Madame Roland and to cry ''O Labour, what vile things are done in thy name!'' '[65]

Although busily occupied with his work at the Herbarium (including the examination and assessment of an old collection of orchid specimens from the Philippines) and his writings (including an article requested for the *Journal of the New York Botanical Garden*), Rupp was forever looking expectantly southwards for the appearance of the first part of W. H. Nicholls's 'magnum opus', which was intended to describe and illustrate every known species of Australian orchid. 'We shall all be in eclipse then!' he cheerfully conceded.[66]

Welcome visitors continued to call. Dr Hugo Flecker of Cairns arrived during a heatwave in January 1946, when Rupp recorded the shade temperature of 102.9°F with humidity at 73 per cent. There were meetings and excursions with other friends, including Walter and Emily Campbell of Hunters Hill, Alick Dockrill of Kogarah, John Young of Murwillumbah, Jean Gemmell of Glen Aplin, and of course Cecil and Pearl Messmer of Lindfield.

Early in 1946, Rupp decided to formalize his intention of donating his personal herbarium of orchid specimens to the National Herbarium of New

South Wales. The decision was reported in the *Sydney Morning Herald*,[67] and there followed an amplified report by 'Waratah' (Peter Joseph Hurley), whose enthusiasm led him to describe Rupp as 'a pioneer, in a land where pioneering was thought to be finished', and the discoverer of 'a subterranean orchid species said to be in the Paterson locality'.[68] These and other exuberant inaccuracies proved too much, and Rupp wrote to the editor refuting and correcting ' ''Waratah's'' kindly-meant but misleading eulogy'. It was Mr Ernest Slater who discovered 'the remarkable . . . orchid' named in his honour, and Rupp further declared 'I am no pioneer in Australian orchidology—its foundations were laid before I was born'.[69]

During the autumn, Rupp checked his herbarium, noting 1,713 mounted specimens or groups of specimens with possibly another one hundred still to be mounted. Over 500 species were represented, 'including about 27 types'. But, 'I am told Nicholls has over 5,000 specimens', he added almost ruefully.[70] A year later Rupp was still transferring his herbarium to the state collection,[71] albeit with some misgiving: 'Sometimes I wonder if I haven't been an ass to give it to an institution which is starved to the bone by the Government'.[72] Like others, Rupp was galled by the official indifference to suggestions and requests regarding such matters as staffing and publishing, and above all, the building of a new herbarium.

Popular fame was growing. In March 1946, a 'charming lass' conducted an interview on behalf of the *Women's Weekly*. An article duly appeared, but Rupp found himself credited with '2,000 species'. 'May as well have said 10,000', he groaned, and he presumed that whatever notes were made at the interview must have been lost on the way back to town.[73] Rather more rewarding were the hours spent in the Mitchell Library 'going through Fitzgerald's unpublished plates, and Dr Rogers's MS notes thereon'.[74] He furnished an immediate report on his findings, and this, with Rogers's notes, is still preserved with the ninety or so plates in question.[75] Further interest was stirred some seven months later when Emily Campbell rang to advise that some more unpublished paintings were in an old trunk at the FitzGerald home.[76]

Rupp began the year 1947 by renewing his lament that there was no 'really satisfactory medium for publishing new species'. A glance at the appended Bibliography may suggest that this was a curious statement to make. However, strikes in the printing and electricity production industries had helped to create a 'great backlog' of material at the Linnean Society, and Rupp himself had certainly kept the editors of the Linnean and many other clubs and societies extremely busy. Now seventy-four, he experienced a sense of urgency, becoming a little tetchy when correspondence was not promptly answered and when papers were not promptly read or published. Although his herbarium was 'getting rather beyond me at home', there was still much to be put on record, despite a clear limit on the time left for doing so. By the end of March, he was almost half-way through the task of transferring his collection,[77] but three weeks later he was dismayed to find that weekend intruders had created 'a frightful mess' in the Herbarium:

Strange to say, the compartments occupied by Misses Vickery and Garden, and myself, escaped unscathed; but everywhere else all the drawers were pulled out, and the contents strewn all over the floors. There was of course nothing in the whole place that could be of value to burglars; and it looked as if, realizing this, they had set themselves in spite to make the place a pandemonium. It will mean a lot of work getting things straight again. The specimen boxes do not appear to have been touched except in Miss Tindale's compartment.[78]

For Rupp, the most notable event of 1947 was ecclesiastical, not botanical. His friend, the Rev. Milton Williams of St Peter's, East Maitland, invited him for the centenary celebrations of the diocese of Newcastle. Rupp gladly caught the train to Newcastle on Saturday, 28 June, arriving to take part in the great pilgrimage procession from Old Bishopscourt, Morpeth, via St James's Church, to the grave of William Tyrrell, first bishop of Newcastle, in Morpeth Cemetery. Bishop Batty, seventh bishop of Newcastle, took a leading part in this pilgrimage in which an estimated 15,000 people participated. Rupp declared 'It was the most impressive function I've ever been to, and everything was admirably organized'. He assisted 'at all services' in his host's church the following day, St Peter's Day, and was taken to Weston where he 'met many old friends' of the Depression days. In all, it proved 'a wonderful time'.[79]

Returning home, Rupp faced a very cold snap. Sydney experienced its coldest Wattle Day (1 August) since 1933 and frosts even occurred in Northbridge. One result was an attack of bronchitis with some 'twinges of arthritis',[80] which may well have provided an appropriate opportunity to remain indoors and to do some writing. Perhaps motivated by Alec Chisholm's *Birds and Green Places* (London 1929) and/or by economic conditions, Rupp had written early in the Depression period, a work he called 'Gardens of Nature'. By early December 1930, George Robertson had had the manuscript 'for almost two years' to no effect, and by March 1933 the Australian Publishing Company had been 'considering' the same work for some time.[81] It is possible, even likely, that 'Gardens of Nature' in its 'improved' form became 'Recollections of an Amateur Botanist', from which some material was published in three instalments of 'A Botanist's Story' in the Melbourne *Australasian* in July 1933,[82] sometimes accompanied by a column by Alec Chisholm.

Now, in August 1947, Rupp raised the matter again with Alec Chisholm:

Do you remember some years ago looking through some MSS of mine, 'Gardens of Nature'? You seemed rather to like it. Lately I have been working on it, improving it I think, and making it into a series of recollections of rambles during the last half-century or more. I have offered it to Angus & Robertson, but don't know yet whether they will take it, or give it the boot.[83]

In the event, Angus & Robertson did 'give it the boot', but Rupp chose to say that they 'gently turned down my MSS, on the ground that owing to increased costs, they are now only accepting books that they think will run to 5,000 copies. Am not doing anything about it at present.'[84] He did, however, send it to Trevor Hunt to read,[85] and by April 1948 hoped that the work might be published in New York, where an editor of Oxford University Press declared it to be 'delightful reading'. Three months later it was clear that, however 'delightful', the offering had been declined.[86] Fortunately, there was ample literary contentment in revising Dr Rogers's article on Orchids for the new *Australian Encylopaedia*.[87] Another absorbing task during mid-1948 was to catalogue 'our Australian orchids at Nat. Herb. . . . 463 species. Of course they are not all good specimens.'[88]

It took an inordinate time for Rupp's pastoral and botanical work to be appropriately recognized. Apparently it did not occur to the ecclesiastical authorities that a faithful parish priest with over forty years of varied service and a renowned reputation in the scientific world might be fittingly honoured with a canonry; no university offered to recognize his work by the bestowal of an honorary degree; no learned society saw fit to award him a fellowship. But there were gratifying compensations.

Some, of course, had early appreciated Rupp's immediate worth and likely achievement. Just as Charles Rupp was delighted to discern his young son's potential as a clergyman when the two began to conduct services together, so was John Bracebridge Wilson to notice his nephew's development as a botanist. Baron von Mueller had been so impressed by the Trinity student's botanical knowledge and activities, that he 'used to insist' that his young friend 'was descended from the German botanist' Heinrich Bernard Ruppius (1688–1719). Some sixty years later Rupp commented, 'It may be so; I dunno . . . we know little of our forbears' in Germany.[89]

The first significant public acknowledgements of Rupp and his work were made in 1924 when J. H. Maiden had one of Rupp's early orchid papers published in the *Australian Naturalist*, and in 1933 when he was included in *Who's Who in Australia*. Then, in the spring of 1944, the Council of the New South Wales Orchid Society elected him an Honorary Life Member. Recognition of a different kind came in August 1948 when the vicar and parishioners of Beeac, at first invited, and then implored their stipendiary reader of 1898, 'to come over and officiate at the church's Golden Jubilee' in September.[90]

This proved to be a gala time. Rupp received 'a very generous cheque covering all expenses . . . from the Vicar and congregation of St Augustine's', and observed: 'Few men, I imagine, can experience the joy of going back to their first Church and presiding at its Jubilee! It's still like a dream to me; but this morning I booked my seat in A.N.A. Skymaster plane for Melbourne on Sept. 13.'[91]

And so Montague Rupp boarded an aircraft for the first time, leaving 'Mascot at 11 . . . and by 1.30 . . . walking on the green sward of Essendon'. The feting began immediately. He was driven on a sightseeing tour of Melbourne, and after dinner taken to a meeting of the Field Naturalists' Club of Victoria at the National Herbarium, South Yarra. There he was asked to recount to the audience of 250 his 'few reminiscences of "The Baron"', whose memorial stamp was issued that day'.

After an exciting day at the Herbarium, Rupp went by train to Colac, to be met by the vicar who provided another local tour *en route* to Beeac. Rupp's participation in the celebrations was already well publicized before he left Sydney. The *Colac Herald* reminded district residents that their visitor was not only a 'noted preacher' but also an accomplished botanist, who had in fact been 'a founder of the St Augustine's Church annual flower show'.[92]

At the social functions and jubilee services held 17–19 September, Rupp did not meet 'many of the old-timers', but one lady of eighty-four 'flung her arms round me and wept on my neck!' and 'boys and girls' of 1898 'introduced their *grand-children* to me!' Groups of former Beeac parishioners even travelled from Geelong and Ballarat to attend the services, and there were further extended excursions around the parish, this time 'bowling along' in the De Soto of his host, Dr M. W. Cave, through country he had last traversed on horseback, often in the teeth of vicious westerlies blowing across Lake Corangamite.

On 21 September Rupp left Beeac (where in fifty years the population had increased from 1,800 to 8,000) after 'a most memorable experience' and returned in Melbourne for yet more nostalgia during 'a very crowded fortnight... visiting relatives, old friends, and the Herbarium, and seeing the sights'.[93] He met James H. Willis for the first time, and telephoned a friend of Trinity days, Dr Edward Field, who took him back to see their old college and the university campus before 'lunch at the Australia'.[94] Then on Monday, 4 October, Rupp returned to Sydney, flying at 11,000 feet 'with a tail wind' which reduced the time to two hours. He found 'all well at home'—and a mass of 'correspondence and specimens' awaiting him.[95]

This excitement had hardly abated when there arrived a letter from the Royal Society of New South Wales advising the award of the Clarke Memorial Medal for 1949. Rupp was astonished as well as elated, for 'this is the highest honour in the gift of the Roy. Soc., and I never dreamed of it coming to me, especially as I have never been a member'.[96] Inundated by congratulatory messages, the bewildered recipient protested that the award could well have been made to 'such outstanding botanists' as C. T. White, C. A. Gardner, W. H. Nicholls, Edith Coleman 'and a certain J.H.W.'[97] His colleagues at the Herbarium thought it was time that a photographic portrait was taken. On 25 November Rupp was hustled off to Howard Harris's Studio, and the pleasing result is still on display in the institution he served so loyally.

Memories of a lifetime were stirred by the trip to Victoria and by the Royal Society's citation. Rupp again lapsed into a strongly reminiscent mood. When Joyce and Archibald Telfer 'with their usual kindness' sent 'a David Jones gift token' for Christmas 1948, he 'bought Henry Handel Richardson's "Myself When Young"' which, however inaccurately, recalled yet more memories of childhood days at Koroit. On Boxing Day, Rupp confided to Gwen Scrivener:

I'll let you into a secret; I'm engaged in writing my own 'Memoirs'... Whether my 'Retrospect' will ever be published I can't say; but I'm going to try. I've known a good many notable Australians, and think I can make it interesting; but it will take months yet.[98]

Rupp worked much more quickly than he had anticipated. On 1 January 1949, he advised Trevor Hunt in similar vein:

Wotjer think I'm doing? Writing my memoirs, with an eye to Angus & Robertson taking 'em on. I've met quite a lot of notable Australians in my time—knew Henry Handel Richardson when she was a brat of a girl; then there are Baron von Mueller, Bracebridge Wilson the great headmaster of Geelong Grammar, James Lister Cuthbertson the poet, Sir Fredk. McCoy, Sir Baldwin Spencer, Sefton Delmer, and lots of others; these sandwiched in with my personal experiences should make a decent story.

He shortly added D. G. Macdougall, Colonel Arthur Lynch, J. H. Maiden, Bishop George Merrick Long, 'Trevor Hunt and many other celebrities'. On 17 January the completed script was delivered to W. G. Cousins of the editorial department of Angus and Robertson.[99] A verdict was promised 'in about three weeks'.

On 3 February, Rupp received £6 6s. 0d. for his revision of Rogers's article on Orchids for the *Australian Encyclopaedia*, but there was no mention of the memoirs. Sadly, Rupp's friend, Walter George Cousins, died in August 1949,[100] and hopes for the publication of 'Retrospect' apparently died with him.

Early in April 1949, Rupp recorded his survival of the 'ordeal at the annual meeting' of the Royal Society to receive the award:

I hadn't anticipated having to do more than make a bow and acknowledge the honour; but was called upon to 'say a few words'—not an easy matter to a fairly large audience of folks who were mostly chemists and physicists and mathematicians! But it's over now... The Medal itself, apart from the honour, is a thing of beauty! What a wonderful old chap Clarke was.[101]

Before the bestowal of similar honours, there was a pause during which Rupp resumed some intensive work on the genera *Prasophyllum* and *Pterostylis*. He came to be worried about possible indebtedness to the Field Naturalists' Club of Victoria through overdue subscriptions or the costs of reprints, for he wished to meet any debts before bowing to economic restraints and terminating his membership at the end of 1952. To its credit, the club's council resolved the matter by electing him an Honorary Life Member,[102] a gesture which was greatly appreciated.

In April 1953, J. H. Willis and Dr M. M. Chattaway nominated Rupp for another award—the Australian Natural History Medallion. Instituted to commemorate the life and work of Baron von Mueller, the award was first presented in 1904. Although not immediately successful, Rupp received a telegram early in June 1955 advising that he had won the Medallion for 1954. This 'came as a great surprise; I had no idea I was in the running', he wrote.[103] A month later, R. H. Anderson 'organized a most delightful afternoon function' at the Gardens. Rupp had been so ill that he thought that the presentation 'may have to be done by proxy'. It proved to be a memorable occasion with a plethora of resounding tributes from Anderson himself, who made the presentation and delivered 'a merited eulogy'; Dr W. R. Brown, secretary of the Linnean Society; H. J. R. Overall, president of the North Shore Fauna and Flora Protection Society; Rev. W. J. Siddens, rector of St Thomas's, North Sydney; A. B. Porter, president of the Orchid Society of New South Wales; H. K. C. Mair, senior botanist at the Herbarium, and Mrs Pearl Messmer, 'a collaborator on occasions' who 'feelingly told of his worth as a friend and botanist'. Once all of this was accomplished, the embarrassed guest of honour rose on his 'hind legs and tried to reply'.

Other guests included Rupp's sister, Florence Monypenny (then ninety years old and 'in great form'), his son Arthur and daughter Eileen, and P. A. Gilbert, E. A. Hamilton and Marjory and Norman Loader, all of the Orchid Society.[104] Regrettably, Florence Rupp and Alec Chisholm were too ill to attend. Rupp noted that an ABC reporter was also present, and that there was 'quite a good report on the 7 o'clock wireless'. On the other hand, 'the S.M. Herald ignored it, I being neither a negro prizefighter nor a dubious jockey'.[105]

While such awards, sparingly granted and greatly prized, brought tremendous satisfaction, the knowledge that plants had been named in his honour must have been even more gratifying—*Acacia ruppii* (1912); *Boronia ruppii* (1927); *Prasophyllum ruppii* (1927). In 1955, Stanley F. Goessling St Cloud named *Cadetia ruppii*, 'a beautiful species, in honour of Rev. H. M. R. Rupp whose valuable assistance and advice have been so generously given to students of North Queensland Orchidaceae', and in 1960 A. D. Hawkes of Florida proposed the name *Dendrobium ruppianum* for the North Queensland Oak Orchid 'in memory of the late Rev. H. M. R. Rupp,

author of many excellent studies on the Orchidaceae of Australia'.[106]

It was inevitable that towards the end of such a long life, Rupp should experience the sadness of losing several ecclesiastical and 'orchidy' colleagues whose friendship he had long treasured. His own vitality seemed to ebb with each loss, especially if the person were younger. Early in his seventieth year (1942) he attended the funeral of Canon John S. Needham, former chairman of the Australian Board of Missions. This was held the day after the 'dreadful broadcast...about the last days on Singapore Island' which had left Rupp 'frantic with horror and fury'. Apart from Bishop Francis de Witt Batty's address, 'which lifted it up a bit', Rupp felt 'rather upset' for 'everything seemed so gloomy and dismal and uninspiring'. He also learnt that Needham was 'two years my junior'.[107] A month later, in March 1942, Rupp lamented, 'Dear old Dr Rogers is dead!' He was 'the leader' of Australian 'orchidians...and a very dear friend'.[108] In the spring came news that the Rev. John Jones, another ABM colleague, had died in Sussex.

Early in June 1948, Rupp was saddened by the death of the Rev. E. Norman McKie, 'a most lovable man', and in August 1950 there came the 'staggering blow' of C. T. White's death, 'sad, sad news, alas!':

We have lost our greatest botanist, and a man who surely won the affectionate regard of all who knew him. For years I have counted it a real privilege to be numbered among his friends... There is no one of equal calibre to take his place... God rest his soul.[109]

A month later, Rupp was still disconsolate: 'I shall miss him very much, he was always so ready to help in any way he could. He was in my opinion Australia's greatest botanist since Mueller and Bailey.'[110]

The first shock of 1951 came in March when W. H. Nicholls died—'Poor old chap; it does seem hard that he should not have been spared to enjoy the fruition of his years of labour'[111]—and a few days later:

Sad news. Poor Nicholls died last Saturday! It is really tragic that he should go just as the publishers are about to issue Part I of his great work... It is an irreparable loss to Australian botanical circles. One can hardly realize that nevermore shall we see those delightful line-drawings appear in constant succession in...the 'Victorian Naturalist'.[112]

On Easter Day (25 March) 1951, Rupp recorded that he was

just back from a glorious Easter service at St. Thomas's, North Sydney; I thought often of Will Nicholls and others of my contemporaries who have passed through the veil of late. I was grieved to hear. . .that Mrs Elma Nubling of Mentone had gone, about the same time as Nicholls. The poor old man will be lost without her; she was a woman of great charm, and looked after him most devotedly. F. W. Schmidt, of Sydney, who has cultivated native orchids for many years, and has collected valuable specimens for us, also passed on last week.[113]

To another correspondent he confessed being 'dreadfully distressed by the death of Mr Nicholls. It is really tragic, that he should pass away on the very brink of the fulfilment of his lifelong ambition.' Against incredible odds, Nicholls had 'stuck to his aim, and at long last the goal seemed in sight'.[114] Two months after his friend's death, Rupp's feeling of loss had hardly changed: 'the passing of poor Nicholls still gives me a pang whenever I look at his autographed photograph above my desk'.[115] While they often disagreed on botanical matters, Rupp and Nicholls never wavered in their mutual respect. A little spirited disagreement was far preferable to empty silence: 'I feel lost without poor old Nicholls. I do not know of any Victorian upon whom his mantle might fall.'[116]

In June there was more sad news from Victoria. Edith Coleman had died: 'I had a letter from her only about ten days ago, chirpy and full of news. . . Dear me, all my old friends are going—I suppose my turn can't be far around the corner.'[117] Shortly after there occurred the death of another 'old friend', the Venerable Keith Single, archdeacon of Newcastle, and in the spring Edwin Cheel died in retirement at Ashfield.

At the beginning of 1952, Rupp was 'shocked' to learn of the death of John McConnell Black (1856–1951), doyen of Australian botanists, who in 1939 to the surprise of some and the inspiration of all, began a revision of his *Flora of South Australia* at the age of eighty-three. He died in North Adelaide on 2 December 1951, but 'nobody over here knew anything about it', Rupp complained, until Miss (later Dr) Mary Tindale saw the news in *The Bulletin*. One would have thought that 'in the case of such an outstanding and remarkable personality' the obituary would have been more widely published.[118]

In April 1952 'another dear good friend. . .passed on. . .Mrs Florence Perrin of Launceston. I shall miss her letters very greatly, as I still do those of Edith Coleman.'[119] 'So many are going before me,' he observed.[120] Early in the following year, Edith

Coleman's daughter, Dorothy, advised Rupp that he had lost yet another 'old friend', Erwin Nubling, whose meticulous work on the Prasophyllums had evoked such admiration.[121] Within three weeks, Rupp was 'deeply distressed by the death. . .of Dr Archie Telfer, who married . .Joyce Scrivener. He was not only one of Sydney's most distinguished medical men, but was a very fine character, and incidentally quite keen on orchids, like his wife. He was only 49.'[122]

The friends who remained were as faithful and supportive as ever, especially during periods of illness which became more frequent after Rupp, in his own words, 'entered the ranks of the "octogeraniums" '.[123] His friends were greatly and fondly appreciated. 'I am extraordinarily fortunate, or rather blessed, in my friends', he told J. H. Willis. Some new friendships, made quite late in life, in part compensated for the losses which were so deplored. Rupp cited as examples Norman and Marjory Loader, then of Castlecrag:

Not many men make new friends at the age of 76 as I did with these dear people. Their generous kindnesses overwhelm me. Most folks their age wouldn't be bothered with an antique like me beyond the bounds of ordinary courtesy; but there seems no limit to what they'll do for me.[124]

The Rupps found the bitter winter of 1949 difficult to endure. Florence Rupp suffered a heart attack and was confined to bed; Rupp himself developed asthma and bronchitis, while being obliged to 'drag wood in from the bush' for warming the house and cooking in the backyard. He felt unable to 'make any comments on miners or Governments' as he would be carried 'too far' into the realm of invective.[125] Rather slow recovery was followed by 'an obscure and horrible form of "flu" (blessed word, covering a multitude of medicoes' don't-knows). . . It was like a combination of dengue, typhoid, bronchitis, and African sleeping sickness!' The resulting enervation caused him to decline an invitation 'from Harvard. . .to serve on the "International Commission on Orchidology" ' and for the same reason he began to transfer 'a lot of my bush-house orchids to the Botanic Gardens' where they would be adequately tended.[126]

Still resilient, Rupp enjoyed early in 1950 'lunch in town. . .with Mrs H. Sinclair, the Baron's niece' whose son Bruce had been the family's physician at Paterson.[127] In the spring, there was the unforgettable time with the Loaders at the Althofers' Nindethana Nursery near Wellington—but only after recovering first from bronchial influenza and then from hepatitis. Then came the uplifting news that, as from 1 July 1950, the Department of Agriculture (after two or three years of badgering by R. H. Anderson) almost doubled the famed honorarium to £50 per annum. Before the end of the year, despite recurring attacks of asthma, Rupp found time during one session at the Herbarium to go across to the Domain to identify West Australian orchids for the organizers of the Red Cross Flower Show.

Rupp's humour concerning industrial matters hardly had cause for improvement during 1951. By May, Northbridge residents were experiencing up to five power blackouts a day, and Rupp was reduced to typing letters by the light of a hurricane lamp, often shivering until nine o'clock at night when the use of radiators was permitted. There were bus strikes, wharf strikes and continued power problems, Rupp reported during one of his now-frequent attacks of bronchitis and cardiac asthma. He was advised to give up his pipe after 'nearly 60 years', and as the hills of Northbridge became more testing, it was decided to move to a house in Neville Street, East Willoughby, 'devoid of our 40 odd steps and precipitous garden'. The move was effected on 25 October.[128]

As he approached his eightieth year, Rupp was sharply aware of his increasing infirmities. Recovering from one bout of illness in May 1951, he told Trevor Hunt, 'I'm afraid my days of activity are drawing to a close—I can't growl'.[129] 'I am getting too old,' he remarked to J. H. Willis, 'memory and power of concentration are failing.'[130] Nevertheless, he valiantly maintained his correspondence and the taxonomic and other work at the Herbarium. Sometimes when he was indisposed, specimens were sent home for an opinion before they should wither. The kind concern of Knowles Mair, Joyce Telfer and Norman and Marjory Loader ensured that there was transport to and from the Herbarium, and some assistance during the time spent there. Mrs Loader still recalls Rupp's clear lessons in the most economic use of gummed paper when mounting specimens.

There were still some occasional excursions. In the spring of 1952, the Loaders took Rupp to Springwood, where in addition to scouring the bush with the aid of Miss Isobel Bowden's expert local knowledge, Rupp preached the sermon on Sunday morning as an appreciative gesture to the party's clerical host. There was another excursion with the Loaders in March 1953, this time to the Hawkesbury River. In the following spring there was an enjoyable

The Rev. H. M. R. Rupp (third from left) and his son, Arthur (extreme left) on an excursion to the Hawkesbury River with Norman and Marjory Loader, March 1953. (Marjory Loader Collection)

day at Narrabeen with Knowles Mair and Fr Bede Lowery, a wonderful three-day trip in November with the Messmers to the Jenolan Caves, and a further excursion to French's Forest with Knowles Mair and Fr Lowery to admire the latter's discovery of a large colony of the elusive Helmet Orchid, *Corybas undulatus* (A. Cunn.) Rupp.[131] By October 1953, however, Rupp conceded, 'I have to recognize that my faculties are running down; I think everyone else down here realizes it'.[132] Correspondents such as Fred Fordham had seen signs of this for some time, for when he moved from Brunswick Heads to Armidale, Rupp suggested in similar terms on at least three occasions that Fordham should contact Dr Gwenda L. Davis of the Botany Department of the University of New England on matters of mutual interest. Rupp had known this 'very keen and capable botanist' since 'she was an infant'; she had produced 'a magnificent piece of work' in her 'beautifully illustrated' monograph on the 'extremely difficult genus' of daisies, *Brachycome*, and was a lady for whom Rupp had 'a profound admiration'.[133]

Rupp also realized that he would soon have to relinquish his Herbarium post, 'though I'm hanging on to it as long as I can be of any use there'.[134] Sundry setbacks continued. In the depths of winter in 1952, Rupp was emerging from another period of cardiac asthma when he gaily quipped: 'Apart from the asthma I am pretty well and active; but I tried a new trick the other day—fell asleep in front of a "Fyrside" heater, and suddenly woke with the toes of my right foot burning!' The blisters were slow to heal.[135] Yet, by the spring, he was able to attend some sessions of the Science Congress at Sydney University, and was receiving generous parcels of specimens for assessment from all over the continent.

There were further asthma attacks until, in January 1954, Rupp's doctors 'lobbed' him into the Mater Misericordiae Hospital at Crows Nest, where he remained for a fortnight. 'I can't sing the praises of the Mater Hospital too much,' he wrote, 'they were SO good to me there!'[136] Just as he had missed Queen Victoria's Jubilee Celebrations of 1887 in the sick bay of Geelong Grammar, so now he missed most of the festivities associated with the visit of Queen Elizabeth II. However, Rupp gleefully noted, 'the nurses took me up to the top floor' to see the fireworks. By July he was 'remarkably well' again despite an influenza epidemic and a spate of colds which caused tram travellers to experience 'barking' which 'rivals a dog-kennel'.[137] Shortly, however, he was back at the Mater Hospital, where he wrote a pithy note to J. H. Willis in characteristic style, referring partly to his cardiac asthma and partly to the inappropriate persistence of one writer using the superseded name *Sarcochilus parviflorus* instead of the more widely accepted *S. australis*. But he added, 'I think my writing days are over'.[138] The stay in hospital was prolonged this time, and R. H. Anderson reassuringly advised that he was 'on 3 months' sick leave'. By October, Rupp was convalescing with his daughter Rachel at Broombee, near Armidale, where he contentedly spent most of his time 'in front of the big log fires'.

In January 1955, he was back at the Herbarium 'once or twice a week', and elated at the news that Fr Bede Lowery and Alick Dockrill had discovered the saprophytic Tongue Orchid, *Cryptostylis hunteriana*, at West Head, Broken Bay, the first record for New South Wales.[139] Then there was yet more asthma, and Rupp thought he was 'going to pass out' but 'there came a sudden and astonishing recovery' and his attention immediately returned to orchids until the next admission to hospital in November 1955. At this point, he sent J. H. Willis his 'swan song', an article which he 'could not manage to illustrate'.[140]

In August 1955, Rupp had noted realistically that because of severe arthritis, 'my wife is more or less of a constant cripple; and I am rarely free from this horrible cardiac asthma. There does not seem to be much prospect of us getting better. We cannot manage on our own, and have to have a permanent housekeeper.' This role was faithfully and effectively filled by Mrs Jessie Croft, who apparently joined the household during the latter half of 1954.[141] Despite Rupp's light-hearted assertion that 'we are a pair of old crocks',[142] he and his wife happily celebrated their golden wedding anniversary at Christmas time 1955, but Rupp was readmitted to hospital for six weeks or so shortly afterwards. In March 1956 he tendered his resignation from the Herbarium, for it was 'not right to accept their honorarium when I am quite incapable of doing any work'.[143] Later in the autumn, Florence Rupp was taken by her daughter Rachel to Armidale for a change, but sadly she suffered another heart attack and died in Armidale Hosptial on 4 May 1956 in her seventy-seventh year.

Family and friends provided welcome support at the funeral service in St Thomas's, North Sydney, but Rupp was desolate. Mrs Croft stayed to run the house, and Rupp declared that she was 'like a ministering angel, and nothing is too much trouble for her'.[144]

The Man of Letters now had more messages than ever to compose, and even he viewed 'with alarm the pile of letters and cards' on his desk. Everyone had been 'very kind, from my dear Bishop of Newcastle through many stages to the humblest of my friends'.[145] Family matters, including news of his ten grandchildren, and of course botanical questions, continued to provide interest, but complete recovery from the loss of his 'dear wife' was not achieved.

The last weeks were characterized by sequences of relapses and rallies as kidney failure and other disorders took their toll. Finally admitted to the Cambridge Private Hospital, Mosman, close to the home of Arthur and Lillian, Montague Rupp died on Sunday night, 2 September 1956, about four months short of his eighty-fourth birthday.

Former parishioners, 'orchidy' friends and members of ecclesiastical and scientific circles joined the family at the funeral service on Tuesday, 4 September, in the beautiful Blacket Church of St Thomas, North Sydney. Rupp had 'felt at home' here, perhaps all the more because the first incumbent of the parish had been the Rev. W. B. Clarke. One of the most poignant tributes came in the form of a wreath composed entirely of orchids, with a message of sympathy from Marjory and Norman Loader.

The cortège proceeded to the Northern Suburbs Crematorium, and Rupp's ashes were later interred beside those of his wife in a columbarium fittingly close to the sclerophyll bushland of the Lane Cove River Park.

Always historically conscious, Rupp had been acutely aware of the wider significance of the national events and personal experiences which had occurred during a long and interesting life begun about half-way through the reign of Queen Victoria and extending into the reign of her fifth successor. He enjoyed reminiscing, and appreciated that others might be similarly interested in his clear and diverse memories. Accordingly he wrote, rewrote and revised his memoirs.

Some brief accounts were published in his lifetime, but although valuable extracts or summaries, they did not have the scope or the detail of the two accounts here published for the first time. Mention has been made of some extracts published in the *Australasian* in July 1933; in May 1941 some of Rupp's schoolday memoirs, 'Fifty Years Ago', were recorded in *The Corian*, the journal of his old school; and in the following month the first part of 'Memories of an Orchid Lover' appeared in the *Australian Orchid Review*—others followed in September 1941 and June 1945. Further botanical reminiscences entitled 'Memories of Victorian Orchids,' featured in the *Victorian Naturalist* in March 1953. We can only be grateful that Rupp did put his memories on record.

The first rector of St Thomas's, North Sydney, 1846–71, the Rev. William Branwhite Clarke, 'the father of Australian geology' whose life and work were commemorated by the medal Rupp was awarded, told members of the budding Royal Society of New South Wales in his Inaugural Address of 9 July 1867:

We must strive to discern clearly, understand fully, and report faithfully; to love truth in things physical as in things moral; to abjure hasty theories and unsupported conjectures; where we are in doubt, not to be positive; to give our brother observer the same measure of credit we take to ourselves; not striving for mastery, but leaving time for the formation of the judgement which will inevitably be given, whether for or against us, by those who come after us; contented if we are able to add but one grain to that enduring pyramid which is now in the course of erection as the testimony of Nature to the truth of Revelation.[146]

Montague Rupp would have liked that. The Orchid Man and the Rock Man would have made ideal penfriends.

PART II

RECOLLECTIONS OF AN AMATEUR BOTANIST

Ampliat aetatis spatium sibi vir bonus.
    Hoc est
Vivere bis, vita posse priore frui.

(The good man prolongs the course of his life-time.
    That is
To be able to enjoy one's former life is to live twice.)

    —Martial (Marcus Valerius Martialis), *Epigrams*, Book X

. . . the memories of youth remain, and are very precious.
I do not believe in sighing for the loss of good things that are beyond
recovery. Better far to remember how good they were, and to thank
God for the good luck that made them ours.

    —H.M.R. Rupp in *The Corian*, May 1941.

It will be obvious to the reader that the writer of these recollections
belongs to the species *Presbyter Anglicanus*, and it may therefore seem
curious that so little is said of the experiences of his vocation. The
omission is deliberate: these are the memoirs of a hobby, and other
matters are intruded only so far as they account for the author's
movements or are necessary for presenting the sequence of events.

    —H.M.R.R.

H. M. R. Rupp's photograph of the Hyacinth Orchid, *Dipodium punctatum,* at East Maitland, *c.* 1933.

Hyacinth Orchid

From the unforgiving white sandy loam
And bracken, from the crackling dry
Debris under the grey-green scythes,
An ensign of mauve flowers, summer
Blown, from orchis flying high;
A phallus standing all alone,
Reach for Apollo in the sky;
Sustaining, spoiling sun.
A mortal child cut off too soon
As sun and wind contrived for love,
Risen as a flower bloom,
The flower petals splotched with blood:
Unless the seed of grief is sown,
Love cannot draw blood out of stone.

—Edwin Wilson, in *Banyan*, Woodbine Press, 1982,
with the author's kind permission.

1 Early Days

I do not know when I began to be a botanist. I am credibly informed that when I was five years old my mother,[1] who was a great gardener and a keen exhibitor at horticultural shows, had to place special guards round her prize flowers, to prevent me from investigating them to their detriment just before show day. It may be that I was merely unconsciously forecasting my choice of a hobby! Be that as it may, before I was eleven I could talk flowers with grave and reverend horticulturists, and knew most of the wild flowers of the district.

My father[2] was incumbent of St Paul's Church, Koroit, in western Victoria, and sometimes he would take me with him on his country rounds, and let me gather flowers while he visited a farmhouse. That the treasures of the bush made more than a fleeting impression on my youthful mind is attested by the fact that fifty years afterwards I can remember what they were, though I have forgotten many of the private and particular names we used to bestow upon them.

Most of the Koroit bush has long ago disappeared before the march of closer settlement. No doubt that is as it should be, but one work of destruction that took place has always seemed to me unpardonable. A mile to the south of Koroit, the Warrnambool road leads to the crest of a gently rising slope. As you approach the crest there is nothing in the landscape to give the impression that you are about to see anything extraordinary. But when the crest is gained, what a glorious scene is suddenly unveiled in all its completeness! Let me try to describe it—not as it is now, but as I knew it in my childhood: as I can see it in memory still. From the summit of the ridge the

ground dropped steeply for several hundred feet to the surface of Tower Hill Lake, an almost circular sheet of water about ten miles round. The lake is one of the most remarkable of those ancient craters of extinct volcanoes with which western Victoria is studded. The out-throw of the volcano was evidently towards the north, for the ridge slopes away on either side from the Koroit approach, and on the south the banks of the lake are only a few feet high. Five islands occupy the central portion of the lake—or they used to: I have heard that they are now merely promontories. I cannot vouch for this, but the Koroit side of the lake was always shallow and reedy.

At one place a rough corduroy causeway led across to the largest island, among whose sugarloaf-peaks are several small crater basins. A mile or two beyond the southern edge of the lake is the ocean,[3] with a long sweep of coastline from the crags of Port Fairy to the sandstone cliffs east of Warrnambool. A marvellous panorama it was—and still is, but now shorn of its crown of glory. For the steep banks of the lake in those days were clothed in virgin forest. Bowers of feathery clematis,[4] groves of fragrant musk-trees,[5] fronds of stately fern-trees,[6] meeting overhead to shade their lesser kinsfolk growing by tumbling cascades that issued from the hillside, and lost themselves in masses of watercress on the edge of the lake. It was a paradise of wild Nature, all flowers and ferns and luscious cool glades. And it is all gone.

When I revisited Koroit twenty-five years afterwards, there was no vestige of this primeval forest round the whole circumference of the lake. The springs had been harnessed for a water supply, and

I was told that the forest had been burnt to get rid of rabbits. It reminded me of the man who burnt his house down to get rid of the mice; and I can still offer no other comment.

When I was eleven, my father became vicar of Coleraine,[7] some sixty miles to the north-west of Koroit, and I was sent to the Geelong Grammar School. John Bracebridge Wilson,[8] a great headmaster of those days to whom the school owes so much of its character, was married to my mother's sister. And now I must hold myself in check: for what old Geelong boy who touches on the subject of the school lacks the impulse to fill pages with happy memories? Not I, for one: but I must remember my role of amateur botanist, and be discreet.

Mr Wilson was not only a great headmaster: he was a scientist of some distinction, and a Fellow of the Linnean Society. He specialized on the Algae of Port Phillip, but was an excellent general botanist, and a friend of Baron von Mueller in Melbourne,[9] to whom he gave me a letter of introduction when I left school.

My botanical tendencies found ample scope for development in the Saturday excursions which have always been an outstanding feature of the school life, and Mr Wilson encouraged them and taught me the elements of botany. There were few other boys interested in the subject, but I remember two. One was R. J. Crooke, with whom I lost touch soon after we left school. The other, C. F. Belcher (now Sir Charles, and chief justice of the island of Trinidad) afterwards became a noted ornithologist, and as the above parenthesis indicates, distinguished himself in the Imperial Service abroad.[10] I have never forgotten his apt designation of a small species of ground-orchid which we used to find in the bush near Geelong—'sitting-down orchid'. Botanists who are familiar with our *Corysanthes* will appreciate the soubriquet![11]

We used to do prodigious walks on those Saturdays. It was nothing unusual to leave the school (of course it was the old school then) at 4 a.m. and walk ten miles or more before breakfast. I suppose we often covered another ten miles rambling about during the day, and then there was the tramp back to the School—we had to be in by 8 p.m. Or if we chose the river for our outing, we sometimes rowed twelve miles to a camp at Campbell's on the Connewarre Lakes, through which the Barwon meanders between Geelong and the Southern Ocean. We breakfasted there, then went on across the lakes

and down another ten miles to Barwon Heads, leaving again for the twenty-two miles spin back to Geelong about 4 p.m.

The river and the lakes, of course, are still there, but the removal of the school to Corio has probably made such long downriver days difficult, and today the You Yangs and the Anakies seem to be the most favoured Saturday resorts.[12] In my time these rocky hills across Corio Bay were a little too far out except for the aristocrats who hired spring carts (motor cars were unknown), but I have been to both. My chief recollections of the You Yangs are the capture of a native bear—who subsequently drowned himself in the school laundry and was the subject of a memorial 'lyric' by C. F. Belcher—and the profusion of a beautiful snow-white *Prostanthera* among the masses of granite.[13]

Orchids were even then my special favourites among the wildflowers. There were no epiphytes— Victoria has but four of these, all in the east[14]—but there were beautiful sun-orchids *(Thelymitra)*, doubletails *(Diuris)*, and 'spiders' *(Caladenia)*, besides smaller fry. I can remember about twenty-five species of orchids collected on those Saturday excursions. Mr Wilson used to lend us von Mueller's *Key to the System of Victorian Plants* to work out the identifications of the flowers we gathered but it was rather heavy going for schoolboys.[15]

My holidays were spent at Coleraine, where I maintained the reputation I gained at school as 'a bit dotty on wild flowers'. I remember an expedition to Mount Dundas, where for the first time I found the quaint *Caleana major*, variously dubbed 'Bee', 'Duck', or 'Cockatoo' Orchid. The second name seems the best—the expanded flower certainly suggests a duck in flight. The Wannon Falls, between Coleraine and Hamilton, were often visited, as much for their flowers as for their scenic beauty. I also collected my plants at Wando Dale, sixteen miles north of Coleraine, a grazing property then owned by the late William Moodie. Mr and Mrs Moodie were hospitality incarnate, and their beautiful home was rarely without visitors. There was a large family of boys and girls, and things were pretty lively sometimes. Some miles down the Wando Valley was another property named Wando Vale, belonging to Mr J. G. Robertson,[16] who made an extensive collection of plants for Baron von Mueller. George Bentham, in his great work *Flora Australiensis*, frequently mentioned Mr Robertson and Wando Vale, but consistently spelt the latter 'Wendu'.

2 The Hobby Progresses

Upon leaving school, I entered Trinity College in the University of Melbourne, as a resident student. The Warden was Dr Alexander Leeper,[1] of whose personal kindness I have the keenest recollections. Of Dr Leeper's reputation as a classical scholar of the highest standing it would be an impertinence for me to say much, and he was an admirable teacher, but from his point of view I was a poor pupil. I recognize the value of a classical education, but if a man has a definite 'pull' in other directions and is not attracted by the classics, he will never make a classical scholar. I did what I was compelled to do in Latin and Greek, and no more. History and Natural Science called me.

I hoped to graduate in Science as well as in Arts, but a serious illness lost me a year and disorganized my plans,[2] and I made the foolish mistake of leaving one compulsory science subject—Chemistry—until it was too late to retrieve the error. I did two years' biology under the late Sir Baldwin Spencer,[3] and thoroughly enjoyed it. Professor Spencer and his chief assistant, the late Dr T. S. Hall,[4] took pains to shew that they were interested in the work of each individual, and we somehow felt that they were not only our teachers, but good pals. I remember being tremendously interested in the wonderful collection of natural history specimens and Aboriginal implements which the professor brought back from the Horn expedition to Central Australia.[5] I did some geology and systematic botany under the late Sir Frederick McCoy, better known as 'Freddie'.[6] McCoy had done notable scientific work in his day, but that day was over, and the old man could not accommodate his mind to new ideas. He had a most profound contempt for the theory of evolution, and it was somewhat disconcerting to pass from his lectures to Baldwin Spencer's: to the one Charles Darwin was anathema, to the other, a kind of inspired apostle or prophet.

I duly presented myself at the residence of the famous Baron von Mueller, armed with the letter of introduction from his friend Bracebridge Wilson. I was not quite sure how to conduct myself in so distinguished a presence, nor whether I should call him 'My Lord'. There was little need to worry! Failing to reach the front door by reason of sundry packing cases and other obstructions, I ventured round to the back, where I was greeted by an exceedingly shabby old man in a dressing-gown and a woollen muffler. He took my letter and, somewhat to my surprise, opened it. Then—'Ah! You are neffew to Meester Vilson? Come in! come in! Anyvon from Meester Vilson is velcome here!' And in we went, to a bare and carpetless room, where I was promptly regaled with biscuits and lemonade.

The dear old Baron! He was an oddity to be sure, but it is almost *lèse majesté* to say so, for he was probably at that time the greatest living botanist of the world. Honours were showered upon him from almost every civilized country. Absurdly oblivious of his personal appearance, he was exceedingly proud of all his titles, degrees and medals, and allowed no one to be blind to the fact.[7] Yet in personal intercourse he was a man of the simplest and humblest of manners, transparently sincere, and brimming over with enthusiasm for his beloved plants. He loved Australia, too, and had great visions of her future destinies: well do I remember how he 'let himself go' in this direction, lapsing into a German phrase now and then, to the bewilderment of his youthful guest. He did not forget me: I still have a few valued fragments of letters signed 'Regardfully yours, Ferd. von Mueller', and he used to send me seeds. I saw him at Bracebridge Wilson's funeral, the woollen muffler still in evidence. He survived his old friend only by a few months.[8]

While I was at Trinity College, I used to spend most of my Saturdays in the bush. Over a great deal of the country I covered in those days, there are now rows of asphalted streets, with electric trams and long lines of shops and villas. I went out to Sandringham one day in 1922, hoping to renew acquaintance with some of the beautiful ground orchids I used to gather, quite close to the railway station, nearly thirty years before. Alas! I had to walk miles along asphalted paths before I could see any remnants of the old bushland. But the orchids were still in those remnants. And I believe one can still find them at Black Rock and Ringwood, as I used to; or along the railway line near Werribee, where the dainty, long-tailed White Diuris used to give me such joy.[9] Occasionally I went as far as Fern Tree Gully, where I developed great enthusiasm over the giant eucalypts,[10] and met my first epiphytic orchid—the little fragrant, jewel-flowered *Sarcochilus parviflorus*.[11] I loved Fern Tree Gully—partly, I think, because the fern-trees and scented musks and little cascades always reminded me of childhood rambles on the banks of Tower Hill Lake. Once or twice I went across the Bay to Portarlington, where I found two fine specimens of a sun-orchid considered rather rare, *Thelymitra epipactoides*. Those specimens, dated October 1897, are still in my herbarium in excellent condition.

Meantime my father had left Coleraine to act as *locum tenens* at Warrnambool, and after nearly eighteen months there he became vicar of Buninyong, near Ballarat.[12] While he was at Warrnambool I spent a long vacation with my only sister,[13] who was married and living at Hay, in the Riverina district of New South Wales. The Riverina country interested me greatly, the vegetation being so different from that of southern Victoria. I was fortunate in the fact that heavy rains had fallen in December, and there was a wonderful carpet of grasses and herbs. I botanized assiduously, and the *Riverine Grazier* published an account of all the plants I found[14]: this was, I think, my first venture into print as a botanist. I am afraid it was dull reading, but I was quite proud of it. There were no orchids on the list: but it was at Hay that I first met with

Baron Ferdinand von Mueller (1825–96), in more formal mood. 'The Baron' happily 'regaled' Rupp 'on biscuits and lemonade' when the young student called to present a letter of introduction in 1892.

The humble 'baronial castle' at 28 Arnold Street, South Yarra, where Rupp visited Baron von Mueller in 1892. The old seed room is at the back right. Building destroyed 14 October 1963. (Photograph courtesy of Dr J. H. Willis, taken 19 August 1959)

one or two of the beautiful Swainsona Peas, and the two attractive native bulbs, *Crinum flaccidum* and *Calostemma luteum*.[15] The former was not in flower, but the *Calostemma* was in great abundance.

Buninyong I prospected fairly thoroughly about 1896, and found it rich in botanical treasures. I had begun to keep records of the fruits of my rambles in 1892, at Coleraine. My Buninyong 'census' for the vacations of 1896 shews 32 species of orchids and 232 of other flowering plants and ferns. The ferns numbered 14—a large number for a district containing no areas of the rainforest type.

The explanation is that the ferns grew down on the clay walls of old mining shafts. It was a ticklish job getting hold of them, and often I failed. Some of the spores must have been carried great distances by the wind, for there were ferns in those shafts which would not be found on the surface of the ground, in my opinion, within fifty miles of Buninyong. It will be remembered that this was the scene of the first discovery of gold in Victoria, and the ground for miles round was riddled with these old shafts, so that one had to keep a sharp lookout while rambling through the bush. On Mount Buninyong, which is an extinct volcano, I found a very large and handsome Greenhood orchid which was not correctly determined and classified for many years afterwards, being confused with another very distinct species. It is now known in Victoria as the Sickle Greenhood, *Pterostylis falcata*.[16]

At the end of 1897 I left Melbourne University, and was appointed a lay reader of the Church of England, preparatory to taking Holy Orders, under the late Rev. John Kirkland, vicar of Colac. To my great regret, Mr Kirkland left shortly afterwards; but his successor, the Rev. H. S. R. Thornton, only son of the then bishop of Ballarat,[17] proved a good and sympathetic friend. I was practically in charge of the northern end of the large parish, with my centre of work at Beeac, and as soon as I received priest's Orders this became an independent parish. There was little to encourage my 'hobby' in the Beeac district, which consists in the main of open grassy plains, closely settled and in large part under cultivation. Seven small ground-orchids were found in five-and-a-half years, and the residue of the indigenous flora was not particularly interesting. During my association with Colac, however, I had several excursions into the Cape Otway forests, which are a rich hunting-ground for botanists. Unfortunately, these trips only fell to my lot in midsummer, when most of the flowerings were over. In 1902 I received an invitation from the bishop of Grafton and Armidale in New South Wales (Dr Cooper),[18] who had been suffragan-bishop of Ballarat and had known me from childhood, to come and work in his diocese. The vision of 'fresh fields and pastures new', both in relation to my work and my hobby, attracted me, and later that year I accepted the curacy of Tamworth under the Ven. Archdeacon T. K. Abbott, and bade farewell to Victoria.

3 Tamworth and Warialda

Before I left Beeac, my conceptions of Northern New South Wales were very hazy. It is astonishing how little some of us know of our own country, and how far from the truth our imaginations of those parts we have never seen often are. I had heard of the New England Tableland, and of some fine rivers along the North Coast. For the rest, I pictured the country as parcelled out among great sheep-stations on vast open plains. As the train panted up over the Ardglen gap, I wondered where these plains were to begin. Quirindi gave promise of them: then came the Peel Range upsetting my anticipations, and after that it was dusk. But the moon was high and clear before we passed Duri, and I looked out of the window and saw the dark outlines of peaked hills against a massive loftier range behind them, and at their feet rows and rows of twinkling lights. The approach to Tamworth on a bright moonlight night is altogether charming. All my imaginations crumbled away. Next morning, when I looked out from West Tamworth across the Peel to the bigger town straddling over the flats and creeping up the hills, I felt that my lot was cast in a goodly heritage. And so it was.

No man of finer calibre than Thomas Kingsmill Abbott ever served the Church in New South Wales,[1] and to work under his leadership was a privilege. I found the people congenial, the work interesting, and the summer climate—well, bad enough, but not as trying as I had been told. Most important of all from the personal point of view, in Tamworth I found my wife,[2] and when I left after fifteen months of a very happy curacy, it was with the archdeacon's benedictions on a newly-wed couple. As to my hobby, with which these recollections are mainly concerned, Tamworth afforded it ample scope for development. If 'T.K.' was no botanist himself, he was keenly interested in my doings and findings, and would invariably bring home for me anything that caught his eye during his own country trips and which he thought I might like to have.

Up to this time, I had nothing from which to identify plants except Baron von Mueller's Victorian *Key*. At Tamworth I came across so many things which did not seem to be described in that book, that before long I wrote to the government botanist in Sydney, the late J. H. Maiden,[3] asking if he could recommend some other book. Mr Maiden replied at once, advising me to get Moore and Betche's *Handbook of the Flora of New South Wales*, offering to help with any plants I could not determine, and asking me to send any specimens I could spare for the National Herbarium. From that time onward to his death twenty-three years later, Mr Maiden was 'guide, counsellor, and friend' to me in all things botanical. Whatever little contributions I may have made towards the botany of New South Wales are really due to him. Except when he was ill or out of the State, I do not think he ever failed to answer personally a single letter that I wrote—and I ought to blush at the thought of the number I did write. Of course I complied with his request to send plant material for the National Herbarium, and have continued to do so ever since.

I do not know that I discovered anything very remarkable among the Tamworth plants. I remember being much impressed by the first large

The Rev. Herman Montague Rucker Rupp, BA, curate of Tamworth, at the time of his marriage, December 1904. (Crown Studios portrait, by courtesy of Mrs Rachel Cox)

Florence Mabel Dowe (1879–1956) at the time of her marriage to the Rev. H. M. R. Rupp in December 1904. (Crown Studios portrait, by courtesy of Mrs Rachel Cox)

epiphytic orchid I had ever seen (except in conservatories)—a Cymbidium which is common on the western and southern slopes of New England.[4] It is the only epiphyte of Tamworth. Strictly, we should not include Cymbidiums among our epiphytes. An epiphyte is a plant which grows *upon* some other plant, i.e. on the surface or bark. The roots can be seen on the bark; they feed upon air and water, and almost certainly on the products of certain fungoid growths. Cymbidiums, however, start life by the germination of the seed in a crack of some tree which is decayed and hollow inside. The roots penetrate into the hollow and grow down (ultimately) as far as it extends, obtaining their food from the humus of the decayed wood. In neither case is the tree itself injured, and those fastidious folks who profess to dislike orchids because 'they are parasites on the poor trees' are wasting their sympathies. A parasitic plant feeds on the sap of its host, and may often thus drain the 'life-blood' of the host till death

ensues. The process may be seen going on all too frequently in our scrubs and open forests, where mistletoes are increasing alarmingly, and destroying valuable timber trees. But to class orchids as parasites is nearly as absurd as to class a man as a parasite because he rides a horse. I admit that the horse may have his own opinions about that, and if the man never got off at all—but there! We'll call him an epiphyte and get back to Tamworth. I found a wattle which had not previously been recorded so far north, and a few plants noted by Mr Maiden and his staff as 'rare'. My Tamworth census gives 270 species including 16 orchids and 10 ferns.

In 1904 my wife and I settled in the vicarage of Warialda, on the far western slopes of northern New England, about a hundred miles north of Tamworth. The extent of the Warialda parish at that time (it has since been divided) may be gauged by the fact that Tamworth was no further from us than was my northern boundary at Boggabilla, on the Queensland

The township of Warialda as the vicar saw it in 1905.

border near Goondiwindi. I had horses and sulky, but most of the long journeys, sometimes occupying over a week, were done without the sulky: i.e., I was, as we have agreed, an epiphyte. I also had the medium of conveyance known vulgarly as a pushbike, and in an evil moment I decided to use it for a trip to Yallaroi, forty miles out and nearly half of them black-soil miles. It rained—after I got there; and I had to tackle the home journey next day. I sold the pushbike that week.

Warialda is subject to severe droughts, but in a normally good season it is a veritable paradise for the botanist. The township lies in the valley of the Warialda Creek, an affluent of the Gwydir. The hills on either side are of sandstone, and there are large tracts of sandy scrubland. Here grew in profusion many of our loveliest wild flowers. Never had I seen such flannel flowers[5]; never in one area such a wealth and variety of beautiful wattles. In the winter months a blue daisy-bush, *Olearia ramosissima*, spread its dense masses of bloom like azure seas for hundreds of yards on either side of the track. The lovely native daphne, *Ricinocarpus bowmanii*, grew in miniature scrubs to the south of the valley.

A few miles above Warialda the creek had cut its way through a granite outcrop known as Cranky Rock, and here were plants that one never found on the sandstone. To the north and the west—the directions in which most of my journeys led—were vast areas of red loam and genuine black soil, with stretches of sand again here and there. For forty or fifty miles the country was undulating, the hills gradually lessening in height till the last of them sloped out on to the great plains. Dense pine scrubs were frequent: towards the plains these gave place to mixed scrubs of pine,[6] belah,[7] and sandalwood,[8]

or 'budda',[9] and there were open forests of ironbark and box in between, with apples, *Angophora intermedia*,[10] near the watercourses. Each class of country had a flora of its own, at least to the extent that there were plants never seen on the sandstone or loam but common on the black soil, and *vice versa*.

Out on the plains towards Boggabilla trees were rare, and usually marked the courses of intermittent creeks. On these open plains *Crinum flaccidum*, here called the Darling Lily, was in its glory. In a good season I have ridden through acres of it—tall stems crowned with their perfumed amaryllis-like flowers, white and pink. My census for Warialda parish, recorded during two years and a half, shews 393 species, 24 of which belonged to the genus *Acacia* (wattles), 16 to the orchid family, and 9 to ferns. The orchids were not very striking: there were one or two Greenhoods of interest,[11] and the big *Cymbidium* (*C. canaliculatum*) extended out to the plains.

When we left Warialda, for reasons concerned with my father's health we went to Victoria for two years, and I took charge of the parish of Yea, in the Goulburn Valley district. I did no botanizing there at all: I do not know why. Returning to New South Wales at the end of that period,[12] I was appointed to Copmanhurst, at the head of the navigable waters of the Clarence River.

Florence Rupp, with daughter Rachel, seated in the grounds of the Warialda vicarage, 1906. The Church of St Simon and St Jude is in the background. (From glass negatives kindly lent by the late Arthur Rupp, Willoughby)

The Vicar of Warialda about to set out for Boggabilla, eighty miles away, 1905. (Photograph by courtesy of Mrs Rachel Cox)

4 The Clarence

New South Wales has many beautiful rivers. Every one of the streams that traverses the coastal belt between the Dividing Range and the Pacific, from the Tweed in the north to the Bega in the south, has its own charms. I have seen them all, and should find it difficult to judge between their merits. I suppose that for sheer loveliness the Tweed, with its kaleidoscopic background of Mount Warning, stands above its rivals. And the Hawkesbury, as we all know, has been dubbed the Rhine of Australia. Yet there is a calm, simple majesty about the Clarence, a regal splendour of stately progress, as it sweeps along its broad highway through the rich lands for fifty miles from Grafton to the heads, which I have found in no other river. Above the environs of Grafton, its character is entirely different. On the thirty-mile run up to Copmanhurst there are times when you might easily imagine yourself on the Upper Hawkesbury.

Beyond Copmanhurst, for miles and miles there are magnificent stretches of broad waters backed by lofty mountains clothed in subtropical forests, and linked up by rollicking, rock-studded rapids. I have stood spellbound before many a scene on those upper waters that might even dare to give challenge to the Tweed. Broad reaches, still and silent, the mighty ramparts of the valley mirrored to the smallest detail on the unrippled surface of their deeps; little creeks of liquid crystal, that have come tumbling and gurgling noisily down through the brushes, steal quietly across the narrow flats to feed the great mother-stream; palm-fronds quiver and rustle, caught in the eddy of a whispering breeze; fern-trees gleam golden-green as a stray sunbeam penetrates through a gap in the canopy overhead. To me the Clarence is the Queen of Rivers.

Copmanhurst is a small settlement at the head of navigation; just above its little wharf are the first of the countless 'falls', as the river folks term the rapids. Eighty miles of navigable waters stretch from here to the heads, though none but small craft come up further than Grafton. In my time, the Copmanhurst parish involved frequent journeys to Cangai, a copper-mining township away in the mountains on what we then called the South River. This is the largest tributary of the Clarence, and throughout its upper course it is better known as the Nymboida, now familiar to the public in connection with its hydro-electric scheme. Some twenty miles above Cangai the Nymboida receives the Mann, which rises far away on the New England plateau beyond Glen Innes, and tears its way, foaming and roaring, through a great granite gorge to the junction. Below this point the stream was known as the South River, but I believe it is now generally called by its original name, the Mitchell.

The term 'South River', I suspect, came into use owing to the uncertainty of the main stream at the Clarence–Mitchell junction. The Mitchell, flowing north past Cangai, joins the Clarence, flowing south from beyond the Queensland border, some twelve miles further down; the junction is so nearly in a straight line, and the two rivers are so nearly of equal volume, that the rightful claim to supremacy was considered doubtful. From this point the Clarence turns eastward towards the Pacific, and almost immediately plunges through The Gorge, where for several miles it is divided into narrow, cliff-girt

Rupp's views of the Clarence River at the Grafton Wharves and at the vehicle and stock punt at Ramornie, about 1910. (From glass negatives kindly lent by the late Arthur Rupp, Willoughby)

With the aid of daughter Rachel, the vicar of Copmanhurst uses his sulky to transport poles from the bush, probably for constructing a greenhouse for rainforest orchids, about 1910. (Photograph by courtesy of Mrs Rachel Cox)

Cangai, the Clarence River copper-mining settlement which Rupp used to visit when he was Rector of Copmanhurst, 1908–11. (Photograph, *c.* 1910, from Bishop Cooper's papers, by courtesy of Mr David Stammer)

channels which rush over two beautiful falls. Copmanhurst is twenty miles further down (probably more than forty by water), and between there and Grafton the river receives another considerable affluent, the Orara, navigable for a few miles above the junction.

If the reader has been able to follow this description, it will be gathered that the Clarence watershed drains a vast area, and this is the secret of its great volume. The main stream rises in the highlands not far from Stanthorpe in Queensland, and runs south for something like a hundred miles, between the Richmond Range and the Great Divide. Then it receives the Mitchell, which through the Nymboida drains the New England Ranges almost as far south as the outskirts of the Dorrigo plateau, and through the Mann provides outlet for several streams of northern New England. Finally the Orara, and below Grafton the Coldstream, drain the Coast Range that runs north and south between the Great Divide and the sea.

Mine buildings on the Mann or Mitchell River, as Rupp saw them about 1910. Note his camera case in the lower right-hand corner. (From a glass negative kindly lent by the late Arthur Rupp, Willoughby)

As might be imagined, the Copmanhurst district provided a fine field for a botanist. It was my first introduction to the subtropical vegetation of the North Coast brush forests—and incidentally, to the scrub ticks and leeches which infest them. An enthusiast must put up with such little unpleasantnesses, but the botanist suffers less from these 'varmints' than the bird man, who must lie still and endure torments from ticks, leeches and the ubiquitous mosquitoes, lest he scare his timid quarry. I doubt if I should ever have made good as an ornithologist, much as I love birds. Bird men may scorn my pusillanimity, but there it is: I simply cannot lie still and do nothing with ticks boring into my back and leeches feasting on my nether limbs!

Epiphytic orchids are, of course, abundant in the Clarence brushes, and as they were nearly all new to me, I revelled in them. North of Copmanhurst is a large area of sandstone country, and the high land above the valley is bordered by sandstone cliffs of varying height, from whose bases a long scarp falls away to the river flats. These cliffs, and the little brushes in the gullies that had fought a way through them, were very rich in orchids. I have always regretted that at that time I had not developed a little more conceit in my own capacity to work out the character of a plant, for I am sure that the lack of it caused me to pass over many variations from type, some of which probably indicated 'new' species. But I believed that Moore and Betche had practically covered the whole ground of the New South Wales flora; that to be dissatisfied with any of their determinations was rank heresy and impertinence, and that apparent departures from their descriptions were probably only the vagaries of variable species. I have grown older and bolder since then!

Thousands of plants of the great rock-lily orchid, *Dendrobium speciosum*, and the daintier pink rock-lily, *D. kingianum*, covered the rocks above the Copmanhurst flats. Pencil orchids hung in masses from the trees,[1] and tongue-orchids[2] often completely covered large portions of trunks and branches.

Twenty miles away, at Coaldale (now, I think, officially called Whiteville to avoid confusion with a South Coast town), there were still a few scrubs of Hoop Pine, *Araucaria cunninghamii*, and these noble trees were hosts to many beautiful little kinds of *Sarcochilus*. Ground-orchids were fairly plentiful, chief among them *Calanthe* of the moist brush-forests,[3] with its tall, stately spikes of snowy bloom above the great, broad leaves. The orchids, however, did not receive my undivided attention, for the flora in general was most fascinating. I was very soon introduced to the notorious Nettle Tree, *Laportea gigas*.[4] I had been warned and was wary of it, but anyone who is given to bush rambling where it grows is bound to be caught napping sooner or later. My encounters were not serious: to be well stung by this beautiful, velvet-leaved monster is no joke, and the effects may last for some months. On the Coaldale road I discovered a graceful and attractive wattle which, after very thorough investigation, was pronounced by Mr Maiden to be 'new', and thereafter it was duly enrolled amongst its many relatives under the name *Acacia ruppii*.[5] A new tea-tree, citron-scented, was also to my credit.[6] Mr J. L. Boorman,[7] then official collector for the National Herbarium, paid us a visit once, and stayed a few days. I kept him very busy examining a large accumulation of specimens, to the disgust of an elderly lay reader employed by the parish, who complained that we had done nothing for a whole day but talk Latin names.

My Copmanhurst census is quite a formidable one—572 species. The orchids number 50, of which 23 are epiphytes. Had I known as much about orchids and their ways then, as I have learnt since, it is quite possible that the total might have been doubled. Ferns number 36, wattles and eucalypts 17 each. The family which includes the wattles (Leguminosae) is very prominent on the Clarence[8]: there are 70 members of it on my list.

We left Copmanhurst early in 1911, as I had been appointed vicar of Barraba, on the western slope of the New England Tableland.

5 Barraba and the Nandewars; and Pickings Here and There

Barraba days are among the happiest recollections of my life—though not all of them, for we had the longest run of sickness and mishaps there one year that we have ever experienced. But from a parsonic point of view (no, I'm not going to moralize) Barraba has always seemed to me an ideal parish, because the laymen never allowed me to worry about finances, and nobody grumbled, and there were no snags. But this is not botany.

The western slopes of New England are very distinct in character from the eastern. On the eastern or coastal side, the plateau for the greater part of its length falls away suddenly into deep ravines and gorges, their flanks covered by dense forests which become more and more luxuriant and subtropical in character as they reach the lower levels. On the western or inland side the descent is far more gradual, the climate is drier and the subtropical element is lacking in the vegetation, and though there are wild ravines and dense scrubs, they are quite unlike the humid gullies and brush forests on the eastern fall.

A few miles to the north of Barraba, which lies among the hills some sixty miles north of Tamworth, a spur from New England runs out westward, terminating in an irregular massive range attaining a height of 5,000 feet on Mount Kaputar, and a few feet less on Mount Lindesay. The spur with its western citadels is the Nandewar Range, more commonly called 'the Nandewars'. There is a tremendous drop on the western side of the Nandewars to the great plains that stretch out to the

Darling and beyond. It is the absence of any intervening hills worthy of the name that gives the Nandewars so imposing an aspect as you approach them from Narrabri on the plains. From the Barraba side, where the whole country is hilly, they are far less spectacular.

Barraba is in the very front rank of the wool-producing districts of Australia; and sheep have no respect for botanists. On the hill country more or less cleared of timber and devoted to the culture of fleeces, I was obliged to confine my researches chiefly to grasses, which I had previously only skimmed. I sent specimens of sixty-two species of grasses and sedges to the National Herbarium. But on the highlands to the east towards New England, and in the ravines and gullies of the Nandewars to the west, there was ample scope for my hobby, and it was tantalizing to realize that so many treasures were hidden there awaiting discovery, while comparatively few opportunities to search for them came my way—I had not the time. Still, I managed a good deal.

Two excursions stand out in my memory—one to the Gwydir Falls on the Horton River, and one to the summit of Mount Lindesay. The Horton rises, I think a few hundred feet below the summit of the Lindesay. Or is it Kaputar?[1] I am not sure, for in my time Lindesay was supposed to be the highest point. Anyhow, I have been there—right to the source in a bog under a charming grove of fern-trees. The little stream emerges from these summit-hills on to a small plateau, at the eastern edge of which it takes a plunge over a 370-foot precipice into a mighty gorge. The country here is exceedingly wild and rugged, but the falls and their gorge are worth going to some trouble to see. I am not sure that they are quite worth the trouble I had that night, though. My sleeping-chamber was about ten feet long by six wide: it had a single small pane of glass for a window, and the only door led into a room occupied by a married couple. Certain un-nameable creatures, which American travellers used to call 'sofa-birds',[2] swarmed over me like ants, and I was not pleasant to look upon next morning. I may add that I felt even worse than I looked.

My visit to the summit of Lindesay was marred by a nasty blow to my pride. Without unduly boasting, I may claim to possess a fairly good 'bush instinct', and have often been able to find the way out of tangled forests where others have lost their bearings. On this occasion I accompanied a boundary rider from Mount Lindesay homestead as far as the source of the Horton, where he left me to follow my own sweet will. I had been so concerned on the way up to take mental notes of gullies and corners that looked good for plants and photographs, that I failed to watch the track. Consequently, after visiting the summit and fossicking round the birthplace of the Horton, I realized that I had no idea where the home track lay. I gave the pony I was riding full permission to shew me, and he took me down into an appalling ravine which I had certainly never seen before, and I had the utmost difficulty in getting out of it. As a matter of fact I got back to the homestead an hour or two before I was expected, for I was so anxious not to have a search party despatched after me that I made desperate efforts to get down on to the plateau somehow and somewhere, and did so sooner than was necessary. But it was some time before I mentioned my bush instinct again.

The orchids of the Barraba district are chiefly ground species, but one *Dendrobium*—the dainty little Tongue Orchid[3]—has crept over from the east on to the granite country of these western slopes, and the *Cymbidium* mentioned in earlier chapters is abundant.[4] Several of the ground orchids were new to me, such as the green-flowering *Dipodium hamiltonianum* and the quaint Greenhood *Pterostylis truncata*. Twenty-nine species of the family were found. Among other plants were a *Boronia* to which Mr Edwin Cheel has given my name,[5] and a *Coprosma* on Mount Lindesay which Mr Maiden told me had not previously been found north of the Kosciusko country,[6] but I have seen it since on Barrington Tops. There were many very beautiful

Mementos of Rupp's unforgettable visit to Mt Kosciusko about 1912. *Above:* 'The Top of Australia 7,256 ft.' *Below:* The Blue Lake. (From glass negatives kindly lent by the late Arthur Rupp, Willoughby)

wattles, especially on the granite. The Mount Lindesay forests more nearly approached the character of coastal brushes than anything else I have seen on the west of the main Divide, and indeed there were trees and shrubs and ferns there that one usually associates with the coastal flora. My census gives 451 flowering plants and ferns for the district.

During my term as vicar of Barraba I had a summer holiday at Kosciusko. I need scarcely say that I took no holiday from botany while I was there, and I found the alpine flora most fascinating. Several ground orchids venture up above 6,000 feet in the summer months, and I added half a dozen species to my herbarium, besides a large number of other plants. I need not inflict my impressions of Kosciusko upon my long-suffering readers. Suffice it to say that if I could go there again tomorrow I should certainly do so, and I say this not as a botanist, but as an ordinary person who likes a good thing.

I left Barraba to join the office staff of the Australian Board of Missions, and for the next five-and-a-half years our home was in Sydney,[7] though a great deal of my time was spent travelling on deputation work in New South Wales, Victoria, Tasmania and Queensland. It would be tedious to go into details of how I occupied spare mornings or afternoons in all kinds of places where my duties led me, making hurried explorations after plants. But many specimens were added to my own collection in this way, and National Herbarium records will testify that I did not forget Mr Maiden. I had little opportunity of becoming acquainted with the Queensland flora, as my time there was monopolized by Brisbane. On one free afternoon I was attracted by the name 'Manly', and spent an hour or two gazing upon mudflats and dodging mosquitoes. But I had several glimpses of what Tasmania could do in the way of wild flowers, and they whetted my appetite for more. I was booked to preach and lecture for the Rev. (now Archdeacon) H. B. Atkinson,[8] then rector of Devonport. As we sat in his study I caught sight of Leonard Rodway's *Tasmanian Flora* (1903) on a shelf, and asked if I might look up something I had got hold of in another district. The discovery of a mutual interest in botany proved the foundation of a very delightful friendship. Mr Atkinson's father had been a botanist of some note,[9] and was an authority on Tasmanian ferns, and I found that 'H.B.' knew a great deal that I was very willing to learn. He and his wife took me for a memorable day's outing up the Forth River, among the beautiful native beeches which Tasmanians will persist in calling 'myrtles'. I cannot bring myself to adopt this misnomer—nor have I ever heard a satisfactory reason for it. I suppose as a matter of fact it is no worse than calling Casuarinas 'oaks'—but some Casuarina timbers *are* remarkably like oak—and besides, 'Casuarina' is rather a mouthful, but who can't say 'beech'?

I paid a second visit to Tasmania later, and found 'H.B.' installed as rector of the fine old Church of the Holy Trinity, Hobart. He was preparing to accompany his bishop to England for the 1920 Lambeth Conference,[10] and he rather took my breath away by inviting me to come and act as *locum tenens* for the eight or nine months during which he would be absent. There did not seem to be any particular reason why I should decline, as after more than five years of work with the ABM I was ready to return to a life less concerned about timetables and motor service cars, and Tasmania attracted me. I agreed that, provided my wife was willing, I would come.

Meantime, during those ABM years, I had of course spent a good deal of time in and around Sydney itself. Saturdays were usually free, and most of them saw me exploring the richly dowered bushlands on the outskirts of the metropolis. We lived for most of the time at Longueville on the Lane Cove River, where all sorts of treasures were to be had then, quite close to hand. I found two of the very rarest of our Prasophyll orchids within a few hundred yards of our house,[11] and waratahs grew just outside our fence. That was less than twenty years ago.[12] Looking at my Lane Cove records, I find fifty-two orchids on the list—all but three were ground-species. Among them was the remarkable *Galeola cassythoides*, which my small son discovered during a visit of 'H.B.' to Sydney. Galeola is a climber, and it is not uncommon on the hills of the coastal area, but is now very rare near Sydney.

R. D. Fitzgerald lived just opposite Longueville, at Hunters Hill,[13] and many of the types which he figured in his unrivalled work, *Australian Orchids*, were collected in the vicinity. Recently, through the kindness of a mutual friend, I was privileged to visit his son's home at Hunters Hill, and we were treated to a kind of private exhibition of interesting 'Fitzgeraldiana'. Ghosts of long-departed orchids of Hunters Hill must have stirred that day, for the company which the great orchidologist's son entertained with relics and tales of his father included, besides two enthusiasts in the persons of Mrs C. A. Messmer of Lindfield and myself,[14] Dr R. S. Rogers or Adelaide,[15] *facile princeps* among living Australian authorities on orchids.

6 Tasmania

A *locum tenens*, whether in the ministry of the Church or in any other responsible position, often has a very difficult task. Sometimes the absent party for whom he is acting is not as capable a man as he is himself, in which case he may quite unintentionally create a spirit of discontent to greet the other man's return. Sometimes the latter may be—justly or unjustly—unpopular, in which case endless temptations to take advantage of the fact will certainly present themselves to his substitute. Sometimes the *locum tenens* does more than meets the approval of his returning principal, sometimes he does less. If dissension or other trouble arises in connection with his responsibilities, he may be at wits' end to know how to cope with it without indiscretion. None of these dire contingencies worried me at Holy Trinity, Hobart. 'H.B.' had long since proved his capabilities; he was not merely popular, but beloved by his people. He told me just what he did and what he did not want me to do and there was never a whisper of dissension or bickering. Small wonder, then, that I have happy memories of Hobart. My duties were not sufficiently heavy to preclude more opportunities of exploring the bush for many miles round than I had anticipated. I met the veteran government botanist, Mr L. Rodway,[1] whose son, Dr F. A. Rodway,[2] had practised at Barraba in our time, and has now for a number of years been a valued botanical correspondent of mine at Nowra, on the N.S.W. South Coast.

The Tasmanian flora is of great interest to the botanist, quite apart from its indisputable merits from the aesthetic point of view. It contains a large number of what I should term transitional forms, which are often puzzling, not to say confusing; but puzzles and confusions to the botanist are there to be solved and straightened out. There are, as one would expect, many plants which are also indigenous to the mainland, and the mountain flora is similar to that of the Australian Alps, though I think superior. But what surprised me most was the number of very distinct species, often in prolific abundance, of which no trace can be found across Bass Straits. For instance, several shrubby species of *Pimelea* are very common along the north coast; two are fairly plentiful in Victoria, but a third—the most abundant—is entirely absent from the mainland.

Of course I explored Mount Wellington—not as extensively as I should have liked to, but sufficiently at all events to make me love it like a Tasmanian himself. The hills round Granton, the Longley heathlands, Brown's River, the Bellerive and Lindisfarne scrubs, Kangaroo Valley (the home of Mr Atkinson's parents and sister) were all visited, and I had one delightful trip down D'Entrecasteaux Channel. Another day's outing was not so delightful. A public holiday came, and we decided to spent it 'somewhere where we haven't been'. A map was consulted, and Shark Point on Pittwater was selected. We were rolled out there in a crazy old train that was probably acquainted with George Stephenson. I believe that if we had picked over the whole of Tasmania, we could not possibly have hit upon any other spot so utterly uninteresting, unlovely, or barren. And we had to wait about five hours for a train back!

When the bishop returned from England he offered me a new Launceston suburban parish,[3] and after some hestitation—afterwards amply justified—I accepted. Here I met a botanical enthusiast and a very loyal friend in Mrs G. E. Perrin,[4] who has lately been rewarded by Professor A. H. S. Lucas for her valuable cooperation in some of his exhaustive work

During his short time at Launceston, 1921–2, Rupp's most memorable botanical excursion was to Mt Barrow, where he 'spent a fortnight's holiday in the foothills'. (From glass negatives by H. M. R. Rupp kindly lent by the late Arthur Rupp, Willoughby)

on Australian algae,[5] by the naming of a Tamar River species, *Nitophyllum perrinae*. My own herbarium owes much to Mrs Perrin's kindness. When she visited Western Australia in the spring of 1922, she collected for me several hundreds of specimens, including some of the most interesting of the western orchids. In July of the same year I had spent a few delightful days at the 'George Perrin holiday cottage' at Low Head, where the moorlands are rich in flowers.

The most notable botanical trip I had from Launceston was to Mount Barrow, where I spent a fortnight's holiday in the foothills. At that time—I cannot say what developments have taken place since—Launceston was almost oblivious of the attractions of this massive sentinel of her northern environs, and was content with the picturesque grandeur of Cataract Gorge and the placid beauty of the Tamar reaches. But Mount Barrow is an asset the possession of which any tourist resort might envy. The panorama from its 4,400-foot summit is magnificent, bounded on the north by the blue hills of far-off Bass Straits islands, on the south by the distant bulk of Mount Wellington and its satellites, on the west by the aptly named Tiers, and on the east by a wild jumble of wooded ranges. Between you and the Tiers lies the silver ribbon of the Tamar, with Launceston nestling at its northern end where the two Esks meet. A few miles to the south-east looms up the long rugged mass of Ben Lomond, Tasmania's highest mountain. And then—the Barrow plateau below the summit—what a galaxy of colour and beauty there is! Citron-scented boronia, delicate gentians, mignonette-spiked *Bellendena*, rice-flowered *Richea*—these are but a few of the floral gems. *Richea* is one of Tasmania's most

interesting groups, belonging to the family of Australian heaths (Epacridaceae).

Only eight species are known, and seven of these are restricted to Tasmania, the eighth occurring also on the Australian Alps. Lower down the mountain below the plateau are tangled groves of beech[6] and Christmas-bush (not the New South Wales kind[7]) and waratah. The Tasmanian waratah,[8] though not equal in splendour to the regal 'national flower' of New South Wales, is neither so stiffly aristocratic. Lower still are giant gums, and beeches now giants too, and fern-tree bowers, and mossy dells decorated with azure mountain-berries. In the foothills grows one of the largest and stateliest of our Greenhood orchids, the Sickle Greenhood already mentioned in connection with the Victorian Mount Buninyong.[9] Below the hills is the winding, wattle-lined stream of Saint Patrick's River.

Cataract Gorge is, of course, the pride of Launceston, and a very wonderful place it is. From a botanical viewpoint, I found its most interesting features in a little scrubby patch of vegetation growing among boulders nearly opposite the tea-rooms at the First Basin. Here is a very curious accumulation of plants otherwise practically alien to the district. My theory was that either they or seeds of their elders had been washed down from distant hills by the fierce floods of the South Esk, and being arrested in their voyage by whirlpool processes at the Basin, had been deposited at this spot, and making the best of a bad job, had decided to accommodate themselves to their new circumstances.

At the end of 1922 I accepted an offer of work from Dr Stephen,[10] Bishop of Newcastle in New South Wales, and said goodbye to Tasmania immediately after Christmas.

7 Bulahdelah

After three months of uneventful relief work,[1] I accepted charge of Bulahdelah on the Myall River, about seventy miles north of Newcastle. The village is scattered along the western base of one of the most remarkable rocky hills in Australia, known as the Alum Mountain. Barely 1,000 feet high, its bold cliffs and rock-masses make it the dominant feature of the landscape for miles along the Myall Valley. I know of few more striking scenes than that which greets the traveller's eye when, climbing to the summit of the range that walls in this valley on the west, the road suddenly curves, and he finds himself looking over a sea of undulating tree-tops to the strangely tinted Bulah Delah—'the Great Rock'—on the far side of the valley. The colour scheme of the Alum Mountain is unique. The alunite rock of which it is composed, when newly fractured, is a delicate pink; but on the worn surfaces of the cliffs and natural monoliths of the upper half of the mountain, the pink has faded into different shades, and these blend with the white-and-brown of lichens and the green of mosses and matted rock-creepers till the general effect is undefinable. If you approach Bulahdelah in the late afternoon and are lucky enough to see a passing shower sweep across the Great Rock, you will never forget the opalescent sheen that suddenly gleams as the rays from the western sun strike the wet cliffs.

The Alum Mountain is naturally the pride of Bulahdelah; but there is something remarkable about the Myall River too. It rises in the ranges to the north, only some thirty miles away at most, and until within a mile or two of Bulahdelah it is a mere creek. Then suddenly its bed broadens, and sinks to a depth of thirty to forty feet, which is maintained (except at one spot where a rock-bar has shallowed it) until the channel runs out into the Myall Lakes fourteen miles lower down. Crossing the south-west corner of the lakes—which expand and stretch northward for twenty-five miles among the hills—the river emerges to flow south for another ten miles before it discharges into Port Stephens. How did that small stream come to carve out such a consistently deep channel?

The Bulahdelah district holds the high-water mark of my botanical activities, for in eighteen months I had collected over 700 species of flowering plants and ferns (with very little attention to grasses and sedges). 'A botanist's paradise' is a term apt to become threadworn, and I have used it before in these recollections: therefore I am almost disposed to class Bulahdelah as the botanist's seventh heaven. Certainly I do not know its equal. It includes almost every type of country except the alpine. Extensive peaty moorlands and heathy flats; swamps, bogs and lagoons; sandhills and rocky ridges; tea-tree scrubs, palm scrubs, bush forests, open forests; deep gullies and more or less open grasslands. Bulahdelah was a busy centre of the timber industry, but though timber-getters still form a large part of the scattered population, the old-time activities are gone—maybe to return some day, maybe not.

I recorded eighty-nine species of orchids from this district (inclusive of a few not collected by myself but sent to me for identification, and excluding two or three doubtful forms). Among them is one—constituting in itself a new genus and species—which may fairly claim to be the most remarkable orchid hitherto found in New South Wales, and which I had

Ernest William Slater, who in November 1931 discovered the subterranean saprophytic orchid which Rupp named *Cryptanthemis slateri.* (Photograph taken at Bathurst, *c.* 1960)

Close-up and enlarged view of the flowering head (capitulum) of the subterranean orchid which Rupp described as a 'wonderful discovery' — *Cryptanthemis slateri* (now *Rhizanthella slateri).* (Photographs by courtesy of E. W. Slater)

Rupp's photograph of the subterranean orchid he named *Cryptanthemis slateri* (now *Rhizanthella slateri).* He noted: 'Plant in situ. Alum Mtn. Bulahdelah 10/10/1933. Soil excavated on one side'. (Marjorie Loader Collection)

the privilege of describing and naming as *Cryptanthemis slateri*[2]—the former ('hidden flower') in allusion to its habit, and the latter in honour of its discoverer.[3] This extraordinary plant 'lives and moves and has its being' *under the ground*, and was accidentally discovered by Mr E. Slater when digging up tubers of the Hyacinth Orchid *(Dipodium).*[4] Even the little flowers, which are clustered in a head at the top of the stem, if not actually produced beneath the surface of the soil (a point not yet quite cleared up), are effectually hidden under accumulations of dead leaves and other debris. An orchid of similar habit was discovered in Western Australia three years earlier, creating quite a sensation in botanical circles. The Bulahdelah plant is too distinct to be included, however, in the W.A. genus named by Dr R. S. Rogers *Rhizanthella.*

Including two Cymbidiums, 22 of the Bulahdelah orchids are epiphytes, the remaining 67 being dwellers upon earth. On the Alum Mountain itself 58 of the total 89 orchids have been recorded. One of these, a tiny Helmet Orchid,[5] puzzled even the redoubtable Dr Rogers when I sent him specimens; but ultimately he found that it had been discovered and named by Allan Cunningham in 1833, and thereafter lost sight of for ninety-one years! It has since turned up on Russel Island in Moreton Bay.

This, however, is not a treatise on orchids, and much as I should like to describe some of the larger

Rupp's caption for this photograph was: 'The Alum Mountain Bulahdelah. *Cryptanthemis slateri* was discovered just behind the trees at the back of the old Church'. (Marjorie Loader Collection)

and more beautiful forms, I must forbear. When I turn to the non-orchidaceous flora of the district, I am in a quandary; for to give even a bare catalogue of the 'things of beauty that are joys for ever' would occupy too much space. Beautiful boronias, gorgeous bottlebrushes, flaming Christmas bells, golden peas as large as a sweet-pea, pink hawthorn-flowered tea-tree of the bogs, citron-scented; soft flannel-flowers, snowy everlastings; white and mauve hibiscus, maroon-throated; so the list runs on. Should anyone desire to see the most glorious garden Nature can produce along the seaboard of the mother State, let me direct them to Smith's Lake (tragically prosaic name for a vision of beauty!), twenty miles from Bulahdelah, in September.

Bulahdelah gave me a new friend in Dr H. Leighton Kesteven,[6] whose osteological and other scientific researches are known far beyond Australia. He was keenly interested in the orchids, and has sent me much valuable material since I left the district. It was a pleasure to be able to attach his name to a new *Dendrobium*,[7] and though recent developments may just possibly reduce this plant from specific to varietal rank, it will remain one of the most attractive Dendrobs we possess.

It was at Bulahdelah also that, through Mr Maiden, I first came into personal touch with our Australian 'orchid chieftain', Dr R. S. Rogers of Adelaide,[8] from whom I have learnt so much and whose many kindnesses I can never repay. It was a joy indeed, after nine years of correspondence, to meet 'The Doctor' face to face at the Sydney Science Congress of 1932, and even to succeed in carrying him off for a night to our temporary home at Collaroy Beach. It will have become evident from what I have written that the development of my hobby was leading me more and more to concentrate upon orchids.

From Bulahdelah I sent Mr Maiden some manuscript notes on orchids which I thought might be of use in the National Herbarium. Without consulting me, he sent them to the *Australian Naturalist*, which published them, and I was led to understand that further contributions would be acceptable. From that time on, notes from me on Australian orchids—especially New South Wales forms—and descriptions of new species, have appeared periodically, first in the journal just mentioned, and later in the *Proceedings of the Linnean Society of New South Wales* (of which I became a member), and in the *Victorian Naturalist*.

An article which I contributed from Bulahdelah

to the *Sydney Morning Herald* on 'The Cult of the Orchid' brought me a charming letter from Blackburn, Victoria, signed 'Edith Coleman'.[9] Mrs Coleman's orchid investigations are too well known, at least in botanical circles, to call for praise from one who has learnt so much from them. The correspondence thus begun has continued ever since, with much pleasure and profit to me, and, I hope, with some to the other party. Letters, indeed, now began to come from so many quarters that I was at this time led to consider whether I should not dispose of all my herbarium plants (numbering several thousands) except the orchids, and confine myself to the latter. This I did a little later.

Among the many letters received was one from Mr W. H. Nicholls,[10] of Melbourne, whose orchid contributions to the *Victorian Naturalist* are of such singular value to the student of these plants. The initial correspondence in this case also has grown to permanence, and we have collaborated in an exhaustive review of the Australian Helmet Orchids *(Corysanthes)* published by the Linnean Society of New South Wales.[11] Mr Nicholls is not only an excellent botanist, but also an accomplished draughtsman both in colour and in black and white. About this time also I came into more personal contact with Mr Edwin Cheel,[12] now curator of the National Herbarium, but then Mr Maiden's chief assistant. This has been another valued friendship, and Mr Cheel was later on kind enough to write the 'Foreword' to my *Guide to the Orchids of New South Wales*.

Mr A. H. Chisholm,[13] the well-known ornitholo-gist and author of what are, I think, the finest nature-study accomplishments in Australian literature, introduced me to yet another correspondent whose letters never fail to charm and to teach Nature-lore; for they seem always to bring with them the very atmosphere, with its scents and songs, of the mountain with its flowers and birds, which she loves so well. This was Miss Hilda Geissmann, now Mrs H. Curtis, but best known to her friends as 'Hilliegei', of Tambourine Mountain, South Queensland.[14] An occasional but very useful correspondence was also begun with Mr C. T. White, Queensland government botanist.[15] The late Adam Forster,[16] whose wonderful collection of more than a thousand faithful and artistic water-colour sketches of Australian flowers should certainly have been (but have not been) secured for one of our national institutes, wrote asking me for specimens, and this established yet another friendship. Two more names I must include here—Mr H. E. Finckh,[17] of aquarium fame, was interested in orchids, and wrote very valuable notes to me about several species. Before he died I had the pleasure of spending an afternoon with him at his delightful Mosman home, in his aquarium and among his birds and orchids. Last but not least of those with whom friendship began during my Bulahdelah days, was that Nestor among New South Wales botanists— friend and staunch disciple of R. D. Fitzgerald himself—Mr A. G. Hamilton,[18] who in 1926 paid a visit to Western Australia and brought me more than a hundred specimens of orchids from that treasure-field of terrestrial forms.

8 Paterson and Barrington Tops

In September 1924 I was appointed rector of Paterson, in the Maitland district. The Paterson River, one of the main affluents of the Hunter, has attained some celebrity in the commercial world— Paterson River oranges are considered to carry with them an unwritten guarantee of high quality. To a smaller but still considerable public, the river is no less celebrated for the quiet charm and beauty of its scenery. 'It's a little bit of England,' said a visitor as we stood on the bank below the old churchyard, and watched the shadows from the glorious willows lengthening over the placid stream, and the reflections of the green hills deepening under the margin opposite. Both river and township have other claims to notice besides oranges and scenery; for the history of settlement in the Paterson valleys teems with romantic interest. That, however, is a subject which cannot be followed up here.

The Paterson rises some sixty miles above the township, high up on the edge of Barrington Tops, a lofty eastern spur of the Dividing Range attaining 5,300 feet on Carey's Peak. Immediately beneath this peak, at the head of a mighty gorge clothed in a splendid forest of the Northern Beech,[1] *Nothofagus moorei*, rises a second stream, the Allyn, and the two valleys run south, approximately parallel, until they unite at the village of Vacy, five miles above Paterson. At the latter place the river emerges from a narrow neck between the hills and flows through rich alluvial flats to join the Hunter near Morpeth.

I had not been long in Paterson before I decided that my hobby must definitely be restricted in its scope. My herbarium was growing almost beyond control and was demanding more time than I was justified in giving it. All sorts of 'problems' in orchids

were calling for concentration on that group, and my ever-increasing circle of naturalist friends was pulling in the same direction. My original intention was to present the herbarium collection (except the orchids) to the museum of Trinity College, Melbourne. The warden, Dr Behan,[2] however, wrote saying that while the council appreciated the offer, it was felt that more useful purposes might be served if the collection were transferred to the Botany School of Melbourne University, where the specimens would be more easily accessible for reference, and probably under more effective supervision. To this I willingly agreed, and the collection is now part of the University Herbarium.[3]

Some of the new botanical acquaintances with whom I became linked up while at Paterson may be mentioned here. Mr Charles Barrett,[4] better known as a 'bird man', is interested in all branches of natural science, and was fittingly chosen some years ago to edit that admirable little journal, the *Victorian Naturalist*. Mr Barrett wrote to ask if I could help him to visit Barrington Tops, and I was able to put him in touch with that prince of naturalist-guides, Mr John Hopson—of whom more anon. He stayed at our rectory before and after the excursion, and brought me some specimens from the plateau which whetted my appetite for a trip to the highlands myself. In the following year, during a holiday visit to Melbourne, I was the pampered guest of Mr and Mrs Barrett in their Elsternwick home, where I was introduced to a bewildering number of fellow-naturalists: Mr E. E. Pescott,[5] author of *The Orchids of Victoria*, Mr W. H. Nicholls, now quite an 'auld acquaintance' by letter, Messrs A. J. Tadgell and T. Green, and others. I had also a few delightful days with Mrs Edith Coleman at her Healesville cottage-in-the-bush, where I met Mrs E. M. V. Eaves, who like Mr Green is a magician in the art of stereoscopic photography of orchids. I spent one or two days in the National Herbarium, where I helped Mr P. F. Morris with some *Caladenia* specimens, and met a very old acquaintance in the person of Professor A. H. S. Lucas of *Algae* fame, who had been a tutor at Trinity College in my student days. I also visited Mr Nicholls at West Footscray, and revelled in specimens of his draughtsmanship and photography.

Several Sydneians come into my catalogue here: Mrs C. A. Messmer, daughter of Mr H. E. Finckh, a born orchid-lover and kindest of hostesses at Lindfield—a born letter-writer too, as I am accused of being myself. Mr E. Nubling of Macleay Street also joined the circle about this time, and I have had much kindness from him and his wife. They are indefatigable orchid-hunters of the National Park and the Blue Mountains, and my own opinion is that if Mr Nubling ceased to hide his light under a bushel, and published his knowledge of our orchids, some of us would have to take a back seat. Mr George Scammell[6] and Mr F. A. Weinthal are two more enthusiasts in the same direction who have been good friends. Finally I must include Mr and Mrs John Tucker of Paterson, whose unique fernery includes so many good things, and who, I verily believe, know every orchid of the Paterson hills!

In 1925, during a visit to Sydney, I went to see Mr Maiden, now living in retirement at Turramurra. Though crippled and more or less of an invalid, he was still hard at work on his great revision of the Eucalypts, and full of genial enthusiasm and kindly encouragement. It was touch-and-go whether I could manage to fit the visit in; but how glad I was that I had done so when, three weeks later, the papers announced his sudden death.[7] Like every man, he has been criticized as well as praised: I can only say that to me, through all those years, he was *amicus benignissimus*.

In the days of long ago, the Paterson flats between the township and the Hunter were covered with cedar forests, the only remnants of which—an occasional cedar that may have been a sapling then—seem as though striving to conceal their identity among the willows lining the stream. Along the flanks of the valleys were dense brushes, now for the most part shrunken to mere patches in the pockets of the hills. Across the ridge opposite the township, however, along the rocky glen of Dunn's Creek, there is still primeval forest; and here, and in those patches of brush in the pockets, are orchids in plenty.

Moonabung Gorge, a wild ravine eight miles to the west, is not easy of access, and I visited it only once; but a botanist who could spend a whole day or two there would, I am sure, be well rewarded. Only on the Clarence have I seen larger accumulations of rock-lily Dendrobs, and there are others in abundance. A tea-tree scrub near Martin's Creek proved to be singularly endowed with ground-orchids, and a tiny-flowered *Prasophyllum* which I sent to Dr Rogers from there proved to be new, and was described by him as *P. ruppii*.[8] It has been found since at Bulahdelah, and on the track to Barrington Tops. Another new 'Prassie' was also in the Martin's Creek scrub, and I came across it a few years later at Port Macquarie. In a good springtime, the railway-line reserve between Paterson and Maitland produces the most wonderful show of the lilac-hued *Diuris punctata* I have ever seen. This has always been one of my favourites, and I well remember the excitement over my first specimens of it in a paddock near Geelong, more than forty years ago.

In 1925 I was introduced to an orchardist and farmer of Eccleston, on the Upper Allyn River, named John Hopson.[9] I was told that he was interested in 'bush things', and wished to meet me. Being no entomologist, and only incidentally interested in insects through their agencies in certain matters of the pollination of flowers, I did not then know that Mr Hopson had long since won his spurs in the field of entomology. But he talked orchids to me in a way that shewed he knew much about their habits, and had been attracted by them. 'I want to take you up to the Tops,' he said, 'there are lots of things up there that nobody knows about.'

Midsummer is the best time of year for exploration of that 5,000-foot plateau, and I could not see my way clear to go during the approaching summer, but in the spring I went on a short holiday at Eccleston, and he took me all about the foothills, and over a high ridge separating the Allyn from the Upper Paterson, and up the steep climb of 4,000 feet to the glorious beech forest of Bald Knob. Many there are who can join me in testifying to the delightful experience of exploring the bush in company with John Hopson. He was a devout member of the Congregational Church with a serious outlook upon life, and even the rollicking, happy-go-lucky youths who sometimes visited the Tops under his guidance voted him 'one of the whitest of white men'. His love of the mountains and every-thing they held of bird, mammal, insect or plant; his amazing knowledge of them (every spot he took or directed me to, proved, as he assured me it would, 'alive with orchids'); his kindly consideration, his ready interest in whatever was found; his quiet humour and gentle toleration of idiosyncrasies; all these qualities endeared the man to everyone who came under the spell of his acquaintance.

When, in January 1928, I was actually accompanying John Hopson on the long-contemplated excursion to the Tops, I asked him, as we skirted the deep gorge of the Allyn on a slippery bridle-track, whether he ever had trouble with inexperienced equestrians, he said, 'Well—I do suffer fools now and then, but'—with a quick twinkle of his eyes—'not gladly, you know.' I am half-afraid I was one of the fools that day, for in my anxiety to scan the top of a rock about on a level with my head, I stood up in the stirrups just as my steed was about to negotiate a fallen tree, with the result that he got his fore-legs over it, but his hind feet stuck in the mud. It was a very awkward spot in which to dismount, and Mr Hopson, who was a long way ahead, had to leave his own packhorse and come back to the rescue.

For me, those were four great days that he and I spent on the plateau. The weather was vile until the last day—rain, sleet and mist—but we simply ignored it. We camped in the old hut—since then burnt down—on the Barrington River, went exploring wherever my mentor thought best, and when we got back dried our clothes before a roaring fire; and we took no harm. The plateau consists mainly of broad moors, swampy in parts, traversed by clear, swift-running streams, the headwaters of several rivers. In January, the moors are vast carpets of flowers. There were orchids galore, many of them of great interest and some very beautiful. A new *Diuris* which I had named from specimens brought by Mr Barrett was in great profusion.[10] Two new 'Prassies' were found, and later on I had the pleasure of naming the larger one after Dr Rogers, and the smaller after my good guide and companion.[11] I had conjectured that on this high tableland plants usually associated with the southern States might occur, and this proved to be the case. In one of the bogs near our camp was a large colony of the fine Sickle Greenhood,[12] which I had last seen in the foothills of Mount Barrow in Tasmania. The Tasmanian Christmas Bush, *Prostanthera lasianthos*, was much in evidence, massed with splendid effect in shallow gullies.

On our last day we visited Carey's Peak, and stood for a long time in silence as we gazed out upon one of the greatest panoramic views in New South Wales. I have visited the Blue Mountains, Bulli and Cambewarra, the Brown Mountain and Kosciusko: Carey's Peak can hold its own with any of these. The Allyn Gorge with its incomparable beech forest provides such an impressive foreground, and the outlook sweeps over so vast an extent of ocean, coastline, and inland mountains and plains, that the scene is almost unique.

A few months later, and John Hopson had taken his last climb to the Tops. He died suddenly from heart failure after visiting a sick brother.[13] His body rests in quiet Eccleston among the hills: I like to hope that, some day, naturalists may carve some simple memorial on the rocks of Carey's Peak.

One other friend I have kept in reserve for a special note here—Mr W. J. Enright of West Maitland.[14] Mr Enright is well-known in geological and anthropological circles, but he is an orchid man too, and I met him at Paterson. He offered to lend me his complete set of Fitzgerald's *Australian Orchids*,[15] the recognized 'classic' on the subject, published by the N.S.W. government on a very elaborate scale in the days when governments did such things. Though far from exhaustive of Australian species, it is almost indispensable as a work of reference for botanists; but as it is long out of print and rarely comes on to the market—and then at high figures—I had been compelled to restrict my use of it to the rare occasions when I visited Sydney. Mr Enright brought out his set of the volumes, and allowed me to retain them for nearly five years, an act of kindness which I can never hope to repay. Indeed, no man, surely, was ever blessed with kinder friends: and in this matter of the great reference work on our orchids, I owe it to two of the others mentioned in these memoirs that I am now the happy possessor of a complete set of 'Fitzgerald'.

9 The 'Guide'; the Coalfields and the Pilliga Scrub

For some years the need of a new book on the subject of the New South Wales flora has been voiced among naturalists, and by people who would not call themselves naturalists but who wanted to know more about the wild flowers and ferns of their country. Moore and Betche's *Handbook*,[1] admirable in its time, was in a sense out of date, by reason of the number of new plants discovered since its last edition (1893),[2] and the many alterations made in classification, etc. Moreover it was also out of print, and only to be had occasionally second-hand; and finally, it was too strictly technical for the average person who is no botanist. Miss Florence Sulman's book for a while seemed to meet the need,[3] and is of course still useful; but it was not intended to be exhaustive, and only deals with part of the flora. There is still need for something more complete: Dixon's curious 'Key' has its value,[4] but is not the book we want.

As governments are not nowadays much concerned about these matters, private publishers must issue such books if we are to have them at all, and they are rarely profitable investments, because by their very character they do not appeal much to the general public. We might gradually get what we want in sections: an attractive little book on the *Ferns* would, I believe, be warmly welcomed. Then we might have one on the *Wattles*, another on the *Eucalypts* (Russell Grimwade's *Anthography* is beyond praise, but too costly),[5] one on the *Boronias* and their allies, and so on. Which, you may say, is all very obviously leading up to the fact that I have written a book on the Orchids of New South Wales myself!

True. It had been urged upon me for some time, but I did not think I could undertake it. However, I had notes extending over many years, and began to put them together to see how far they would answer the purpose, when a friend gave me a hint that Messrs Angus and Robertson, the well-known publishers, were looking for someone who might undertake a popular handbook of our orchids, and advised me to get in touch with them. I did so, and the result was the publication in 1930 of the *Guide to the Orchids of New South Wales*. From the publishing firm, and in particular from Mr George Robertson,[6] I have received generous treatment on a scale which I fear will never be repaid by the sales of the book! For although, if one may judge by its reception in the press and by the many letters received, it is serving its purpose—that of a key to identification which, while suitable for popular use shall yet be of some value to botanists—it naturally appeals only to people interested in the subject.

Mr Robertson, though 'an all-wool Australian' who has done far more for the country of his adoption than he will ever get credit for in this forgetful world, is also a Scot and a son of the manse. Whether he is of any value in support of the allegation that the Scot lacks humour, may be gauged by this little tale, which I hope that he will forgive: Circumstances delayed the publication of the *Guide* for a few months, and when it came out he sent me the first copy inscribed thus: 'George Robertson to H. M. R. Rupp. Proverbs xiii.12.' You can look up the reference.[7]

We left Paterson at the close of 1929 and went to Weston, only twenty-four miles away, but in its conditions utterly different, for it is an industrial centre in the heart of the South Maitland coalfields. Many of my friends commiserated with me on this move, seeing no possible connection between orchids and coal. As a matter of fact, I think the great interest of the coalfield orchids kept me going longer than would otherwise have been the case. The intense and increasing industrial depression, with all that it implied to a wholly industrial population dependent on the mines, together with the consequent uncertainty of my position, induced a severe strain upon nerves. But quite close by were miles of bushland, with the heaviest 'crop' of ground orchids I have ever met with. Greenhoods in particular were literally in countless myriads, a new species among them.[8] So it was sometimes possible to get away for a few hours and forget the worries of life in the field of Nature.

After two years, however, the strain proved too much, and I was advised to take three months' rest. These months were spent at Collaroy Beach, and it so happened that the Sydney Science Congress took place in that period, so that besides unfettered rambles about the Collaroy and Narrabeen hills, I had the pleasure, as I have mentioned above, of meeting Dr Rogers, who was the 1932 president of the Botany Section. The Congress meetings, too, were full of interest, and I made new and renewed old acquaintances from many parts. By August I had fully recovered; but a return to Weston was not desirable, and the outlook was not cheerful. The bishop of Armidale[9] offered me four months of 'outback' work in the Pilliga Scrub on the N.S.W. north-western plains, and I accepted it. I found this area of very great interest botanically.

The spring was a dry one, and I was disappointed at the absence of ground orchids; but to my surprise—for 'tree orchids' are rare west of the Dividing Range—there was a *Cymbidium* which, when it flowered, proved by far the most beautiful Australian member of the group I have seen: the flowers, borne in great profusion, are golden yellowish-green flaked with red.[10] But there were plenty of trees and shrubs the interest of which atoned

for the absence of orchids, and Pilliga post office scored well in parcels and letters from the vicarage! I should not have cared to remain there permanently: the parish involved heavy travelling over bad tracks, and I had 'done my bit' in that type of work twenty-five years before. Nonetheless we enjoyed our four months: the people were kind and hospitable, and the climate until mid-December was nearly perfect. We left immediately after Christmas; and with the great plains receding into the distance, this rambling record of bush rambles must end.

I ought not, however, in these botanical recollections, to omit mention of a New Zealand friend whom I have never met, but whose specimens enrich my orchid herbarium and whose letters—now, alas! rare through failing eyesight—have given much valuable information. Frequent allusions to New Zealand orchids in books and journals made me desirous of seeing some, as it appeared to me that there must be many forms closely allied to our own. I wrote to the Auckland Museum, and received a reply advising me to get in touch with Mr H. B. Matthews of Remuera and Kaitaia. I did so, and Mr Matthews, who is a descendant of one of Samuel Marsden's missionaries, responded by sending me a splendid assortment of specimens which have been invaluable. His son is the 'Dick' of that excellent tale of adventurous travel from Sydney to Cape York, *We and the Baby*,[11] and the two travellers called in at Paterson Rectory on their way north.

Finally (I know it looks rather like 'lastly, brethren'), I should be ungrateful not to acknowledge a Western Australian and a North Queensland correspondent, both of whom have sent me generous supplies of specimens. The former is Lieutenant-Colonel B. T. Goadby,[12] whose enthusiasm for orchids is known eastward as well as westward. The northerner is Mr W. F. Tierney of Cairns, and as for his enthusiasm—well, his letters make me shudder to think of the leeches and ticks I should have to endure in those tropic forests if he ever got his hands on me! What knowledge I am gaining of the many attractive North Queensland epiphytes, I owe very largely to his kindness in sending living plants and flowers.

10 Cui Bono?

Yes—I have often heard that asked. 'What's the good of it?' Well—what's the good of any hobby, for that matter? Of course, if a man thinks he will take up a hobby for the sake of the cash profit he can make out of it, I bid him good-day. His ways and mine don't meet. I know there are such people; and that they look upon a man who takes up a hobby without a single thought of its relation to money as a fool. But since he thinks the same of them, only more so, it doesn't matter in the least. My idea of a hobby is an occupation to which I can always turn with pleasure when I want relaxation from the frictions of routine or the strain of worries—and every man has worries, whether he hides them or keeps them in the window.

My hobby must not be anything which I take up with money profits as its objective, for in that case it will always have a door open inviting worry to enter. For the same reason it must not be an expensive hobby. I have no quarrel with the wealthy man who cultivates an expensive hobby: if he can afford it, good luck to him—probably the more he spends, the better for the country. Again, I have conscience enough to stipulate with myself that my hobby shall work harm to no man, and that if possible its value shall be shared by others besides myself. Now *for me*, botany is a hobby which fulfils all these conditions. A man may say he doesn't know what I can see in it: well, if he *wants* to know I am always ready to tell him; and if he doesn't, let him leave it at that. I may not know what he can see in

his postage stamps, or old china, or Buff Orpingtons, or in hitting a little white ball with an expensive club; but I have enough sense to realize that he sees something good in them, and I wish him luck with them. It doesn't in the least follow that what you or I can't see is no good.

Some people are willing to admit the value of agricultural botany, but insist that the investigation of wild flowers is nothing better than a useless fad. My dear dogmatist—for that is what you are—there could never have been any agricultural botany if the wild things hadn't been investigated first: and isn't it all to the good, even from your point of view, that much of this botanical spadework should be done by the voluntary labour of private individuals, leaving your Agriculture Department experts all the more free for the work you recognize as valuable? Besides, you are ignoring scientific values.

'The meanest flower that blows' has other services to perform than that of bringing to the poetical mind thoughts that lie too deep for tears. Its life history, its habits, its structure, its response to its environment, are all in some relation to those of other plants, and all have some definite scientific value. Therefore I am well content to have been admitted into the goodly company of amateur botanists, and my life is the richer for the many good pals I have gained through my hobby. If I have been able to contribute ever so little to the sum of our knowledge of the world of Nature, then I am satisfied that, for me, no better hobby could have been chosen.

Part III

Retrospect

Phaius Orchid*

In swamps from Stradbroke to Siam
A yellow moon, convex on water,
Draws the *Phaius* into bloom;
Along a shaft of pure titanium,
As if to name is to command.
Grandfather Forbes in revelation,
Wrote to the Herbarium: God's honest
Broker's *magnum opus*: Rupp, on p.110.

From a pontoon dredge in an eating
Lake, a metal snake feeds on the
Sedge; each grain of sand from
That pungent bog to fight the Reds,
Or win the space-race to the moon:
But who will make the *Phaius* bloom?

> —Edwin Wilson in *The Dragon Tree*, Woodbine
> Press, 1985, with the author's kind permission.

* The spectacular *Phaius* or 'Swamp Orchid' once grew in swampy areas on the north coast of New South Wales. In 1920 John Copeland Forbes, wheelwright of Mullumbimby, sent a specimen to the National Herbarium, where it remains with other material which was for long in Rupp's care. Over-collection and the destruction of habitats through coastal sandmining during the 1950s and 1960s are believed to have rendered the species virtually extinct in New South Wales.

H. M. R. Rupp's photograph of *Phaius grandifolius* (now *P. australis*) cultivated in the bush house at the Paterson rectory, 1925.

1 Early Days

I was born at Port Fairy, Victoria, on 27 December 1872. My father was the Anglican incumbent (a most unpleasant designation, but too widely used to be discarded) of that little seaside township, which in those days was called Belfast. My mother, whose maiden name was Rowcroft, died at my birth[1]; and soon afterwards my father was removed to Learmonth, near Ballarat. His only other child, my sister, Florence Emily Marie, was then about seven years old; at the time of writing she has attained the dignity of an octogenarian,[2] and lives at Lindfield, a well-known Sydney suburb. At Learmonth my father married Rachel Emma Tillery Kirkpatrick, when I was still an infant. There were no children of the second marriage; but my stepmother, at least so far as I was concerned, left nothing to be desired in the way of devoted affection. When I was four years old we moved to Koroit, only twelve miles from Port Fairy, and a little nearer to Warrnambool, which was—and still is—the largest town in that district of Victoria.

My paternal grandfather had been a German schoolmaster,[3] and in 1846[4] he decided to emigrate to Australia with his wife and three children—my father, aged eight, his sister, aged six, and an infant boy. During the voyage both parents and the baby died.[5] Happily for my father and his sister, there were family friends on board, and the children were adopted by Mr W. F. A. Rucker, a well-known Melbourne merchant of the early days, whose name appears very frequently on the pages of that interesting little publication, John Pascoe Fawkner's *The Melbourne Advertiser*.

My father was educated at a private school in Melbourne, and ultimately decided to train for the Anglican ministry. For that purpose he attended Moore College, Sydney,[6] where his name is still to be seen on the roll of students, 1862–67. He was ordained by Dr Perry, first bishop of Melbourne, and served in that diocese till 1875, when the western portion of Victoria became the diocese of Ballarat, in which he remained until his retirement in 1905. His devotion to duty and sincerity of purpose, coupled with pastoral gifts of no mean order, endeared him to his parishioners wherever he went. His first wife (my mother) was a daughter of General Horatio Rowcroft, who served in the Indian Mutiny. Her sister Oriana married John Bracebridge Wilson, for thirty-four years headmaster of Geelong Grammar School.

My recollections of Koroit, which I left seventy-two years ago,[7] are very clear. I remember the old galvanized-iron parsonage, to which my father added several weatherboard rooms; the log fences along the lanes, over which clematis showed its starry blossoms; and above all, the glories of Tower Hill Lake. Those glories, by the vandalism of man, have been almost completely destroyed.[8] Words are powerless to convey any real picture of the lake as I knew it in my childhood; yet I must try to describe it as it was then. It is a volcanic crater-lake about twelve miles in circumference, on its southern side only two miles from the ocean. On its northern side, a mile away, lay the village of Koroit. From there the road sloped gently upwards, till suddenly you found yourself on the brink of the steep banks of the lake, several hundred feet above the water. Those banks were densely clothed in virgin forest (of which

St Paul's Church, Koroit (dedicated 1890), is now served from Port Fairy. Note the chimney of the old vicarage on the extreme right. (Photograph: L.A.G., 14 September 1988)

today not a vestige remains), and rough tracks led down past bubbling springs of clear, cold water, into glens of musk and blackwood and eucalypt, adorned with all kinds of ferns; tree ferns and lady-ferns and staghorns and what others I forget. With what awe I used to peer cautiously into the huge burrows of wombats as we scrambled past!

The greater part of the lake was comparatively shallow, dotted with clumps of reeds; but on the Port Fairy side there was an area of clear water which was alleged to be 'unfathomable'—no doubt this was the original large crater. On the ocean side of the lake the banks were relatively low. In the middle were five islands, the largest of which had the characteristic conical peaks of old volcanoes. The island vegetation was quite different from that of the mainland, the principal trees being Casuarinas ('she-oaks'), and the ferns were restricted to one or two species which did not occur elsewhere in the neighbourhood. There were several small crater 'basins' on the islands, and one of the joys of children—old and young—was to loosen boulders at the top and watch them go bounding down till they leaped with a great 'plop' into the water at the bottom.

A crazy old corduroy causeway connected the largest of the islands with the mainland near the road from Koroit to Warrnambool. Most children were forbidden to cross unless accompanied by their elders; but two boys set off one Saturday afternoon, and having successfully mastered the causeway, proceeded to one of the crater basins and enjoyed themselves with glee, until they realized that night was falling. How they got back is beyond my power to relate, though I was one of them; not only did they get back, but they meanly evaded a search party which had gone to look for them, and it was not till the small hours of the morning that these unfortunate people learned of the return of the truants. My companion in this escapade, Graeme MacDougall,[9] passed away only a year or two ago; he became well known in New South Wales as the very capable secretary of the Sydney sesquicentenary celebrations. We had not met for over sixty years, yet a portrait in the *Sydney Morning Herald* of Mr D. G. MacDougall at the time of these celebrations caught my attention instantly, and induced me, during a visit to Sydney, to call at his office. Sure enough, he proved to be the mate of my childhood. He was astonished at my

The revegetating of much of the reserve at Tower Hill, Koroit, carried out since 1961, would have greatly pleased Rupp, who was rightly and strongly critical of the previous devastation of the area. (Photograph: L.A.G., 14 September 1988)

recognition of his portrait, and invited me to dine at his Potts Points home,[10] where we spent the evening recalling the adventures and sins of our youth—for the escapade I have described was not the only one of which we were guilty.

My sister not long ago called my attention to the fact that in our Koroit days we were well acquainted with a little girl who became one of the most distinguished figures in the field of Australian literature. Mrs J. G. Robertson, far better known, of course, under her pseudonym of Henry Handel Richardson, was a daughter of the Koroit post-mistress, and she and her sister Lily and their mother, Mrs Richardson, were frequent visitors at the parsonage. We knew the girls as 'Ettie and Lily'—the former dark and rather thin, the latter plump and, as befitted her name, fair. I have recently been reading Henry Handel Richardson's last book, *Myself When Young*, an unfinished autobiography of her juvenile years; she died before she could complete it.

Reluctant as I am to presume to criticize this 'swan song' of so gifted a writer—particularly as she can make no reply—I cannot refrain from commenting on her account of her Koroit days. The kindest interpretation I can put upon this portion of her narrative is that her memory completely failed her, and that she drew upon her ever-vivid imagination to fill the gap which memory refused to bridge. For the stories she tells of our father beating his daughter almost into insensibility, of our stepmother hating us children and being hated by us, and of her chasing me over the furniture with a riding-whip, are sheer fiction of an extremely unpleasant kind. My sister still possesses an old photograph of 'Ettie and Lily', and one of their mother; our recollections of them are very clear, and neither of us can explain H.H.R.'s extraordinary tale except by some hypothesis such as I have suggested. It would have been entirely alien to our father's character to exercise brute force against anyone, least of all to his own beloved daughter.

As for our stepmother (who had no children of her own), she admitted to me on her deathbed that she had often been lacking in sympathy and understanding with my sister; but I am sure this was not made obvious to outsiders. Upon me she lavished such indiscriminate affection that I was quite justly dubbed 'a spoilt brat'; and the story of the riding-whip would be ludicrous, but that it is unjust to the memory of a woman who gave me all the love that might have been bestowed on a child of her own. That love never faltered through all the years; and the pangs of death were softened for her by the knowledge that I was present to administer the Holy Sacrament to her and my father, as she lay dying. Thus it will be understood why I cannot allow H.H.R.'s imaginative tale to pass without protest. Possibly 'Ettie', who describes me as a 'monkey', felt that I deserved to be chased with a whip; and later on, through the medium of her self-confessed habit of 'making-up' came to believe that I had been!

But her whole picture of Koroit is warped. The country, though it might be termed flat when one got away from the lovely panorama of Tower Hill, was certainly not 'treeless' in those days; there was plenty of beautiful bushland within easy distance of the village; and my earliest recollections of those fascinating ground-orchids, the Spiders and Doubletails,[11] came from the open forests of Koroit in the late 'seventies. Let me add in conclusion that I yield to no one in my sincere admiration of the charm and quality of Henry Handel Richardson's literary work. *The Fortunes of Richard Mahony* alone would entitle her to a very high place, if not the highest, in the ranks of Australian writers. She shall have at least that tribute from the 'tiresome monkey' of her childhood days.

My botanical inclinations made their first appearance at Koroit. I fear that their beginnings did not create a good impression. 'Mother' was a great gardener; and devoted as she was to her small stepson, even she was not pleased when he came in clutching some of her choicest blooms in his grubby fingers. One of our best friends was Mr Frank Norman, who had a wonderful garden not far from the parsonage. He used to tease me by giving me the long botanical names of his treasures. So one day I hunted through a book of my mother's for the longest name I could find. With considerable difficulty I mastered it, and committed it to memory. When I was quite sure of it, I sauntered down to 'Mr Frank's' garden. He began his usual game with me; but instead of nibbling at the bait, I retorted, 'Ha! ha! you haven't got *Amaranthus hypochondriacus* in your garden!' He looked at me in astonishment for a moment, then threw back his head and broke into the roar of laughter for which he was famous in the village. I was not quite sure whether I had triumphed or not.

The Irish element was very strong at Koroit. I remember little whitewashed stone cottages, whose occupants betrayed their origin directly they spoke. Pigs and fowls, at least in the daytime, strolled in and out without let or hindrance. Next door to the parsonage was a little wooden cottage with a thatched roof, where dwelt an old dame named Biddy Maxfield. She was a small woman with a very wrinkled face, and a voice like a trombone. I was frankly terrified of her, though the poor old soul was harmless enough. Her skirts were as short as those which became fashionable after the First World War forty years later, and she used them as a receptacle for dried cow-manure from the lanes round about. Her husband, 'Old Tom', was a Protestant of sorts, and suffered considerably from the lashings of her tongue. He cultivated vegetables, and used to feed me surreptitiously through the fence on spring onions. Biddy could produce at will a dreadful cough, and she used to go the rounds begging for cast-off clothing, fortifying her pleas with the most alarming noises from her throat, to show how badly she needed warm clothes. If the lady of the house was kind, the blessings of all the saints of Ireland were invoked upon her; if she hesitated, Biddy threatened to expire on the spot, a martyr to 'Prathestant' lack of charity. Years afterwards I heard that when she died, more than three hundred dresses were found stored away in her cottage. Biddy and Tom were childless.

Camperdown, some sixty or seventy miles away, was the nearest railhead in those days. A visit to Melbourne—which was a very rare event for us—meant travelling by sea from Warrnambool or Belfast. The little tubs that did duty as passenger steamers were also cargo boats, and as the cargo usually included hordes of squealing pigs, these trips were not conducive to a love of the sea—particularly if there chanced to be a south-westerly gale blowing round Cape Otway. The attractions of Melbourne were many, but in the mind of a small boy there was always the dread of the inevitable return trip on the *Dawn* or the *Julia Percy*.

At the beginning of 1884 my father became vicar of Coleraine, some sixty miles to the north-west of Koroit, and a few miles from the Wannon River, chief affluent of the Glenelg. The church here was a picturesque stone building, ivy-clad, and possessing some fine stained-glass windows. In my father's time a peal of eight carillon bells, the first of their kind in Victoria, was installed. My acquaintance with

Coleraine was limited to holiday times, for soon after my father was settled there I was sent to school at Geelong. My sister, after attending one or two schools in Western Victoria, had become a boarder at St Catherine's School for daughters of the clergy at Waverley in New South Wales; and before she left there, she had made the acquaintance of the man who was afterwards to be her husband, Arthur A. Monypenny, then an officer in the Lands Department.

Geelong Grammar School was to be my real 'home' for the next eight years. It was then, of course, situated within the town boundaries; the removal of the school to Corio, across the bay of that name, and the erection of the present magnificent buildings, were not begun until 1913. But it had a fine reputation even in earlier days; and this was due chiefly to the character and work of its great headmaster, John Bracebridge Wilson, whose memory is deeply revered and honoured by the dwindling band of old boys who still survive from the years when he held office. During 1884 I resided at the Junior Grammar School, which was only unofficially connected with the Senior School across the street, and in those days was managed by Mrs Wilson's brother, Mr A. P. Rowcroft. Next year, however, I was duly enrolled as a member of the big school itself. I had always been a delicate child, and for some time I was not allowed to take any active part in sports; perhaps I may be pardoned for recording that I lived to carry off the school athletic championship for two years in succession. I may be further forgiven for adding that this feat was duplicated, many years afterwards, by my son at The Armidale School in New South Wales.

Mr Wilson aimed at the building up of character in his boys above all else. But he was a splendid teacher and a great scholar. His Sunday lectures on the Greek Testament, and his weekday classes in physiology, left indelible impressions on the minds of those who attended them. He was a strict disciplinarian, but an eminently just one. He was a distinguished naturalist, and was made a fellow of the Linnean Society of London in connection with his work on Australian algae. He gave generous encouragement to any boy who showed an inclination towards natural history; to him I owe the beginnings of what became a lifelong hobby—the study of Australian native plants. He had published, under the title *Florula Corioensis*, a small volume recording the plants of the Geelong district,[12] and this was followed later by another on the algae of Port Phillip.[13] He was an enthusiastic yachtsman, and

most of his holidays were spent cruising in his own boat, the *Archdale*, in the neighbourhood of Sorrento, near Port Phillip Heads, dredging for 'seaweeds'. He was an intimate friend of Baron von Mueller and Professor Sir Baldwin Spencer.

Among the staff of masters at the school in those days was one whose verses find a place in every anthology of Australian poetry—James Lister Cuthbertson.[14] Cuthbertson was an old Oxford Blue, and a very fine classical scholar, who could—and did (to the joy of his boys, who thus evaded many awkward questions)—quote Euripides or Aristophanes 'by the yard'. He was passionately devoted to the school, and like his 'Chief', he set first in his aims the building up of character. Self must be controlled for the welfare of the school community. So he wrote in 'Our Motto':

'Not self, but side,' our schoolboy pride
    Is thus to win the day;
And if defeat we chance to meet,
    'We played the game', we say.
So in the strife of later life
    The battle we may fight,
Not win applause, but aid our cause,
    Defend the side of right.
Then, Lighter Blue, whate'er we do
    Let this our motto be,
'Not self, but side,' whate'er betide,
    Shall win the victory.

That Cuthbertson was a genuine poet, capable of far greater things than his popular school ballads, is beyond question. A memorial volume was published by Melville and Mullen in 1912, under the auspices of the Old Geelong Grammarians, with the title *Barwon Ballads*. I quote a couple of stanzas from 'At Cape Schank':

Over the mirror of azure
    The purple shadows crept,
League upon league of rollers
    Landward evermore swept,
And burst upon gleaming basalt,
    And foamed in cranny and crack,
And mounted in sheets of silver,
    And hurried reluctant back.

And the sea, so calm out yonder,
    Wherever we turned our eyes,
Like the blast of an angel's trumpet
    Rang out to the earth and skies,
Till the reefs and the rocky ramparts
    Throbbed to the giant fray,
And the gullies and jutting headlands
    Were bathed in a misty spray.

He was a man of moods, irritable at times, and unable to conceal the fact that some boys were his favourites, though I think he was unconscious of this. But most of us loved him, because we knew that his whole life was wrapped up in the welfare of the school, and that he hated anything mean or underhand. For many years he coached the school crews; and woe betide the boy who failed to pay due attention to him! I cannot refrain from quoting two more verses, from 'The Independent Oar':

The Coach's voice is ringing rough
    From off the southern shore—
Of 'time', and 'feather', and such stuff
    I daily hear him roar:
'Your feet against the stretcher jam,'
    'Your hands away,' and more.
I care not for the Coach—I am
    The Independent Oar.

My vigour—when I give my mind—
    Shall time and pace restore,
And this, to their dismay, shall find
    The crew who row before.
We may be thrashed—such little blows
    The truly wise ignore—
They do not stir from his repose
    The Independent Oar.

Cuthbertson left the School a little more than a year after Bracebridge Wilson's death in October 1895, and after an extended visit to England, returned to Geelong, manifesting in many ways his unfaltering love for the old school. After a visit to Queensland, and several changes of residence, in the summer of 1909 he went down to the mouth of the Glenelg River, as he had done for many years, to fish for mulloway. He was taken ill, and died at Mount Gambier in January 1910.

Another master of my time was Lieutenant-Colonel A. F. Garrard,[15] who married Mr Wilson's elder daughter. He died not long ago at Heidelberg, near Melbourne, at the age of ninety. His wife is still living in Heidelberg.

The mathematical master in my earlier years at the school was named Lynch—a quiet man, generally well-liked. No one would have guessed that some years after he left he was to become famous as the redoubtable Irish 'rebel', Colonel Arthur Lynch![16] But so it was.

The School Chaplain was the Rev. Canon George Goodman,[17] vicar of Christ Church, Geelong—a position which he held for no less than fifty years. I cannot remember that we boys ever saw him except on Sundays, and in the course of confirmation classes. He was an old-fashioned Evangelical; a learned man, with a fine delivery; but I fear it must be said that 'Goodie' exercised no influence among us. He was examining chaplain to the bishop of Melbourne when my father was ordained in 1862, and still held that office when my father retired from active work in 1905!

In the minds of the boys at least, in those old days, Saturday was easily the greatest day of the week. This was not just because we were free from lessons, but because we were allowed to go out into the bush, or down the Barwon River, in parties of not less than three, for the whole day. We could leave the School as early as 4 a.m., and had to report 'Back' by 8 p.m. To go down the river, of course, one had to qualify for membership in the School Boat Club; the test for this was to swim eighty yards in boating costume. We performed prodigious feats on those Saturdays. I regretted very much to see a hint in a recent number of the School magazine, *The Corian*, to the effect that the school Saturday excursions appear to be gradually dying out.

Barwon Heads, twenty-two miles away from the old School boatsheds, was often the goal, though there were other attractive camping-places in between. A favourite resort only five miles downstream was The Willows, on the St Albans estate. The name indicates the main feature of the river bank here; behind the willows were extensive swamps, where budding ornithologists could seek the nests of water-birds. Across the water were acres of lignum bushes, tangled and untidy, but beloved by small birds. But when we were bound for Barwon Heads we would usually row twelve miles down to the Connewarre Lakes before breakfast. The river enters these shallow tidal lakes by a narrow reed-bordered channel which was known as the Gut. Black swans frequented the lakes literally in thousands; and in my memory I can still hear, as the boat emerged from the Gut, the flapping of innumerable great wings and the plaintive music of the birds.

At Campbell's Point, a promontory crowned by a tall Norfolk Island pine, we would enjoy a well-earned breakfast, grilling chops on twisted wire and of course boiling the billy for tea or coffee. Then we did some botanizing or bird-nesting or fishing or just loafing, according to inclination, until the tide was high enough to guide the boat along the tortuous, narrow channel through the lakes to the lower river. Arrived at the heads, there was a dip in the breakers, a midday meal, some fishing perhaps, or more rambling, and then came the homeward journey. I do not say that we rowed the twenty-two miles without stopping, but there was keen rivalry between crews as to who would be first home. In the tidal waters of the river from the Connewarre Lakes downward, there was always the chance that the coxswain would miscalulate his course at a bend and run the boat on to a slimy mudbank, thus incurring the wrath of his crew, some of whom would have to get out and wade up to their waists in mud in order to move the boat back into the channel. Great joy prevailed, of course, if this happened to one of the rival crews, and not to your own.

But the river, much as we loved it, was not our only rendezvous on those Saturdays. Bream Creek, Spring Creek (now a fashionable seaside resort graced with the name of Torquay), Waurn Ponds on the Colac Road, Grub Lane on the Queenscliff Road, the beautiful Moorabool Valley near Batesford; even the You Yangs across Corio Bay, called us, and we answered. We often walked more than forty miles in the day. There were plutocrats among us who hired spring carts and *drove* out to one or other of these places; but most of us walked. And what friendships were formed and fostered by those Saturday parties!

Most of my contemporaries, I fear, have now passed away. Of those who, to my knowledge, still survive, I may mention Sir Charles Belcher, formerly chief justice consecutively of Kenya, Cyprus and Trinidad, who now lives in retirement in Kenya, and still writes to me; Mr Sidney Austin, well-known New South Wales pastoralist; Mr A. Brownscombe, of Rose Bay, Sydney (who was 'best man' at my wedding in Tamworth forty-six years ago),[18] Mr H. Leahy, surveyor, whom I met quite recently during a visit to Beeac, Victoria; Mr K. C. McCormick of Killara, Sydney (affectionately known as 'Tim'); and Mr W. T. Menck, of North Carlton, Melbourne. No doubt there are others; but memory plays tricks at my age, and they are omitted only because I cannot remember hearing that they are still alive.

I went up from Geelong Grammar School to Trinity College, Melbourne University, at the beginning of 1892. But before I deal with that stage of my life we must hark back to Coleraine, of which Adam Lindsay Gordon had sung—

On the fields of Coleraine there'll be labour in vain
Before the Great Western is ended!
The nags will have toiled, and the silks will be soiled,
And the rails will require to be mended.[19]

The 'Great Western' was a famous steeplechase which used to be run at Coleraine. Gordon was often there; and when my father became vicar there was

a local hotel-keeper named Trainor who had known the poet well. The parish had for some years been the home of another literary celebrity, Ada Cambridge[20]; as a matter of fact she was the wife of my father's predecessor, the Rev. G. F. Cross. Some of her poems and novels—few of which, I am afraid, have survived the judgement of the passing years— were written at Coleraine.

There were some very fine pastoral estates in the district, and some beautiful homes. Chief among these perhaps was Murndal, the residence of the Hon. S. Winter Cooke, MLC.[21] I have never been in England, but Murndal was more suggestive of pictures of an old English manor house than anything else I have seen. Built of bluestone, with white facings, on level ground facing the flats of the Wannon Valley, it was approached through noble avenues of elm and oak planted in the early days, and in front were lovely lawns and gardens, on one side of which was a splendid orchard. My outstanding memory here is of two magnificent English mulberry trees on the lawns; I have never seen their equals elsewhere. Needless to say, a schoolboy was always ready to accompany his father when parochial duties called him out to Murndal. On the way, occasionally we called in at a homestead owned by an eccentric old man of patriarchal appearance named Arden, who was a descendant of the family from which William Shakespeare's father chose Mary Arden for his wife.

A mile or two north of Coleraine was another beautiful home, Konongwootong Creek, originally owned by the McConochie family. One of them, Walter McConochie, died a few years ago at Wahroonga, Sydney, and I think his widow still lives there. But Konongwootong Creek passed into other hands. I remember it best as the home of a Mrs Stanley, who added a second storey to the house, which was built of the same peculiar and attractive local stone as Coleraine Church. She entertained on a rather lavish scale, and was pleased to give a large dinner party in honour of the twenty-first birthday of the vicar's son.[22] A famous Melbourne elocutionist, Miss Nellie Veitch, proposed my health, and I wished myself a hundred miles away as I rose and stammered out a few words in response. Later on, as Mr Walter McConochie told me when a friend took me to visit him, the place changed hands again, and the beautiful homestead was completely demolished, the stones being carted to Coleraine for the erection of other buildings. *Sic transit gloria mundi.*

But clearest of all my memories of Coleraine are those associated with yet another pastoral homestead, sixteen miles out on the Harrow road. This was Wando Dale, where dwelt Mr and Mrs William Moodie with their family of twelve. Wando Dale was a synonym of unfailing and generous hospitality. Mr Moodie was the nephew of a distinguished early Victorian botanist, J. G. Robertson[23] (no connection, as far as I know, of Henry Handel Richardson's husband),[24] who is mentioned so frequently in Bentham and Mueller's *Flora Australiensis*. Robertson's home was Wando Vale, some miles down the river from Wando Dale. He was a friend of the Tasmanian botanist Ronald Gunn,[25] who paid him a visit at Wando Vale. The Wando was a small tributary of the Glenelg; Bentham consistently misspelt it 'Wendu'. Several of the Moodie family were approximately of my age; and as an invitation was always forthcoming when my holidays were due, I think I spent nearly as much time there as at the Coleraine vicarage. The boys were full of mischief, and I was never loth to follow suit. A favourite prank was to fasten a few dead crows, or a couple of well-seasoned sheeps' heads, on the back of a departing visitor's vehicle, hoping that the impedimenta would hang on at least till they had been dragged through the Coleraine streets. My father, who drove a pair of horses in a hooded buggy, 'got wise' to this trick, and as soon as Wando Dale was out of sight, used to look through the little window of the hood to see if there were any undesirable appendages behind.

Mrs Moodie was an enthusiastic gardener, and always had lovely flowers. Some of the family shared my love of wild flowers, and we used to comb the hills for Spider Orchids and other treasures. Only a few of that delightful family now remain; among them are the two youngest, Miss Grace Moodie, floral artist of Mosman, Sydney, and her brother Murray, who grows her flowers, has a fine collection of orchids, and is an accomplished landscape artist. It has been a pleasure to renew, through this brother and sister, my old friendship with the family. Murray will perhaps forgive me for recording my earliest recollection of him. We were at dinner at Wando Dale when one of the younger members of the family rushed in, exclaiming, 'Mother! Mother! Murray's trying to pull a snake out of the rockery by the tail!' And he was, too. He was barely two years old then.

All visitors to the Melbourne Art Gallery know Buvelot's beautiful picture, 'Waterpool at Coleraine'.[26] I remember that pool well. It was about half a mile from the town, and was really one of a series of billabongs formed by overflows from Bryant's Creek. Children used to bathe in them.

In April 1890 my sister Florence was married in Holy Trinity Church, Coleraine, and went to live with her husband at Hay, N.S.W. Shortly afterwards our parents left on a twelve months' tour of England, Scotland, France, Switzerland, Italy, Egypt and Palestine. During that year most of my holidays were spent with the Rucker family at Kew, Melbourne. At the end of 1891 I bade goodbye to Geelong Grammar.

2 University Days

I went up to Trinity College, Melbourne, in the first term of 1892, having won the Mary Armytage Scholarship from Geelong Grammar School. The warden of Trinity was then Dr Alexander Leeper, whose leanness earned for him among irreverent students the soubriquet of 'Bones'. Of Dr Leeper my personal recollections are of a man who was exceedingly kind to me through all my Trinity days. But it would be idle to pretend that he was popular. He was not tactful in dealing with high-spirited undergraduates; and his intense preoccupation with the Classics, in which he was a really distinguished scholar, gave the impression that he had little sympathy with students of other branches of knowledge. In 1891 there was serious friction between warden and students, and forty men left the College *en masse*.[1] Thus when I went up in the following year there were only thirty-nine resident students—the Thirty-nine Articles, as we were dubbed.[2]

Despite our small numbers, it was decided to enter a crew for the Intercollegiate Boat Race. Our rivals were the big Presbyterian College, Ormond, and the Methodist, Queen's. (Newman, the Roman Catholic College, was not then in existence.) There were three Old Geelong Grammarians in the crew; Harry McWilliams and Phil Parsons, medical students, and myself, a theological. Our stroke was Alec Thynne, son of a well-known Queensland politician, and an old boy of St Ignatius's College, Riverview, New South Wales. Others in the crew were Frederick Sefton Delmer, in after years to become something of an international celebrity in Berlin; his younger brother Harold; Clive Gaunt from the Melbourne

Grammar; and Harold Jackson from Scotch College. The coxswain was Eustace Jarrett, who died in the college the following year. We were a scratch lot; and our coach, Mr George Upward,[3] of international rowing fame, and for many years sergeant-at-arms in the Commonwealth parliament after Federation, had gloomy views of our impending fate. We were tipped to come a bad last. But Thynne set us going at thirty-two, and kept it up; we forged ahead, and won by two lengths from Ormond. Rumour said that the grave and dignified warden, on the official launch, threw up his top-hat, and lost it in the murky waters of the Lower Yarra! However that might be, no restrictions were imposed on our celebrations that night. 'Twas a famous victory, indeed.

My university course was not very satisfactory to anybody concerned. I was ambitious to do Arts and Science simultaneously, but I fear it must be said that I lacked steady application; and I needed counsel and advice which I did not get. In 1893 I secured the Cusack Russell Scholarship from the diocese of Ballarat; and in 1896 I won the Wyselaskie Scholarship in Natural Science at the university. But I lost a whole year through illness[4]; and in the end I left with only the BA degree.[5] Had I been able to return for another year the coveted BSc would have been easy; but circumstances prevented this. During the year in which illness intervened, I became the 'publisher' of a cyclostyled college periodical designated *The Crab-Catcher*, from which it may be inferred that rowing men were originally responsible for its appearance. In its columns the contributors libelled each other and their fellow-students in the most unscrupulous terms imaginable; nevertheless

Trinity College crew, winners of the Intercollegiate Boat Race, 5 May 1894. *Standing (l. to r.):* J. M. Semmers (Coach), (7) A. H. Parker. *Sitting:* (5) G. E. Broughton, (4) H. C. Delmer, (Stroke) H. E. Bullivant, (6) H. M. R. Rupp, (2) H. South. *Front:* (3) A. H. Bullivant, (Cox) J. Lang, (Bow) P. H. Parsons. (From a photograph kindly lent by the late Arthur Rupp of Willoughby)

it was very popular, and its early demise was regretted. I cannot remember who were associated with me on the 'Editorial Board'; but I recollect that Andrew Peacock wrote a 'leader' on the record reign of Queen Victoria.

Of my contemporaries at Trinity, however, I have many other clear recollections. A number of them became well-known figures in public life. T. Slaney Poole,[6] H. H. Henchman,[7] D. J. Bevan,[8] and C. F. Belcher[9] all reached the Judicial bench; Poole in South Australia, Henchman in Queensland, Bevan in the Northern Territory, and 'Charlie' Belcher's distinguished career in the Imperial Civil Service has already been alluded to. 'Dave' Bevan, a son of the minister of the Collins Street Independent Church, Dr Llewellyn Bevan, was one day walking along what we called the Cloisters, outside the studies on the ground floor of the Clarke Wing of the College. Someone threw a log of firewood out of a window. 'Dave' instantly picked it up, and not realizing that the culprit had closed the window, hurled it back—

through the plate glass! There was a glorious smash; luckily no one was hurt. A. A. Uthwatt,[10] probably one of the most brilliant men on the roll of Trinity, has risen high indeed; for he is now one of the Law Lords on the Judicial Committee of the Privy Council. Andrew Peacock,[11] senior student in my last year, was naturally dubbed 'Chuck'. His lovable and transparently sincere character exercised great influence in the College, and later on in the diocese of Ballarat. He was not a robust man, and died at Warrnambool (where he was archdeacon) in 1912, deeply mourned by all who knew him. He was a brother of Sir Alexander Peacock,[12] for some years premier of Victoria. Need I say much at this stage (for we shall meet him again later on) of George Merrick Long,[13] to be widely known as bishop, statesman and scholar, whose sudden death at the Lambeth Conference of 1930 came as such a grievous shock to Church and nation? Dr Edward Field, in those days affectionately known as 'Teddie', still practises in Melbourne.[14] To my great joy, I

H. M. R. Rupp in his undergraduate days at Trinity College. Photographic portrait by James Meek, Hamilton, about 1895. (By courtesy of the Archivist, Geelong Grammar School)

discovered him during a recent visit to the southern capital; he drove me to Trinity to see what changes time had wrought, and then took me off to lunch in the city. He and I share one sad memory of the Long Ago—the loss of Herbert Appleton Palmer, who was a friend beloved to both of us, and my study-mate for several years. He went as a medical officer to the Boer War, and was killed in the disaster to the Victorian forces at Wilmansrust.

Hugh Bullivant is still well-known in pastoral circles; his elder brother, Arthur, lives in Melbourne. Hugh stroked the Trinity crew of 1894, in which Phil Parsons was bow, Harold South ('Caesar') 2, Arthur Bullivant 3, Harold Delmer 4, George Broughton 5, myself 6, and A. H. Parker 7. John Lang was cox. Broughton and I were the heaviest men in the boat, at 10 stone 12 pounds. Most critics said we were far too light a crew to stand the racket of the race; but we romped away and won by six lengths! Phil Parsons died in London many years ago. Harold South built up a good medical practice in Boonah,

Queensland, and his comparatively early death there was deeply regretted by the community. Harold Delmer became a master at Sydney Grammar, and died there. I do not know what became of Broughton and Parker. John Lang and his brother Pat, who were nephews of Andrew Lang of literary fame, came up from Geelong Grammar with Stan Elder in 1893. Pat became a doctor. John is best remembered as having been the prime mover in the establishment of the Melbourne aquatic carnival, Henley-on-Yarra. Stan Elder, a well-known Melbourne solicitor, died in 1948. The Langs, too, have gone. So has Heber Green, who, although not very popular in his early days at the College, did a brilliant science course, and was very highly esteemed by members of the Victorian Field Naturalists' Club.

Russell and Frank (now Sir Frank) Clarke,[15] Leo and Clive Miller and their cousin 'Tommy', were members of prominent Melbourne families; our paths have not crossed since Trinity days. Nor have I heard anything of J. B. Kiddle,[16] a giant of (I think) 6 feet 8 inches. I remember being near him in the dense crowds that packed Swanston Street during the illuminations celebrating the Diamond Jubilee of Queen Victoria in 1897. A small urchin, unable to see much of what was going on, gazed up at the towering form of 'Dinny' Kiddle, and yelled out, 'Hey! reach us down a star, mister, will yer?' Clive Shields, a medical student in those days, I met unexpectedly in Sydney a few years ago at a wedding. The bridegroom, Dr H. Leighton Kesteven (whom we shall meet in a later chapter at Bulahdelah) had asked me to act as best man (he was then a widower); and Clive Shields had come over from Melbourne to give the bride away!

Nearly all my theological contemporaries have passed on. Besides Andrew Peacock and George Long, I may mention A. B. Rowed, Frank Lynch, Alfred Gates, L. Townsend, H. T. Fowler, J. S. Wells, J. H. Frewin, A. E. J. Ross, John Forster, and Charlton Brazier. I do not know if the last-named is still living. John Forster became archdeacon of Armidale, and died a year or two ago.[17] A few others remain: Tom Langley, until recently dean of Melbourne; Canon Kitchen; the Rev. T. Keyran Pitt of Tasmania; and Bishop E. N. Wilton, who has recently closed a long period of service as a colleague of the rector of St Thomas's, North Sydney, where he is loved by pastor and people alike. He was a fine cricketer in his younger days, and captained the Trinity eleven in 1895, in one match against Ormond scoring 128 not out. In the same match he took four wickets for 42.

Trinity days, 1895. Rupp, the central (sixth) figure, played John Pym in Robert Browning's *Strafford. (Alma Mater,* June 1895)

In 1896 the warden persuaded us to stage Browning's *Strafford* at the St Kilda Town Hall.[18] The newspaper comments on our performance were for the most part kindly, but guarded. *Strafford*, though containing some fine passages, is a very difficult play, avoided by professional actors and actresses; and I daresay we laid ourselves open to the old saw of fools rushing in where angels fear to tread. Slaney Poole took the part of King Charles; Dave Bevan was Strafford; Miss M. A. Bartrop was Queen Henrietta; Peacock was John Hampden; I was John Pym; and so on. One journal, which had been omitted when the complimentary tickets were handed out, avenged the insult thus:

The acting of the Trinity students is not worth a *denarius*. Their performance of 'Strafford' must be condemned as a wrong committed against Browning, and an affront to dramatic art... Mr Bevan can't act a cent, and never will be able to act... Mr T. Slaney Poole's Charles I was like a trick Pomeranian poodle, and a very indifferent one at that... Pym was played by Mr H. M. R. Rupp, and was played with ludicrous feebleness. Indeed, all the fiery Puritan malcontents were as namby-pamby as a collection of vegetarian curates.

Well, I hope the critic felt better after getting that off his chest! The *Sun* was more playful:

To put it in cricket parlance, these greatly gifted ones have had a splendid innings, and as there was nobody to bowl or field against them, each player has scored gloriously off his own bat. Therefore does

The beaming Dr Leeper,
As he winks each glistening peeper,
Rub his hands with all the fervour of unqualified delight,
While Poole, Peacock, Rupp, and Bevan
Almost taste the joys of heaven,
In recalling all the glories of that most eventful night.

Townsend, Bullivant, and Frewin,
That triumphant scene renewing,
Shout rejoicings o'er the spreading of the cult that they adore,
While Pitt, Palmer, and a'Beckett,
To the end that nought shall check it,
Are conspiring for occasion to repeat the dose once more.

But each poor unhappy buffer
They condemned to sit and suffer,
Thinks this blessed Browning business is the saddest sort
of jest.
And the cry of bitter anguish,
As in shrinking dread they languish,
Is—We'll take it *all* on credit if you'll let us off the rest.

I don't think the stage gained any recruits as a result
of our performance at St Kilda.

When I went up to Trinity, I took with me a letter
of introduction from John Bracebridge Wilson to
Australia's most distinguished botanist, Baron
Ferdinard von Mueller. For some considerable time
I was shy about presenting it. I was not accustomed
to paying calls at baronial castles; and I wondered,
too, whether a man whose fame I knew to be world-
wide would resent the impertinence of a youngster
like me daring to intrude. However, I felt that my
uncle would be annoyed if I made no use of his letter,
so one day I summoned up courage and went out
to the address given in a South Yarra street[19]—to
find that the 'castle' was only a very ordinary and
modest cottage.

I could not reach the front door, for the verandah
was completely occupied by large and small cases of
specimen plants; so I went round to the back, not
quite sure whether I had mistaken the address. A
boy, certainly not clad in baronial livery, opened the
door, and in reply to my queries said yes, this was
Baron von Mueller's house, and he was at home. I
sent in my letter of introduction. Presently there
came to the door an elderly, bearded man, in an old
dressing-gown, and with a woollen muffler round his
throat. With a strong German accent which I shall
not attempt to reproduce here,[20] he said, extending
his hand, 'You are from my dear friend Mr Wilson?
Come in! Come in! Anybody from Mr Wilson is
welcome here!'

He led me into a barely furnished room, with a
coverless table strewn with more specimens. For a
moment or two he seemed undecided as to what the
next procedure should be; then he shuffled out of the
room, returning presently with a bottle of lemonade
and a jar of biscuits, determined that this young
fellow should be suitably entertained. He questioned
me for a few minutes about Mr Wilson, and about
my botanical interests; and then he began to talk.
And that talk threw a great light on the man's
character. That a world-famous botanist, busy day
and night over the innumerable specimens that came
into his hands, should spare more than an hour
discussing the wonders of our Australian flora with
a callow undergraduate, was surely remarkable; but

the dear old Baron was, of course, a remarkable man.
He did his utmost to inspire me with something of
his own enthusiasm; nor did he fail. When I left he
loaded me with packets of seeds, and invited me to
write to him if I found anything that seemed to be
of special interest. I took him at his word; and I still
possess two or three of his letters, which always
ended, 'Regardfully yours, Ferd. von Mueller'.

I remember that Mr Wilson always insisted on
alluding to the classic *Flora Australiensis* as 'Bentham
and Mueller's'; and in my opinion this is preferable
to the common practice of using only Bentham's
name. For it is unlikely that Bentham, great botanist
as he was, could ever have produced this work
without Mueller's cooperation, so freely given in
both material and manuscript.

The baron estimated that he had spent £20,000
which he had saved out of his official income as
Victorian government botanist, in promoting the
cause of science. He died a comparatively poor man,
leaving only a little over £1,000 in his will.[21]

On the Professorial Board of the University I
remember best Edward Ellis Morris (English),[22]
T. G. Tucker (Classics),[23] Henry Laurie (Logic and
Philosophy),[24] J. S. Elkington (History),[25] Sir
Frederick McCoy (Natural Science),[26] and Sir
Baldwin Spencer (Biology).[27] Morris was not
popular, nor were his lectures appreciated as at least
one of his old students thinks they should have been.
I have always been grateful to him for teaching me
how to enjoy the English poets; particularly was I
impressed by his handling of Browning and Landor.
Tucker I did not know well, as I only took first year
Latin and Greek; but I need scarcely add that he was
a classical scholar with a world-wide reputation.
Henry Laurie was beloved by all his students. One
of his Moral Philosophy lectures impressed his classes
so deeply that he was persuaded to publish it, under
the title *Some Thoughts on Immortality*. (I still possess
a copy.) It was ably reviewed in the Melbourne *Age*
by David Syme himself; but what the famous editor
and proprietor of the *Age* did to the unfortunate
printer of his review we never heard. In setting up
the title he had omitted the first 't' in 'Immortality'!

I came in contact with McCoy and Baldwin
Spencer, of course, in connection with science
subjects. Sir Frederick McCoy was born in 1823,[28]
and died in 1899. In his time he was a great zoologist
and palaeontologist, with an international reputation.
But 'his time' belonged rather to the 'fifties and
'sixties of his century than to the 'nineties. He was
a pre-Darwinian, and anything savouring of the
theory of evolution was anathema to him. In one of

Alexander Leeper, classicist and warden of Trinity College, 'a man of scholarly taste and reputation . . . a product of the combined culture of Dublin and Oxford'. *(Alma Mater,* July 1898)

Sir Walter Baldwin Spencer (1860–1929), anthropologist, explorer, and professor of Biology, University of Melbourne, 1887–1919. Rupp considered him 'one of the finest characters' he had known. *(Alma Mater,* June 1897)

his palaeontology lectures he would unroll, year after year, a huge wall-picture, depicting a fearsome fight in a prehistoric swamp between an *Ichthyosaurus* and a *Plesiosaurus*, antediluvian monsters as unlike one another as any two creatures could be. Then, beaming with triumph, he would turn to the class and remark scornfully, 'Now, gentlemen, let these progressive development theorists prove their case. Let them tell us, Which of these two monsters was developed from the other?' There was nothing more to be said; evolution was squashed. Nevertheless

Sir Frederick McCoy (1817–99), palaeontologist, museum director, professor of Natural History, University of Melbourne, 1855–99. Rupp appreciated that he was 'a pre-Darwinian' to whom 'the theory of Evolution was anathema'. *(Alma Mater,* August 1895)

Victoria is deeply indebted to McCoy, for he laid the real foundations of the magnificent collection which now constitutes the National Museum. To Baldwin Spencer is due the chief credit for that great institution as we now know it; but he was the first to acknowledge how much he owed to McCoy's spadework.

Spencer, as I knew him in the 'nineties, was the antithesis of McCoy; comparatively young, he was one of the latest products of English scientific study, and was a pronounced evolutionist. To attend his lectures and those of McCoy simultaneously was at times somewhat bewildering. Spencer was one of the finest characters I have met, and was a man of singularly lovable disposition. So far was he from jibing at the views of his older colleague, that he said to me one day in the biological laboratory, 'For all I know, we may all come back to the old boy's ideas yet, Rupp; science never stands still, and sometimes seems to move round in circles!' In later years, his great work as an anthropologist in connection with the Australian Aborigines quite overshadowed his biological research. He resigned in 1919, but continued his scientific activities to the end; in 1929 he went on an expedition to Patagonia, and died there at the age of sixty-nine.

Canon Robert Potter was chaplain of Trinity College in my time.[29] He was vicar of St Mary's, North Melbourne; and it goes without saying that the vicar of a big industrial suburb cannot have much time for the duties of a college chaplain. He was a black-bearded man with a thin, squeaky voice which at first repelled listeners; but when you became used to it, you forgot its defects; for he was a man of great intellectual calibre, and I do not think I ever heard a sermon or a lecture of his which I would not gladly have heard again. He had a rather caustic vein of humour, and was noted for his quick repartees. One of the clergy said to him one day, 'Canon, I don't know how you manage to produce all those witty little remarks so quickly'. 'Ah, my boy,' was the instant response, 'you wouldn't say that if you knew how many I keep to myself!' We saw little of the canon at Trinity, however, except on Sunday mornings in chapel, and at his theological lectures.

The lecturer in Economics at the college was the Rev. Reginald Stephen, MA.[30] He subsequently became successively dean of Melbourne, bishop of Tasmania, and bishop of Newcastle. He was recognized in and beyond university circles as a real authority on economic subjects, and I don't think he ever had cause of complain of absentees from his lectures. They were thoroughly enjoyed, and I think absorbed; he spoke fast enough to sustain our interest, but not too fast for most of us to take copious notes. He was an old Trinity man himself; and was also an old Geelong Grammar boy. He resigned the see of Newcastle in 1928, but is still living in retirement near Melbourne.

Botany had by now definitely become my hobby; and during my years at Trinity most of my available Saturdays were spent in the bush on the outskirts of Melbourne. I was more interested in the taxonomic side of the subject than in any other; I wanted to learn the identity of every plant I found, and something about its relation to its nearest allies.

Orchids always had a special fascination for me, though I did not begin to concentrate on them until a little over twenty years ago.[31] If I could get a companion for my rambles, so much the better; if not, I went alone. Many of the areas over which I used to wander are now densely populated.

In those times one could collect beautiful 'Spider Orchids' within a stone's throw of Sandringham railway station. Beaumaris, Black Rock, Mordialloc, Oakleigh, Spring Vale, Ringwood, were all within easy reach; occasionally I got as far as Fern Tree Gully, and twice, I think, to Healesville. On the basalt country to the north and west of Melbourne there was not much 'bush'; but the grassy plains had their own flora, and all was fish that came into my net. There was a curiously interesting spot on the Moonee Ponds Creek, not very far from the Melbourne Zoo. Here grew several plants— *Billardiera cymosa* is the only one whose name I can recall now—which were not found anywhere else within a hundred miles or so of the metropolis.

Alfred Gates, another theological student, and also another budding botanist, used to go out with me sometimes; his name reminds me of an incident which is perhaps worth recording. Like me, he was very weak in mathematics; and he had engaged a private tutor from Carlton to come and give him lessons. This gentleman was rather too fond of the cup that cheers and inebriates. At the time I speak of, Gates was my study-mate. Some of my relatives were coming to have afternoon tea with us, and Gates discovered that he had an appointment with his tutor about an hour before they were due. He did not turn up; and Gates went off to a university lecture, intending to join the party later. I went down to the Sydney Road gate to meet my visitors, who were a little late. Leading them upstairs, I opened the door to usher them in, remarking, 'This is our room; Mr Gates will be here presently'. An aunt was the first to enter; as she did so she glanced behind the door, and screamed! The errant tutor had arrived during our absence, very much the worse for Johnny Walker. Stumbling in, he had fallen into the big wicker firewood basket behind the door, and had promptly gone to sleep there, with his legs sticking up in the air. I had to enlist the help of neighbours to get rid of him; and I doubt whether my aunt's suspicions were ever entirely dissipated.

One of my vacations was spent at Hay in New South Wales, with my sister and brother-in-law. I had never been in the Riverina before, and found the botanical features of the great inland plains very interesting; it happened to be a good season, and there were many flowers quite new to me. I made my debut in print at Hay; the *Riverine Grazier* published a list of the native plants I had found in the district.[32] At that time my father had left Coleraine, and was acting as *locum tenens* for Archdeacon Beamish at Warrnambool.[33] As he and my mother were living in the vicarage with Mrs Beamish, they could not accommodate me just then. Later on the archdeacon resigned, but my father remained there for nearly two years, and was then appointed to Buninyong, near Ballarat. I found the Buninyong district a paradise for a young botanist, and always enjoyed my vacations there. One of the most interesting botanical features of the district was the number of beautiful ferns which grew in old deserted mine shafts, relics of the era of gold in Victoria. The subject has been dealt with by Mr R. W. Bond in the *Victorian Naturalist* for December 1942, but he does not mention Buninyong. Many ferns grew in these shafts which were not to be seen within a hundred miles on the surface of the ground. The spores were probably carried by wind. Plants were usually out of reach, but sometimes it was possible to rake a few up. Among the Buninyong orchids I remember best a colony of giant Greenhoods just below the summit of Mount Buninyong[34]; the musky-scented *Caladenia angustata*; and the lovely little perfumed lemon-yellow Sun Orchid, known in Victoria as 'Rabbit's Ears', from the curious appendages in the centre of the flower.[35]

3 Ordination — and After

It will be obvious from what has been said in the last chapter that I had decided to follow in my father's steps, and seek Holy Orders in the Church of England. This decision had been reached before we left Coleraine, where, as a layman, I conducted my first service in the local church during my father's absence for a Sunday at Portland. From various causes, I was unable to prepare for ordination examinations before leaving Trinity College, and after some consideration it was decided that I should serve for a while as a stipendiary lay reader and continue until I passed the examination for deacon's orders. Actually, this set me back for at least a year. My father had hoped that I would work with him; but owing to unforeseen circumstances we were not able to arrange for this. So it was decided that I should go to the Rev. John Kirkland, vicar of Colac, and should take charge, under his supervision, of Beeac, at the north end of his extensive parish.

Unexpected difficulties confronted me soon after my arrival, into which I need not enter here. John Kirkland was a man of great natural ability, who had started his working life as a pit-boy in a mine. It can be imagined that he faced no ordinary difficulties in responding to the call which he felt to seek Holy Orders; but he was one of those men who believe that difficulties are created to be overcome. When I knew him he was a scholarly priest and an eloquent preacher; moreover his gifts as a good man of business were well known, and proved of great value at Colac, where the parish had experienced grave financial difficulties. But when I went to Beeac in 1898 his health was unsatisfactory, and before long he resolved on a trip to England. This was rather upsetting for me, because his *locum tenens* was not likely to understand the difficulties to which I have referred. Colac is in the diocese of Ballarat, whose first bishop, Dr Thornton,[1] had been there for twenty-three years. His only son, the Rev. H. S. R. Thornton,[2] was appointed to take Mr Kirkland's place during his absence abroad.

Harry Thornton had been at Geelong Grammar School for a few years, and was then sent 'home' to Harrow, whence he proceeded to Oxford. He was a man of most unselfish character and lovable disposition; but regrettably these qualities were offset by eccentricities which were very disconcerting to anyone working under his supervision. When Mr Kirkland returned from England he resigned, and became vicar of Port Fairy; and Harry Thornton became vicar of Colac.

In September 1898 we built the first Anglican church in Beeac; it was dedicated in honour of St Augustine of Canterbury. The dedication service was performed by the suffragan bishop of Ballarat, Dr Cooper.[3] Fifty years afterwards, living in retirement in a Sydney suburb, I was astonished to receive an invitation from the vicar and parish council of Beeac to come over and officiate at the jubilee celebrations of St Augustine's! As I am well over the allotted span of three score years and ten, and had retired on a small pension, I was very dubious about accepting, though I appreciated the wonderful compliment of the invitation more than I can say. But a further letter arrived, urging me to come, and enclosing a cheque to cover all expenses. How could I refuse after that? So it came about that I had my first air flight, over to Melbourne (and loved it), and went down to Colac by train. There I was met and driven out to Beeac, where I spent what was to me a wonderful week.

Of course in fifty years there had been many changes. The township itself had not grown very much, but the streets and buildings were greatly improved. The church was just as it had been when we built it in 1898, except for internal improvements; and when strolling down a back street one day I discovered the identical little weatherboard cottage where I had lived. Dr M. W. Cave, whose guest I was, drove me out on some of his rounds, and one day took me over to Lake Corangamite. This great sheet of water (nearly a hundred miles round, and so salt that no fish, and only a tiny mollusc, can live in it) I remember only too well. On many a cold winter day I had ridden for miles on muddy tracks along its inhospitable shores, with a bitter westerly gale blowing across its troubled waters. We drove along that old route of mine, on a perfect road, in the doctor's car. There was a welcome social evening in the public hall at Beeac. It was a happy gathering, yet tinged with sadness for me as I realized how few of the 'old-timers' were left. But there were many present whom I remembered as children, and they brought their own children (and in some cases grand-children) to meet me. The church was crowded for the Jubilee services. Altogether this was a unique experience in my life, for which I felt more thankful than I could say.

During the five and a half years of my ministry at Beeac, I was ordained deacon in St Paul's Cathedral, Melbourne, and priest in St Peter's Church, Ballarat. The district became an independent parish just before Harry Thornton (who died quite recently in British Honduras) resigned Colac, and went to join his father (who had retired from the see of Ballarat) in England. He was succeeded by my old Trinity friend, Andrew Peacock, and it was very pleasant to have as my neighbour one for whom I felt such affection and esteem. My father by this time was beginning to feel the burden of his years, and had accepted an invitation to return to the parish of Learmonth, which he had left twenty-five years before. Here our stepmother died[4]—in the same vicarage to which she had come as a bride. Her passing was slow and painful; and after her death my father asked me to accompany him on a visit to my sister, who was then living at Forbes, on the edge of the Central Western Slopes of New South Wales. Afterwards we went on to Sydney, which I saw for the first time. That, of course, was long before people could ask visitors,

'Have you seen the Harbour Bridge?' but they could and did ask the question with the omission of the word 'Bridge'—and needless to say, we saw as much of the glories of Port Jackson as we could. I little thought that before long I should see Sydney again, not as a casual visitor, but on my way to become a country parson in northern New South Wales.

One more incident in connection with Beeac I think I should relate. I felt the need there for a parish Mission, to rouse people to a keener sense of their responsibilities. But I did not want any fanatical 'revivalist' Mission; it must be something that went deeper. Somehow my thoughts were led back to George Merrick Long, of Trinity days. 'Long is the man I want,' I decided, and I wrote to him. He was then curate of a country parish in Gippsland. He replied promptly, saying that he did not think he had the gifts for this kind of work, and in any case he lacked the experience. He recommended several names to me, but I felt that he *was* the man I needed. I urged him to come; at last he put my letter before his archdeacon (Hindley),[5] and asked his advice. The archdeacon said, 'Go'. So he came; and the memory of those ten days he spent with us is bright and clear today, nearly fifty years afterwards. He made a profound and lasting impression on the parish; and many years later I heard how people still spoke with gratitude of 'Mr Long's Mission'. I think no one was less surprised than I when, some ten years afterwards, I read the announcement that Canon George Merrick Long had been elected bishop of Bathurst.[6]

I remained for another year at Beeac, and then I felt that the time had come when I should move on. Unexpectedly I received a letter from Dr Cooper, who had become bishop of Grafton and Armidale in the north of New South Wales, asking whether I felt disposed to come over and accept work in his diocese. I had never thought of leaving Ballarat; but I had known Bishop Cooper since I was a child; and his offer of work in that great area of New South Wales which constituted his diocese, and which was *terra incognita* to me, was very alluring. But I was doubtful about going so far away from my ageing father. However, he urged me not to hesitate on his account if I felt I should go; he had an excellent housekeeper who was unlikely to leave him, and he meant to remain at Learmonth as long as he was able to fulfil his duties. The bishop of Ballarat, Dr Green,[7] wrote to say that he was watching for an opportunity to transfer me to another parish; but by that time my decision was made. I accepted the senior curacy of Tamworth, N.S.W., under its vicar, Archdeacon

Thomas Kingsmill Abbott.[8] The good folks of Beeac gave me an affectionate and generous send-off, and a new stage in my life began early in 1903, in fresh fields and pastures new.

'T.K.', as he was generally called, was a man of commanding presence and forceful personality. He was not an easy man to know, and was rather inclined to use a gift of quiet but biting sarcasm a little too freely, especially with strangers. When I had been with him for a week I felt more than doubtful of the wisdom of the step I had taken. But by the end of the first month all doubts were dissolved; for I knew that in my vicar I had not only a fine leader, but a staunch and generous friend. Ten years later, when I was vicar of Barraba, I was deeply shocked by the news of his sudden death on the tennis courts of The Armidale School, of which he had been appointed headmaster. He was an ideal man for such a post, for young people everywhere were readily influenced by his gifts of leadership and his fine character. By his devotion to duty he had literally worn himself out prematurely, and a very severe attack of typhoid fever had affected his heart. He was only forty-eight when he died.

When I joined him as curate he was still a bachelor. The staff numbered four—the archdeacon, two curates, and a lay reader who helped chiefly with religious instruction in the schools. We all lived together in the old vicarage at West Tamworth; and we were a very happy family, save for the brief intrusion of one lay reader who shall be nameless, but who was one of the most accomplished rascals I have ever encountered. When unmasked, his interview with T.K. was short and to the point; he disappeared from our circle, and we saw him no more. The junior curate was the Rev. Thomas Caine, a fair-haired, rosy-cheeked Englishman who came straight from a similar position at Hawarden in Cheshire. Very unkindly I was wont to remark that it was small wonder he decided to come out to Australia; for in an ungarded moment he confided to us that the name of his colleague at Hawarden was *Abel*! Tom Caine was a good fellow, and our friendship lasted beyond the days of our Tamworth association. He became vicar of Collarenebri, 'way out west on the Darling, but after a few years returned to England and then went to South Africa, where he died of some obscure ailment while still quite a young man.

It was in Tamworth that I met and won my wife. Never shall I forget T.K.'s reception of my announcement of our engagement. He was not in a very approachable mood when I knocked at his study door, and was very busy over a pile of papers on his desk. 'Come in—what do you want?' he growled. 'Er—I've got something to tell you, Archdeacon,' murmured the curate. He looked round. 'Well—I'm here; why don't you tell it?' he asked. This was not exactly encouraging, but I blurted out the great fact. He snorted, and turned back to his papers. 'Oh, is that all? I thought it might be something important.' Then, 'Who is it?' he asked. I told him. 'Just like your hide,' he said, 'taking the best teacher I've got in the parish!' Then he relaxed, and held out his hand. 'Oh well,' he said, 'perhaps you won't be in such a hurry to leave me now?' (My curacy was for twelve months, and he wanted me to stay longer, which as a matter of fact I did.) His 'best teacher' was Miss Florence M. Dowe, eldest daughter of Richard Andrew Dowe, a well-known Tamworth solicitor. It was said to me by someone that 'you couldn't go round a corner in Tamworth without running into a Dowe'. The family connection was certainly an extensive one. My wife's grandfather, Dr Dowe, had been a familiar figure from the Liverpool Range to the Nandewars. Talk of the wonderful journeys of 'the flying doctors' of today—yes, they are wonderful, and an incalculable boon to the lonely settlements out-back—but after all, can they compare with the journeys of pioneer medical men like old Dr Dowe, decades before aeroplanes, or even motor cars, were dreamed of? Journeys sometimes of over a hundred miles, by gig or on horseback or even afoot, with maybe flooded river to cross, or drought-stricken areas to traverse, or bushfires to be dealt with?

The Dowes were of Irish stock. Dr Dowe married Sarah Loder, through whom my wife is descended from John Howe the explorer,[9] who discovered the Upper Hunter, and named Singleton and Jerry's Plains. The doctor's widow was still living when we were married in December 1904—a handsome and gracious old lady, but withal a bit of a martinet in her way, who required assurance that a visitor's boots had been thoroughly wiped on the doormat ere they trod her spotless floors. She bore the doctor seven sons and one daughter. Most of the sons had large families of their own. My wife has two brothers and four sisters, all of whom are still living at the time of writing. Their mother's maiden name was Bloomfield; and if all men had mothers-in-law as lovable and wise as mine was, there would be no point in the time-worn jokes of the conventional humorists. To which I will add, that I am sure my daughters' husbands and my son's wife would pay the same tribute to *their* mother-in-law, my partner now for forty-four years through the joys and sorrows of life's journey.[10]

We were married by Archdeacon Abbott, Tom Caine assisting, in St John's Church, Tamworth. My 'best man' was Alfred Brownscombe, an old school-fellow of Geelong Grammar days, who is still living at Rose Bay, Sydney. The honeymoon was to be spent with my father at Learmonth in Victoria. We left in terrific heat, and looked forward eagerly to the cooling breezes of the southern State. Alas! they were not forthcoming. It was a record hot summer in Victoria, and even on the highlands around Ballarat there was no relief. None came, in fact, until we made a short stay in Sydney on our way to our new home. This was to be Warialda, on the north-western slopes of New England, where I had been appointed vicar. It was more than a thousand miles by rail from Learmonth, and very thankful the newlywed couple were to reach the end of their travels.

When I began my Tamworth curacy, I found much that was strange to me in the native flora of the district. Though no botanist himself, 'T.K.' was very sympathetic towards my hobby, and gave me every possible opportunity, consistent with the carrying out of my duties, of following it up. Being unable, however, to identify many of the plants I came across in the outlying areas of the parish, I wrote to the government botanist in Sydney for advice. And for twenty years thereafter, until his death in 1925, Joseph Henry Maiden was my guide, counsellor, and friend in all things botanical.[11] Fitting tributes have long since been paid to Maiden's tireless energy and great accomplishments, not the least of which was his establishment on a sound basis of the National Herbarium of New South Wales. My correspondence with him was voluminous; and I think I am correct in saying that except on a few occasions when he was too ill to attend to business, every letter I wrote was answered by him personally. He came to see us at Warialda; and later on we were his guests for a few days at the Sydney Botanic Gardens, where Mrs Maiden dispensed gracious hospitality. He urged me to send down specimens from wherever I might be stationed, for the National Herbarium, and this I did for many years.

Tamworth and Warialda are both situated in that portion of New South Wales known as the North-Western Slopes; but the latter town lies about a hundred miles north of the former. Within the bounds of the Tamworth parish one traversed the rich loam of the Liverpool Plains, the granite of the Moonbi Range leading up to the New England plateau, and the limestone country about Attunga. Each of these areas had its own peculiar botanical features; but all alike were of great interest to the young amateur botanist from Western Victoria. A much richer field for exploration, however, lay before me at Warialda. In the first place, a far greater expanse of country had to be traversed. The area of the Tamworth parish, worked by three clergy and a lay reader, was perhaps 2,000 square miles; that of Warialda, worked by the vicar and a lay reader,[12] was approximately 7,000 square miles. This, remember, was some years before the era of 'the horseless carriage'.

Warialda lay close to the southern boundary of the parish; the only other township was Boggabilla, on the Macintyre River opposite the Queensland town of Goondiwindi, a hundred miles from Warialda by the route I had to travel. In between were sheep and cattle stations, with a fair sprinkling of 'selectors' scattered here and there. At first I was rather overawed by the long distances; in fact I remember that on the occasion of my first visit to Boggabilla, I wired to my wife to let her know I had got there! But one grows used to anything in time. Most of my journeys were done on horseback. The longest single day's ride I ever did was seventy-six miles, with one change of horses.

Occasionally I used a sulky or gig but, if one got caught in a vehicle on the black soil when it rained, it was advisable to have a saddle strapped on behind, for the wheels soon became sticky blocks of mud and refused to function.

4 1905–1914

The extensive parish of Warialda included country of the most varied character. I do not know enough geology to describe it from the point of view of that science; but there were sandstone hills and sandy slopes, black soil hills merging into the great plains along the Macintyre River, miles of rich loam, an outcrop of granite on Warialda Creek a few miles from the township, and farther on in the direction of Inverell, a patch of ironstone. In a good season, the sandstone hills surrounding the township were a botanist's paradise. There were acres of soft white starry flannel-flowers,[1] larger than any I have ever seen near the coast; more than twenty species of wattles; and hosts of lovely flowering shrubs and herbs, far too numerous to specify here. There were great patches of a beautiful bright blue daisy-bush *(Olearia)* which flowered in winter and early spring[2]; and clumps of an exquisite little shrub known locally as native daphne *(Ricinocarpus)*,[3] with clusters of pink blossoms resembling that garden favourite, but lacking its perfume. I found about sixteen species of orchids, all terrestrials; the only epiphyte, a *Cymbidium* which grew in great clumps on trees providing hollow trunks for its root-masses,[4] belonged to the country with heavier soils. Among the trees of the sandstone area, the most interesting to me was the Thready-barked Oak *(Casuarina inophloia)*,[5] which I have never seen anywhere else. Most Casuarinas have a hard, rough bark; but in this little tree it is more like that of a coarse stringy-bark eucalypt, and can be pulled out by the handful.

Cypress-pine scrubs abounded except on the plains towards Boggabilla. I have never seen a cypress-pine struck by lightning, and have often wondered whether this was merely a coincidence. Violent storms were frequent in summertime, and many other trees were seen shattered into huge splinters. It was not a pleasant experience to be caught in these storms. On one occasion I was riding to Yetman, fifty-three miles north of Warialda. The track for the last ten miles ran through a dense scrub of belah,[6] buddah,[7] and small eucalypts. I had almost reached this when a terrific storm blew up from the west. Rain pelted down, branches began to fall, and I turned my horse's head and galloped back to the nearest settler's home; both my horse and I were hit, but not by anything big. Next morning I went on to Yetman, and counted twenty-seven trees down across the track in the scrub! A similar storm struck Yallaroi homestead one day. I was coming down from Boggabilla, and just missed it, but when I reached Yallaroi it looked as though there had been an earthquake. The verandah and most of the roof had gone; half of the woolshed was blown clean away, the other half left intact. One of the men was lying down when he heard the crash at the woolshed; he got up to see what had happened, and just as he did so a huge beam tore through the roof of the room, and went right through the bed on which he had been resting.

The botanical features of the black-soil country were very different from those of the sandstone. In good seasons the growth of herbage and grasses were almost incredible. I have ridden through areas where I could not see over the top of the 'big kangaroo grass'[8] from my horse's back. In some areas one of the Darling Peas *(Swainsona galegifolia)* was very prevalent.[9] Stock-owners hate this plant for its fatal effects on animals which develop a taste for it; and one had to be careful not to praise the unquestionable beauty of the flowers when visiting the station people; as well try to persuade the proverbial bull that the red rag is really very pretty!

On the plains towards Boggabilla there were often extensive patches of the lovely Darling Lily *(Crinum flaccidum)*,[10] and of its smaller relative *Calostemma*, which I always called Native Jonquil.[11] On the black soil hills the principal tree was the Silver-leaf Ironbark *(Eucalyptus melanophloia)*; this petered out towards the plains, where there were groves of brigalow, yarran, and myall.[12] These all belong to the wattle family. The myall, which grew on nearly all soils, is very attractive, with its drooping, silvery foliage; it is also one of the most valuable fodder trees. The timber is beautifully grained, and when freshly cut has the perfume of violets. Unfortunately, like most of our wattles, it is very subject to the attacks of borers.

We had been at Warialda for nearly a year when our first child, Rachel, was born. She is now Mrs L. C. Cox, of Broombee, near Armidale. Towards the end of the next year (1906) alarming reports came of my father's health. He had been compelled to retire, and was living in Melbourne, with his faithful housekeeper, Miss Sarah Eastwood, in attendance. But the reports we received suggested that he was likely to be a chronic invalid; and I felt that if possible I should move nearer to him, since it was out of the question for my sister to do so. Accordingly I applied to Bishop Armstrong of Wangaratta,[13] and was offered the position of rector of Yea, about sixty miles from Melbourne. So in 1907 we left New South Wales. The move to Yea, however, was not a happy one. From a scenic point of view the district was beautiful; but the climate did not agree with us, church affairs were at a low ebb, and the project of a huge irrigation dam on the Goulburn River, which would have submerged Yea to a depth of forty feet, did not improve matters. (The scheme was ultimately abandoned, but that was after we had left.) Quite unexpectedly, my father's health took a turn for the better; and he decided to accept a long-standing invitation from my sister and brother-in-law, that he should go and live with them at Forbes. My wife and I were more than willing to return to New South Wales. Bishop Cooper was away in England; but his vicar-general, Archdeacon Moxon of Grafton, offered me the parish of Copmanhurst, and I accepted. Before we left Yea, however, our only son, Arthur Richard, was born.

Copmanhurst is a small village on the Clarence

White-flowering *Dendrobium kingianum* Bidw. from Lismore, 1926, photographed by Rupp when it flowered at Raymond Terrace ten years later. (From the Fordham Album, by courtesy of Mrs J. Trotter)

A New Zealand species, *Dendrobium cunninghamii* Lindl., photographed by Rupp at Woy Woy, N.S.W., in November 1933. (From the Fordham Album, by courtesy of Mrs J. Trotter)

Family Group, probably photographed in 1907 by H. M. R. Rupp during a visit to his father, then living in retirement. *From left:* Rev. C. L. H. Rupp, holding Rachel Mary Rupp; Miss Sarah Eastwood (housekeeper, standing); Marie Sarah Augusta Monypenny (H. M. R. Rupp's niece); and Florence Mabel Rupp. (Photograph by courtesy of Mrs R. Cox)

River, thirty miles above Grafton, and at the head of navigable waters. The North Coast of New South Wales was new country to us, and we were both deeply impressed by its glorious scenery. The Clarence is the finest river in the state. Rising in the highlands of southern Queensland, it flows southward among the eastern foothills of the New England plateau, to meet its chief affluent, the Mitchell or South River, some twenty-odd miles above Copmanhurst. The Mitchell drains a large area of high country through its two main tributaries, the Mann and the Nymboida. At its junction with the Clarence, the combined waters turn eastward through the Clarence Gorge, where they have cut their way along narrow channels between rugged cliffs, tumbling over roaring cataracts and two more imposing falls. Sir Earle Page,[14] the doughty medical knight of Grafton who has represented his native district for so long in the Federal parliament, is

determined that some day the Clarence Gorge shall be the scene of the establishment of the greatest hydro-electric scheme in Australia; and well it might be.

Below the Gorge the great river flows past Camelback Mountain, broad and deep reaches alternating with shallow stony rapids, till it comes to the head of navigation at Copmanhurst. Then it winds majestically along the thirty miles to the city of Grafton, receiving on the way the beautiful Orara River, which drains the Coast Range. At Grafton the Clarence is a mile wide, and there are ninety-nine islands between the city and Clarence Heads, fifty miles downstream, where the river is three miles across. Some of the islands contain farming communities with their own little villages, and on one—Harwood—is a mill of the Colonial Sugar Refining Company.

From the ecclesiastical point of view, I have no

desire to recall too many memories of Copmanhurst. It was a difficult parish; the travelling was not without its dangers, owing to the numerous fords on the Clarence and the Mitchell which had to be negotiated—there were no bridges. I was once 'bushed' on horseback at ten o'clock at night in the middle of the river, two miles from any habitation. It was raining, and I could not find a certain rock round which one must turn to get safely across. I did not realize that there was a 'fresh' in the river, due to rain away up on the New England tableland, and the rock was covered. However, here I am to tell the tale; so I did not drown.

My chief outlying centre of work was at the Cangai Copper Mine field, thirty miles away over two steep ranges. The mine, I understand, is now defunct, but in my time there were a thousand people there. My first church service at Cangai was held in a small slab 'hall', alongside a more pretentious building just erected, in which roller skating was in full swing! On the other side of me, in the bright moonlight, a big two-up school was going merrily. The atmosphere was hardly conducive to devotions.

Forty miles above Copmanhurst, on the main river, stood a palatial residence known as Yulgilbar Castle, the home of Mr and Mrs C. A. G. Lillingston. There I met their daughter, who will perhaps pardon me, if this ever meets her eye, for saying that her father, a rigid Conservative of the old school, did not altogether approve of the ideas on politics and the social order which were developing in his daughter's mind at Sydney University. Beyond the revelation of her present identity, she needs no further introduction to my readers; for Mrs Jessie Street is well known in Australia and beyond it, as a leading feminist and a staunch supporter of Labour.[15] Our paths have not crossed again since those far-off days at Yulgilbar.

My worst troubles at Copmanhurst were not due to the difficulties of travelling, nor yet to certain disagreements with a few parishioners. The latter trial comes to every parson sooner or later if he has any convictions of his own worth battling for. But in this connection I cannot omit reference to one man, whose loyal and generous support carried me over many an obstacle. This was Mr Charles Tindal,[16] owner of Ramornie station, and the Ramornie Meatworks on the Orara River. The mere fact that a man is wealthy is sufficient to damn him in the eyes of small-minded, envious people who brag about their own love of the poor, but take good care that it doesn't cost them much. I hold no brief for the rich. I am a pretty good 'mixer', and in the course of my ministry could feel equally at home in the boundary-rider's cottage or the station homestead or the farmhouse. I have had dear friends in all three. Charles Tindal was a rich man; but I think he came nearer to my ideal of a humble-minded Christian layman than any other man I have known. The Scriptural maxim, 'When thou doest alms, let not thy left hand know what thy right hand doeth', seemed to be always in his mind, for he hated ostentation.

To get back to my worst troubles at Copmanhurst; they came through the lay readers I was compelled to employ. I had three, one after the other; and my experiences of them were so devastating that I refused to have a No. 4, and carried on as best I could by myself. The lay-reader system was a bad one. Bishop Cooper used to uphold it on the ground that if you got a man who was no good, he could be dispensed with before he actually entered the ministry. That, however, ignored the point that he might do irreparable mischief before his unlucky vicar could get rid of him. And it was a bad system for the readers themselves, too, as I knew from experience. Very few men could read steadily and consistently for theological examinations while they were required at the same time to travel about the parish conducting services and giving lessons in the public schools. Times have changed for the better now, thanks to the establishment of theological colleges where men can be tried out and trained.

It is time I made some reference to the botanical riches of Copmanhurst; but there is so much that I *could* say, that I am afraid of boring my readers. It was, as I have already hinted, my first experience of the subtropical vegetation of the New South Wales North Coast. There were beautiful palm scrubs and glorious brush forests,[17] usually with the noble river close at hand, and the great scarps of the New England Range for a background. Closer in to Copmanhurst there were rocky sandstone hills not unlike those to the north of Port Jackson. Every part of the district was rich in botanical interest. Some of the palm scrubs and brushes were badly infested with leeches; ticks were abundant, and of course there were mosquitoes; but no naturalist allows trifles like these to interfere with his objectives! I did not come across many snakes; 'snake-infested jungles' are largely an invention of imaginative writers. The pretty green tree-snake, which is harmless, was fairly common, but I doubt if I met more than a dozen venomous snakes in three years. To attempt an enumeration of the trees, shrubs and herbaceous plants observed or collected is impossible in such a narrative as this. I had my first encounter with the notorious Nettle Tree *(Laportea gigas)* not long after we arrived[18]; but I had been warned about it, and inspected it with due caution. In appearance it is truly a magnificent tree, often well over a hundred feet in height, with beautiful velvety heart-shaped leaves up to a foot in length; but it is useless for any economic purposes, and the stinging hairs of those alluring leaves can inflict intense pain.

Near Copmanhurst I discovered a tea-tree *(Leptospermum)* with a pleasant citron scent. It proved to be a new species, of considerable economic value, and was named by Edwin Cheel *L. citratum.*[19] Some miles away to the north I found a very beautiful wattle, with drooping branches, rather large golden-yellow flower-heads, and leaves of both kinds—the broad 'phyllodes' and the true feather leaves. This also proved to be new, and was described by Maiden and Betche as *Acacia ruppii.*[20] Fifty-one species of orchids were collected; had I known as much about the habits of these plants then as I learnt in after years, I think that number might have been doubled. My records show that I collected 572 species of flowering plants and ferns during our three years on the Clarence.

Our next move came in 1911, when I was nominated and elected vicar of Barraba. This town lies about midway between Tamworth and Warialda, so we were back again on the western slopes of New England. The parish was fairly extensive. A few miles to the north of the town the Nandewar Range breaks away from the New England plateau and runs westward for about twenty miles, then forming a T of which one arm runs north and the other south. This is the highest portion of 'The Nandewars', reaching a height of over 5,000 feet on Mount Kaputar, and a little less on Mount Lindesay. Snow often fell on Kaputar and Lindesay in winter months. Barraba, at an altitude of 1,700 feet, is situated on the Manilla River, a small tributary of the Namoi. The great western stream of New South Wales, the Darling, receives a good deal of its water from the Nandewars by affluents of the Namoi and the Gwydir. The chief tributary of the latter, the Horton, rises about 500 feet below the summit of Kaputar, and after plunging over a grand fall of 370 feet, winds through a deep gorge into a broad valley running north for thirty miles to join the Gwydir about twelve miles from Warialda.

To the east of Barraba the country rises steeply by tiers of shale and serpentine to the granite of New England near Bald Rock Mountain, an immense mass of granite some 4,000 feet high, surrounded by

The fine Church of St Lawrence, Barraba, as the vicar saw it about 1912.

densely wooded ravines. Except along the river valleys, there is no level land in the whole district. Magnificent panoramic views are unfolded from many of the higher points in this jumble of hills and mountains. The valleys of the Manilla and Horton Rivers have been cleared of most of their original timber, and are noted for the quality of their wool clips.

We were very happy at Barraba, where in 1913 our youngest child, another daughter, Eileen, was born. She married Mr A. C. ('Don') Cox, a brother of her sister's husband, and lives near Murrurundi, on the Liverpool Range.[21] In Barraba I was entirely free from worries like those which had hampered me at Copmanhurst. The fine parish church of St Lawrence had been built under the vigorous leadership of my immediate predecessor, the Rev. G. P. M. Ware,[22] who died in Sydney a year or two ago. We had a splendid body of church workers, and made many good friends. Among these were Dr F. A. Rodway, a son of the noted Tasmanian government botanist, Leonard Rodway. Botany must run in the veins of the Rodway family; for the doctor, who has now lived for many years at Nowra, on the South Coast below Sydney, has accumulated a most valuable herbarium, and has contributed many specimens to the famous Kew Herbarium and to other collections, including my own; while his younger daughter, Mrs G. L. Davis, is lecturer in Botany at the New England University College in Armidale.[23] Dr Rodway and I have met only once since Barraba days; but we have corresponded for over thirty years.

The botanical features of the Barraba district were extremely interesting. On the pasture lands of the

Manilla and Horton valleys I collected many grasses and sedges for the National Herbarium; but I loved best to get into the ravines of 'The Granite' towards New England, and, at the other end of the parish, the gorges of the Nandewars about the Upper Horton and Mount Kaputar. Just below the summit of the latter, at the very head of the Horton, there was a beautiful grove of tree-ferns;[24] the whole character of the vegetation there was more suggestive of a coastal forest than of the far western slopes of the Northern Tableland. The panorama from Kaputar is magnificent; for the range on its western side drops abruptly 4,000 feet to the edge of the great inland plains that stretch out to the Darling.

Eastward you look up the valley of the Manilla River and beyond it to the heights of New England. Up in the hills east of Barraba, about ten miles from the town, is a tract of serpentine country traversed by Ironbark Creek. The flora of this tract was quite distinct from that of the shale on one side and the granite on the other. It was here that I found yet another new species; a pretty little shrub with pink flowers and strongly odorous leaves, which received from Mr Cheel the name *Boronia ruppii*.[25] Edwin Cheel was Maiden's chief assistant at the National Herbarium, and became the director of that institution on the latter's retirement.[26] He, too, proved a good friend, and gave me much help in my botanical hunting.

Unfortunately we had a good deal of sickness at the Barraba vicarage, though the climate was considered to be a very healthy one. During our third year there the parish was visited by the Rev. John Jones,[27] general secretary of the Australian Board of Missions (commonly called ABM). Mr Jones, a man of outstanding ability who had declined nomination to a bishopric, was greatly in need of an assistant secretary to relieve him of much deputation work. He urged me to apply for the position. After due consideration I did so, and was accepted a few months after the outbreak of the first Great War in 1914. So we left Barraba and came to Sydney, living first at Abbotsford, and then at Longueville on the Lane Cove River.

Before *Cactoblastis* — Rupp's view of part of the Prickly Pear problem at Gravesend on the Gwydir River, about 1905. (From glass negatives kindly lent by the late Arthur Rupp of Willoughby)

5 The First World War Years — and After

I thought a good deal about the question whether I should volunteer for an army chaplaincy. I knew, however, that a good many of the clergy who did go proved to be square pegs in round holes; they did not fit. I did not feel that I had the necessary qualifications myself; but I consulted my 'Chief', Mr Jones, who said 'Unless you feel that you have a definite vocation for army work, you should not go. You have a wife and a young family; there are plenty of unmarried clergy available. You are doing very useful work where you are, and I think you should stick to it.' So I stayed. I remained with the ABM for five and a half years, first as assistant-secretary, and then, after the reconstitution of the Board and the appointment of John Jones as chairman,[1] I became secretary. During the war years both my wife and I lost our fathers. Mr Dowe died at Tamworth from typhoid fever;[2] my father passed away in my sister's home (then at Glen Innes) at the age of seventy-nine.[3]

My work involved almost incessant travelling, chiefly over the greater part of New South Wales—once into Queensland, once into Victoria, and twice to Tasmania, organizing, preaching and lecturing on behalf of the board's work in New Guinea, Melanesia, among the Australian Aborigines, and elsewhere. In the course of my duties I met many notable men and women missionaries, such as Bishops Stone-Wigg[4] and Sharp[5] of New Guinea, Copland King[6] and the Tomlinsons[7] from the same field, Bishop Gilbert White of Carpentaria,[8] Ernest Gribble of Yarrabah,[9] and others. In Brisbane I was

the guest of Archbishop Donaldson,[10] and met his domestic chaplain, the Rev. F. de Witt Batty,[11] who in after years was to be my bishop in the diocese of Newcastle. I stayed with my old friend Bishop Long at Bathurst, and addressed his clergy assembled in Synod. I renewed my acquaintance with Hay for a similar meeting of the Riverina synod under Bishop Anderson.[12] My old diocesan, Bishop Cooper, presided at a lecture I gave in the Glen Innes parish. He died about a year later,[13] having lived to see the fulfilment of his desire to divide the great unwieldy diocese of Grafton–Armidale into the independent sees of Grafton and Armidale. Bishop Druitt was at Grafton[14]; among the parishes I visited there was Copmanhurst, which had not altered much except for the erection of a new church. In the Goulburn diocese, then under Bishop Radford,[15] I paid my first—and only—visit to Canberra, which at that time consisted mainly of plans for future buildings. St John's church, of course, was there, and the vicar, the Rev. Fred Ward, explained the 'lay-out' of the city that was to be. At Holbrook I met an old contemporary of Geelong Grammar School days, Jack Bowler; while at Woodstock in the Bathurst diocese I was the guest of another old schoolmate, Billy Whitney of Waugoola.

I could relate many curious and interesting incidents of my travel during those ABM years, but they would take up too much space. One must suffice; it was an unpleasant experience at the time, but I could look back and laugh at it later. I had had a tiresome and bitterly cold tour in the Goulburn diocese in midwinter, and after sundry mishaps and delays, arrived at Young in pouring rain. I was to lecture there that night (Saturday), and to preach in the beautiful church of St John next day. The rector was the late Stanley Champion, a fine man and greatly beloved by his people, who turned out to the lecture in force despite the weather. Sunday proved just as wet, but we had good services.

I was the guest of one of the bank managers. I had been away from home for nearly six weeks; I was very weary of travelling under unpleasant conditions, and decided to catch a train at 3 a.m. which connected with the main Sydney line at Harden. In those days the electric power at Young was turned off at midnight, but the good folks at the bank provided me with a roaring fire, plenty of candles and books, and a sumptuous supper. I was to let myself out by a side door; my luggage had already gone to the station, so my host and hostess bade me goodbye and went to bed. At half-past two I clicked the latch of the side door and descended into the garden. It was raining in torrents, and pitch dark, and for some time I could not find the gate. I fell into muddy flower-beds, tore myself on invisible rose bushes, and at last reached the street. I knew where I had to go; but when I came to the street corner, I could see no sign of the station lights; actually that cross-street had a slight bend in it which I had not allowed for. Blundering about, I next found myself lying full length in the gutter, which was full and overflowing! When at last I saw the station lights, I could not find the entrance. Fearful of missing the train, I climbed a barbed-wire fence, tore my clothes, and stumbled into a great pool of water. The train had been signalled when at last I reached the platform. My appearance prompted pardonable grins from the few people about; but a porter put me into an empty compartment, and as I had plenty of spare clothing in my cases, I succeeded in making myself fairly presentable by the time we arrived at Harden. My troubles were not quite over, however; there had been a derailment on the main line, and the express was delayed for nearly three hours, necessitating a wire to my wife and a long wait in the cold.

On my first visit to Tasmania I met at Devonport the Rev. (now Archdeacon) H. B. Atkinson.[16] I think he viewed me with some suspicion at first, for he told me later that he had had one or two missionary deputations which proved rather futile. But in his study I caught sight of Leonard Rodway's *Flora of Tasmania* among his books, and immediately pounced on it. The discovery that we were both botanically inclined thawed the atmosphere at once, and opened the way to a friendship which has lasted for thirty years. On my second visit to the island state I found him rector of the fine old Church of the Holy Trinity at Hobart. The bishop (Dr Hay) was going to England for the Lambeth Conference of 1920, and Atkinson was going with him. He asked me to come and be his *locum tenens* while he was away. I had now been more than five years with the ABM, and though I enjoyed the work, I was beginning to feel tired of the constant travelling and the lack of any real home life. So my wife and I discussed the matter on my return to Sydney, and it was decided to accept the Hobart offer. I was the more disposed to take this step because my much-loved chief, John Jones, had been offered the parish of All Saints', St Kilda, in Melbourne, and had accepted. A few years later he followed his great friend, Archbishop Donaldson, who had been elected bishop of Salisbury in England; he became a canon of Salisbury Cathedral, and died a few years ago.[17]

We had a very happy year in Hobart. The path of a *locum tenens* is not always strewn with roses. Disgruntled parishioners are apt to seize the opportunity of the rector's absence to make trouble; and cases have occurred where his substitute has been tempted to make capital out of this. But even had I been so disposed (which I certainly was not), there was no risk of such happenings at Holy Trinity; the rector was a 'man greatly beloved', his people were intensely loyal to him, and all went well. Of course we climbed Mount Wellington, and visited as many of the nearby scenic resorts as time and opportunity permitted.

I found the Tasmanian flora very fascinating; there were so many plants which I had never seen or heard of before; so many endemic forms, and so many that appeared to me to be 'intermediate' and possibly on their way to evolve into new species. My favourite hunting-grounds were Mount Wellington, Mount Nelson, Knocklofty, and the country behind Bellerive; but the hills near Bridgewater, Brown's River, and the Channel received attention, too. Among the most interesting Tasmanian plants are the Richeas, a group belonging to the 'Australian heath' family (Epacridaceae). Only eight species are known and seven of these are strictly confined to Tasmania. The eighth is there too, but it is also found on the Australian Alps. Several of them are very beautiful when in full bloom. One, restricted to Tasmania's western highlands, is a small tree; the rest are somewhat rigid shrubs. The Common Heath *(Epacris impressa)* abounds in many parts of the island,

and is a universal favourite, with its massed blossoms of white, pink, or crimson. Of orchids, Tasmania is now known to possess more than eighty species, all except two being terrestrial plants. The two exceptions are found only in the northern half of the state; a little rock-loving *Dendrobium*,[18] and a small but exquisite *Sarcochilus* growing on trees.[19] Both are also found in Victoria and New South Wales. Among the terrestrials, pride of place perhaps belongs to the 'Spiders' (*Caladenia patersonii* and allied forms), named by Robert Brown in honour of Lieutenant-Colonel William Paterson, who played so large a part in the early affairs of Australian settlement, and was the founder of Launceston. The 'Spiders' have a wide range of colour, and flowers have been measured nearly ten inches across the long, spidery segments. Several species are common on the mainland of Australia; but they do not extend into the tropics.

When my friend Atkinson returned from England, the bishop of Tasmania offered me the new suburban parish of St Aidan's, East Launceston. I knew something about it; it had been portion of the old parish of St John's, and its traditions were alien to those of the school of thought in the Church to which I belonged. I was therefore very reluctant to accept the offer. I was never a 'party man', and I knew that St Aidan's had been controlled by a party with which I had little sympathy, and which had no tolerance for any views but its own. I have worked amicably with 'High', 'Broad', and 'Low' churchmen; but I cannot work with the man who will tolerate no ideas except his own. However, my objections were pronounced unnecessary, I was persuaded to accept, and to Launceston we went. It is a beautiful city, set in beautiful surroundings; but to it I owe the unhappiest period of my life. So much I must say; but will say no more. Of course there were bright patches; we made some good and loyal friends; and the botanical treasures of the district were a joy, and a blessed relief from nerve-strain. But at the close of 1922 I wrote to the bishop of Newcastle in New South Wales, Dr Stephen, who had known me long ago, and who also understood the situation, for he had been bishop of Tasmania himself. He replied 'Come', and I resigned. So once more we made our way northward.

After three months as *locum tenens* at Branxton, between Maitland and Singleton, I was offered the parish of Bulahdelah, some twelve miles north of Port Stephens. The bishop told me, however, that he did not intend to leave me there, as he had other plans in view, which had not quite matured. But we never

regretted going to Bulahdelah. It was a sparsely populated, struggling parish, but the people (mostly engaged in timber-getting, dairying, or fishing) were kind and considerate, the scenery was very beautiful, and as for the wildflowers—well, if Warialda was a botanist's paradise, Bulahdelah might almost be termed a botanist's seventh heaven.

The outstanding feature of Bulahdelah is the massive bluff known as the Alum Mountain. It is not quite 1,000 feet high, but it dominates the valley of the Myall River with its crags of alunite, rose-pink in the sunshine after a passing shower of rain. The Myall River, too, is a stream of no small interest. It rises in the ranges above Bulahdelah, and is scarcely more than a small creek until it reaches the village. There it suddenly broadens out, and reaches a depth of forty feet, which is maintained for most of its remaining journey to the Myall Lakes, through which it flows, emerging below the Broadwater finally to empty into Port Stephens. How did such a small stream suddenly carve out such a great channel? The Myall Lakes themselves are very lovely. But only a mile from their northern end is a smaller but still lovelier lake, which I always call the Lake of Flowers. Alas! its official name is just 'Smith's Lake'. I do not wish to cast any reflections upon the honourable name of Smith; but it seems hardly appropriate for such a beautiful lake as this; moreover, I could never find out who or what the Smith was, whose name was thus immortalized.[20] I described my Lake of Flowers many years ago in the Melbourne *Australasian*, but have mislaid the cutting, so I cannot give the reference.[21]

Smith's Lake (Rupp's 'Lake of Flowers') as the rector of Bulahdelah saw it in January 1924. (From glass negatives kindly lent by the late Arthur Rupp, Willoughby)

The Myall Lakes as seen by Rupp from a vantage point at Bungwahl, 1923.

It was at Bulahdelah that I first made contact (through the agency of Mr Maiden) with Dr Richard Sanders Rogers of Adelaide,[22] for so many years *facile princeps* among Australian orchidologists, and a very distinguished citizen of South Australia. My tribute to his character and accomplishments will be found in the 'In Memoriam' notice in the *Australian Orchid Review* for June 1942. We did not meet until he came to Sydney as president of the Botany Section of the 1932 meeting of the Australian and New Zealand Association for the Advancement of Science; but our correspondence was prodigious, and I still treasure his letters. How he managed to write them all, amid his professional duties and those which fell to his lot through the public offices which he held, with scores of other correspondents in and beyond Australia seeking his opinion on orchid problems, I do not know—he never used a typewriter. Probably his widow, Mrs Jean Scott Rogers, who shared his enthusiasm, and who still writes to me, could let us into the secret. He was the most generous of friends;

to him I owe my eight volumes of Mueller's *Fragmenta*,[23] about 150 loose plates of Fitzgerald's classic *Australian Orchids*, and other valuable publications, besides specimens of a majority of South Australian orchids.

Here I may remark that I have been singularly fortunate in the number cf good friends who have enriched my orchid herbarium (now part of the National Herbarium at Sydney) by sending specimens from various parts of Australia and New Zealand. Prominent among these are Mrs Florence Perrin and Mr Neil Burrows of Launceston, Tasmania; Mrs Edith Coleman and W. H. Nichols of Melbourne (of whom more anon); Mr H. Goldsack of Adelaide; the late Lieutenant-Colonel B. T. Goadby of Perth; the Rev. Fr B. W. Haydon, formerly of Wagga Wagga; Mrs Pearl Messmer of Lindfield; Mr A. W. Dockrill of Carlton, N.S.W.; Mr E. Nubling, formerly of Normanhurst, N.S.W., but now of Mentone, Victoria; Miss I. Bowden, of Beecroft and Woodford, N.S.W.; Mrs Joyce Telfer and her sister Miss G. I. Scrivener of Mt Irvine in the Blue Mountains; the late Captain J. D. McComish of Wahroonga; Mr M. W. Nichols of Kurri Kurri; Messrs G. W. and P. Althofer of Dripstone; Mr Fred Fordham, formerly of Brunswick Heads but now of Armidale; the late Rev. E. Norman McKie of Guyra; Mr Trevor Hunt of Ipswich, Queensland; the Misses Gemmell of Glen Aplin, Queensland; Dr C. P. Ledward, of Burleigh Heads, Queensland; Mr Warren Abell, of Durong, Queensland, Mr Kenneth MacPherson of Proserpine, Queensland; Dr H. Flecker and Mr W. F. Tierney of Cairns, Queensland; Mr W. W. Mason, of Cape Tribulation, Queensland; the late Mr H. B. Matthews of Remeura, N.Z.; Mr E. D. Hatch of Laingholm, N.Z.; Mr Cedric Smith of Stewart Island, N.Z. The list could easily be lengthened; but these, I think, have made the largest and most valuable contributions to my collection; and I have previously acknowledged the assistance given by Dr. F. A. Rodway of Nowra.

Besides Dr Rogers, two other noted students of our orchid flora began correspondence with me when we were at Bulahdelah. These were Mrs Edith Coleman[24] and Mr W. H. Nicholls[25] of Melbourne. Mrs Coleman is well known for her many contributions to the study of natural history in Australia; botany, entomology and zoology alike have been illuminated by her facile pen; her articles are usually enriched by first-rate photographs, or admirable drawings by her daughter, Miss Dorothy Coleman. She has written much that is of great value on our orchids, especially in connection with methods of pollination. Mr W. H. Nicholls has long been recognized as the leading orchidologist of Victoria; and every reader of that excellent journal, *The Victorian Naturalist*, is familiar with his beautiful line-drawings, which he seems to run off as easily as you or I can write a,b,c. Years ago he embarked upon the ambitious project of emulating R. D. Fitzgerald; for he hopes to illustrate in colour every procurable species of Australian orchid, with full details of floral structure, etc. May the day not be far distant when this great work will be completed and published![26]

Bulahdelah gave us new friends in Dr H. Leighton Kesteven and his wife and family. The doctor, who holds the degrees of MD and DSc, is well known in scientific circles for his numerous articles, in Australian and overseas journals, on osteology and kindred subjects.[27] He proved, however, to be quite a keen orchid hunter, too; and we had many pleasant rambles together. Our wives became close friends, while the younger Rupp daughter and a Kesteven lass of about the same age were very fond of one another.

The Bulahdelah bushlands were very beautiful. The brush forests were not quite such a riot of jungle as those of Copmanhurst, but there were palm scrubs, and a wealth of fine trees. Outside the brushes there were large tracts of sandy country, interspersed with peat or bog; here grew many glorious flowering shrubs and herbaceous plants. Boronias were in great abundance, among them the deep rose-coloured, bog-loving *B. falcifolia*. About the lakes the magnificent North Coast Christmas Bells (*Blandfordia flammea*) grew in great profusion. Gorgeous Bottle-brushes (*Callistemon*) were plentiful, and a lovely pink-flowering Tea-tree (*Leptospermum liversidgei*) decked the boglands with masses of bloom. I recorded more than a hundred species of orchids for the district, including about twenty-three epiphytes. Among the terrestrial species the most remarkable is one which was discovered some years after we had left—the subterranean *Cryptanthemis slateri*.[28] The first specimens were sent to me as curiosities; ultimately I was able to describe and name this extraordinary plant as a new genus and species. It lives and blooms *underground*. It was described in the *Proceedings of the Linnean Society of New South Wales*; but a more popular description, with two full-page photographic reproductions, appeared in the *Australian Orchid Review* for June 1938. Here I may say that it was during our Bulahdelah days that I first began to contribute articles on orchids to the Linnean *Proceedings* and to the *Victorian Naturalist*. This habit once begun, has continued ever since, and has been extended to the columns of various other journals. Not long ago I attempted to trace the number of my orchid papers published since 1923, and was rather taken aback to find that there were more than 150. It seems high time, therefore, that I should retire in favour of younger enthusiasts.

In 1924 Bishop Stephen appointed me rector of Paterson. We remained there for almost six years.

6 Paterson and the South Maitland Coalfields

Paterson, on its willow-fringed river of the same name, about twelve miles north-east of Maitland, was named in honour of that same Lieutenant-Colonel William Paterson who was the founder of Launceston, Tasmania. He was sent by Governor King in 1801 to explore the Hunter River, into which the Paterson flows at Morpeth. At first the Paterson was mistaken for the main stream, and Paterson's name was attached to the Hunter itself above the junction, but the error was rectified later. The Paterson rises high up in the lofty Mount Royal Range, not very far from the 5000-foot plateau of Barrington Tops. Winding down its deep valley to the village of Gresford, it there receives its chief affluent, the Allyn, which has its source at the foot of Carey's Peak, the highest point of 'The Tops', and runs swiftly down a magnificent gorge whose steep flanks are covered by a dense forest of the noble Negrohead Beech *(Nothofagus moorei).*

Below Gresford the Paterson flows past Vacy and Paterson, where its valley opens out into the rich flats opposite Morpeth. In the early days these flats were cedar forests, but of them no trace is left beyond an occasional odd cedar on the river bank. It is a beautiful little river, and the whole district is very attractive from a scenic point of view. Years ago the Paterson Valley was famous for its oranges, and there are still some orchards left; but in the main the chief industry is now dairying. The village of Paterson lies on the river bank immediately opposite a wooded spur of the valley known as Hungry Hill. For an Australian area, this valley is very rich in historical interest; and in the course of time I came to learn the story of every one of the old estates (now mostly cut up into farms) from the Barrington foothills to the Hunter. The parish is now a small one, with only three centres of population to be served—Paterson, Vacy and Martin's Creek, each about five miles from the other two.

The old stone church of St Paul at Paterson, which celebrated its centenary a few years ago, is not attractive externally, but its history is quite fascinating; and it fell to my lot to bring to light the hidden or forgotten things of the past. A considerable sum had been spent on repairing the roof, which had been neglected almost to the point of danger. But even with this improvement, the church was in a bad state. The interior was plastered, and the plaster on one side had collapsed, leaving a most unsightly wall of stone and rubble. Stowed away behind a cupboard in the vestry I found a small marble tablet, which had fallen when the plaster collapsed. It bore the inscription: 'In memory of the Reverend John Jennings Smith, M.A., of Catherine Hall, Cambridge, First Incumbent of St Paul's Church, Paterson. Died September 8th, 1846, aged 62.' Jennings Smith's tomb was just outside the east end of the church, and it was in almost as bad condition as the church itself. I wondered whether, if I could find out any particulars about this pioneer priest of Paterson, they might serve for launching an appeal for the restoration of the church and tomb. The churchwardens, however, threw cold water on my suggestion, saying that the parish had spent all it

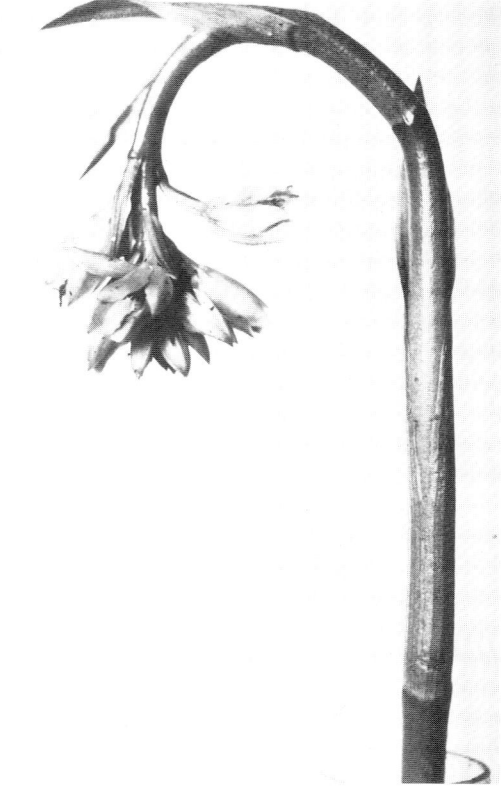

Rupp's photograph of the North Queensland epiphyte, *Saccolabiopsis armitii* (F. Muell.) Dockr., which he described in 1934 as *Cleisostoma orbiculare.* (From the Fordham Album by courtesy of Mrs J. Trotter)

The Pink Nodding Orchid, *Geodorum densiflorum* (Lam.) Schlechter (long known as *Geodorum pictum* (R.Br.) Lindl.), photographed by Rupp in January 1934. (From the Fordham Album by courtesy of Mrs J. Trotter)

could afford on the roof. That same week, a letter reached me from a lady in Brisbane, asking if I could send her a 'snapshot' of St Paul's Church, which had been built by her great-grandfather, the Rev. John Jennings Smith. I sent the picture, told her of the condition of the church and tomb, and asked whether she could give me any information about her great-grandfather. She replied advising me to write to a relative in Melbourne, Miss E. Selwyn Smith, who, she understood, had the old man's diary and other papers concerned with him and his family.

Miss Selwyn Smith proved to be a veritable mine of information. John Jennings Smith was born in 1784. In his earlier years he was a schoolmaster; he is said to have been a partner of R. H. Barham, author of the *Ingoldsby Legends*. When nearly forty years old—a married man with a family—he went up to St Catherine's Hall, Cambridge, with a view

to Holy Orders. He took his degree, was ordained, and held a curacy at Hartpury, near Gloucester. In 1831 he was removed to London, and came under the notice of the Duchess of Kent, mother of Princess Victoria, who appointed him one of the future Queen's preceptors. Two years after Victoria's accession he accepted a colonial chaplaincy, and at the age of fifty-six, with his wife and ten children, came out to Australia, whose first bishop, Dr William Grant Broughton, offered him the archdeaconry of Hobart. He replied, however, that he had come out to do pioneer work; and thus it was that he was sent to minister to the new settlements in the Paterson and Allyn Valleys.

The family travelled by schooner from Sydney to Morpeth, whence the men walked, and the women were accommodated in drays, for the fourteen miles through the cedar scrubs to Paterson. They arrived at 10 p.m.; and that same night the youngest child was born! She ultimately became Mrs Edward Milner Stephen, of Sydney.

There was not a church building in the whole area. But within the brief seven years of Jennings Smith's pastorate there were erected stone churches at Paterson and Allynbrook, a stone schoolhouse at the former place, and a slab church at Lostock on the Upper Paterson River.

The story of the pretty little church of St Mary's, Allynbrook, though it falls within the Jennings Smith period, chiefly concerns the romance of William Boydell and Phoebe Broughton. Boydell's elder brother Charles was the pioneer of Gresford, where he built his home, Camyr Allyn. William came out from Wales to join him, and travelled by the same ship as the newly consecrated Bishop of Australia, Dr Broughton. On the voyage he and Mary Phoebe, the bishop's elder daughter, fell in love. That was in 1836. On hearing whither the young man was bound, Dr Broughton withheld his consent to the marriage until such time as there should be a church within reach of Camyr Allyn. Jacob served for Rachel no more faithfully than did William for his Phoebe; nine years were to pass before St Mary's was consecrated near Caergwrle, the northern part of the Boydell estate, which Charles had made over to his brother. Then the faithful lovers were married, and at Caergwrle their happy lives were thereafter spent. Both are buried just near the door of St Mary's.

Jennings Smith died at Maitland in 1846, as the result of an accident; he was thrown from his gig while on his way from Paterson to visit his eldest daughter and her husband, Captain Horsley, at Hexham. His descendants—Jennings Smiths, Selwyn Smiths, Grey Smiths, Milner Stephens, Shands, Curries and others, are now widely distributed through the eastern states. Both he and his wife are commemorated by mural tablets in St Andrew's Cathedral, Sydney.

Having obtained all the information which is briefly outlined above, I felt that I could launch an appeal (for funds to restore the church and tomb) which would reach far beyond the limits of the present Paterson parish. Accordingly I wrote to the *Brisbane Courier*, the *Sydney Morning Herald*, and the Melbourne *Argus*, telling the story of this grand old pioneer and asking the assistance of his descendants in the work I had planned. The response far exceeded my hopes. Generous contributions flowed in from all three states. Not only were we able to restore the church and to renovate the tomb in strict keeping

Rupp's photograph of St Paul's Church, Paterson, taken for the eighty-fourth anniversary celebrations, 1929.

The Christmas Orchid, *Calanthe triplicata* (Willemet) Ames, flowering in the Paterson rectory bush house, December 1926. (From glass negatives kindly lent by the late Arthur Rupp of Willoughby)

with the original design, but also a second tablet was erected under the old one, recording the actual work of Jennings Smith in the district. Furthermore, a window of beautiful stained glass, bearing the Jennings Smith coat of arms, which had been the property of his second son, was offered by a grandson, the late Mr M. Selwyn Smith of Beaudesert, Queensland, and was duly built in alongside the pulpit of St Paul's.

On 27 November 1929, the eighty-fourth anniversary of the consecration of the church by Bishop Broughton, we had our grand day for the dedication of the restorations and renovations by our own bishop of Newcastle; and it was an additional joy to me that he was none other than my old friend George Merrick Long. For in 1928, Dr Stephen had resigned, and Dr Long, then bishop of Bathurst, had been elected to succeed him. To the celebrations there came Jennings Smith descendants from Queensland, New South Wales and Victoria, to swell the large congregation from the district iself; and of course there were clergy from neighbouring parishes. The service was followed by a 'high tea' in the public hall, at which interesting speeches were made by some of the visitors, including one by Miss Thea Milner Stephen, a grand-daughter of the youngest child of the Jennings Smith family, and at that time the leader of the Women's Auxiliary of the ABM in New South Wales. This day was the 'high light' of our years at Paterson.

It is time that I put in a word or two about the education of our children. Before we left for Tasmania, our elder girl, Rachel, had been enrolled as a boarder at St Gabriel's School, Waverley. She came to Tasmania with us, but went back to St Gabriel's before we returned to the mainland. From there she proceeded to Sydney University, where she took her BA degree, and subsequently taught in several of the leading girls' schools. While on the staff of the New England Girls' School at Armidale, she met her future husband, Leslie Clarendon Cox of Broombee, Armidale. They were married in St John's Church, Raymond Terrace, where I was rector from 1936 to 1939. Our son Arthur, after attending the Friends' High School in Hobart and the Launceston Grammar School, upon our return to New South Wales won a scholarship to The Armidale School. Subsequently he obtained his BA at Sydney University, and returned to Armidale as a member of the teaching staff. After several years there he was appointed to the staff of the Sydney Church of England Grammar School, better known as 'Shore', where he still is. Allusion to his experiences in the Second World War belongs to a later chapter in these memoirs. He was married in Shore school chapel, in Christmas week, 1948, to Lillian Elizabeth Williams. Our younger daughter, Eileen, like her sister, went to St Gabriel's School at Waverley, but did not go on to the university, preferring a business college course, and afterwards held a position with the Queensland Insurance Company in Sydney. At her sister's wedding she met the younger brother of the bridegroom, Ashley Clarendon Cox, and before long we understood that it would be a case of the two sisters marrying two brothers. The younger couple were also married in St John's, Raymond Terrace, and now live near Murrurundi[1]. We have at the time of writing nine grandchildren; three girls in the Armidale family, and three boys and three girls in that of Murrurundi.

As related above, Dr Stephen had resigned the see of Newcastle in 1928, and was succeeded by Dr Long. The latter had been consecrated bishop of Bathurst in 1911, and the story of his sixteen years in that office is a story of incessant hard work and triumphant achievements; but the strain of those years was very severe, and undoubtedly was a contributory cause of his comparatively early death at the age of fifty-five. He was a born leader of men, and was recognized as such far beyond the bounds of his native country. I do not hesitate to call him a 'great man', but his intellectual calibre and commanding personality never led him to set himself up on a pedestal 'above the common crowd'; he could always make himself perfectly at home with all sorts and conditions of people, without any suggestion of conscious superiority. His great gifts inevitably led to many calls being made upon him for work outside the diocese. He was the first national president of the Church of England Men's Society; while he was in constant demand in connection with the drafting of a new constitution for the Church of England in Australia—a project which, alas! owing to the obstructive tactics of certain 'irreconcilables', is still incomplete.[2] But his greatest work of an extra-diocesan character was due to his appointment as director of education in the AIF during the First World War. He enlisted as an ordinary chaplain, but was soon selected for the difficult task of organizing vocational and civil training for 200,000 men. As the bishop of Ballarat points out in his memoir of George Merrick Long (written when the author was dean of Newcastle), 'the full scheme was not carried out owing to the unexpected collapse of the Germans and the signing of the armistice in 1918. Nevertheless a work of immense value was achieved.' Upon undertaking this work Bishop Long was given the rank of brigadier-general, and later the honour of CBE. The Archbishop of Canterbury conferred upon him the degree of DD, and the Universities of Cambridge and Manchester that of LLD. In 1919 he returned to Bathurst, and threw himself into his work with all his old energy, though friends felt some anxiety about his health. Eight years later the Rt Rev. A. L. Wylde was appointed to assist him as coadjutor-bishop; and in the following year the call came to Newcastle.

The diocese of Newcastle has been singularly fortunate in its bishops. It was founded in 1847, and then included practically the whole of eastern Australia north of the diocese of Sydney. The Queensland area was severed by the formation of the Brisbane diocese in 1859, and in 1867 the diocese of Grafton and Armidale was established. The first bishop of Newcastle, William Tyrrell, was a man of far-seeing vision and statesmanlike policies. During the centenary celebrations of the Diocese in 1947, I was privileged to take part in 'the Morpeth Pilgrimage', when some 12,000 people assembled to honour Bishop Tyrrell's memory at his tomb in Morpeth cemetery. He died in 1879, and was succeeded by Dr J. B. Pearson, who resigned after ten years. Then came George Henry Stanton, translated from the diocese of North Queensland. Bishop Stanton was a fine preacher, and a man with a broad missionary outlook and a bright and cheerful disposition. Curiously, my only personal recollection of him dates back to Trinity College days. Arthur Vincent Green, archdeacon of Ballarat and formerly a brilliant student at Trinity, had been elected bishop of Grafton and Armidale; and members of the college presented him with his episcopal ring.

At the presentation in the college dining hall, Bishop Stanton was a visitor, and made a delightful speech; but the only actual words I can remember are—'You are going to be the shepherd of a vast area, in which you will encounter many difficulties; but you will find that your people up there have hearts as big as. . .*pumpkins*!' Dr Stanton died in 1905, and was followed at Newcastle by Dr John Francis Stretch. 'Jack' Stretch, as he was irreverently called in his earlier years, was another of the not inconsiderable band of brilliant Trinity College men. He was an orator of the first rank; a handsome man of fine physique, scholarly, energetic to the point of restlessness, very unconventional (when he was dean of Ballarat he used to shock Bishop Thornton by sitting on the kerbstone garbed in an old cassock, smoking and yarning with the 'cabbies'), and gifted

with a keen sense of humour. Henry Handel Richardson confesses in *Myself When Young* that she fell violently in love with him when he was vicar of Maldon, whither her mother had been removed from Koroit. She was then only fourteen, but she describes the affair as having been quite a serious one for her, particularly as the object of her affections married, and brought his wife to Maldon! She remarks unconvincingly that his eloquence was 'wasted' on a country parish; but admits that later on, she acknowledged fate's wisdom in baulking her young desires: 'As the wife of a Bishop I should indeed have been a misfit.'

Dr Stretch was bishop of Newcastle for thirteen years, resigning in 1919 owing to increasing deafness and failing health. He died shortly after his retirement. His only surviving son, the Rev. Carlos Stretch, is rector of St Paul's, West Maitland. The next two bishops were also alumni of Trinity College, Melbourne. Reginald Stephen, as I have noted above, was lecturer in Economics there in my time. He became dean of Melbourne, and then was elected bishop of Tasmania, whence he was translated to Newcastle on the resignation of Bishop Stretch. No two men could have been more unlike. Dr Stephen, tall and rather gaunt, was quiet and reserved; not an easy man to know, and by those who failed to know him considered cold and even inconsiderate. He was nothing of the kind; but in his younger days he had been very shy, and could never quite rid himself of the habits of a shy man. He disliked having to attend social gatherings, and I remember that at Paterson he was missed from a 'welcome social' in his honour in the parish Sunday School, and was discovered solacing himself with a pipe in the dark street outside! He was a man of great intellectual gifts, and notwithstanding any appearances to the contrary, possessed a quiet vein of humour which could be very effective. He rather loved to tell a story against himself; and it lost nothing of its effectiveness by being delivered without the vestige of a smile on the episcopal face until the *denouement* was over. It was in his time at Newcastle that the diocesan children's homes were established, the Broughton School for boys was founded, and the Provincial College of St John for the training of candidates for the ministry was removed from Armidale to Morpeth.

Then came George Merrick Long. It was a great joy to me to be able to look up to my old friend as my bishop and Father-in-God. But during his first visit to Paterson I felt seriously concerned at the state of his health. He was obviously far from well; and matters were not improved by the bogging of his car on the way back to Newcastle at night (it was a very wet season), so that he had to return to us by the courtesy of a passing motorist, and remain for the night. Some time after the Jennings Smith celebrations, I told him that I felt I had been long enough at Paterson. The parish was now on a sound financial footing, and we had built a comely new church at Martin's Creek. The bishop asked whether I would undertake work on the South Maitland coalfields. The parish of Weston was about to become vacant, and he would nominate me if I cared to go. The finances, he said, were not very satisfactory, but he added, 'You know you've got me behind you'. That was enough for me. I had never worked among coalminers, but saw no reason why I should not get on with them. So we pulled up stakes at Paterson, and moved to St Mary's rectory at Weston, about twenty-five miles away, and in the heart of the coalfields.

The dingy, smoke-grimed cottages of Weston were not very attractive. The church was a wooden building, rather small but quite nicely furnished; the rectory was a fairly comfortable weatherboard house alongside. I had not much travelling; there was a church at Abermain, two miles away, and another at Neath, a little further on towards the 'big town' of the coalfields, Cessnock. Congregations at Weston were always good; but I was unable to raise any enthusiasm for the church at Abermain, and Neath was crippled by debt. Indeed, I found that the bishop had put it mildly when he described the finances of the parish as 'not very satisfactory'; but diocesan assistance was coming; I had the bishop's pledge to back me, and I was prepared to do my best.

At this stage the bishop left for the 1930 Lambeth Conference. He had been with us barely two years, but had won the hearts of his people as he had done at Bathurst. There was considerable concern about his health; but everyone was glad he was going to Lambeth, as it was felt the voyage and the respite from diocesan duties would do him good. Little we thought that we were never to see him again! Scarcely had he left, when a long-threatened general strike on the coalfields broke out. The resulting depression made my position precarious, and shortly afterwards I had an acute attack of sciatica, necessitating hospital treatment. Then the blow fell. We learned from our morning papers that the bishop had been taken ill at the Lambeth Conference, and had died next day from cerebral haemorrhage. Newcastle was stunned; we had lost our great leader. As for me, I could scarcely bring myself to believe the news. But it was

true; and we knew that if he could speak to us, his message would be, 'Carry on'. A special meeting of the diocesan synod was summoned, at which the administrator, the Venerable Archdeacon Henry Alexander Woodd, spoke with deep feeling of the loss sustained by not only the diocese, but also the nation. Synod then proceeded to the election of Dr Long's successor, and chose the Rt Rev. Francis de Witt Batty, coadjutor bishop of Brisbane. I had met him, as I have mentioned, when he was domestic chaplain to Archbishop Donaldson. I felt we had made a good choice. Happily Bishop Batty still presides, after nearly eighteen years, over the affairs of Newcastle diocese. To me he has ever been a true Father-in-God and a kind friend.

Weston was proving too much for me. I had something in the way of a nervous breakdown, and the doctor ordered three months' rest, adding that he did not think it would be wise for me to return. We stored our furniture, and went to Collaroy Beach near Sydney. I sent in my resignation to the bishop, and we awaited the turn of events. I was most anxious not to leave the diocese of Newcastle; but the great depression of the early 'thirties had set in, and when my health was restored, the bishop, who was extremely kind to me all through this period, had literally nothing available for me. Meanwhile, during my temporary retirement in 1932, I enjoyed the delight of meeting at last with Dr R. S. Rogers, who had come over from Adelaide as president of the Botany Section of the session of the Australian and New Zealand Science Association in Sydney. We even coaxed him down to Collaroy Beach to spend a night with us.

Another good friend, whom I met pretty often during those three months, was the late George Robertson,[3] for so many years head of Sydney's famous publishers and booksellers, Angus and Robertson. When he heard the reason for my presence at Collaroy, he promptly put me down as an honorary member of the Sydney Book Club for the period of my stay. While we were at Paterson Mr Robertson had wished to publish a book on the native orchids of New South Wales, and someone recommended me for the job of writing it. He wrote asking me to come down and see him. He wanted a very small book that could be slipped into a pocket; but I suggested that the illustrations would then have to be so small that they would not be of much use for purposes of identification. He replied, 'Give me your own idea, then'. The result was the publication in 1930 of the *Guide to the Orchids of New South Wales*. It was very attractively produced, and received favourable notices both in Australia and overseas; nevertheless truth compels me to record with regret that it was a financial loss to the firm. Yet my larger book, *The Orchids of New South Wales*, published by the National Herbarium in 1943, although it was far more technical, and also more expensive, paid its way and showed a small profit.

During the preparation of the *Guide*, I had to come down occasionally to see Mr Robertson; and every time I left his office I was the richer for books which he thought I 'might be able to carry home'. There was some delay over the actual publication of the *Guide*, and when the first copy came out he sent it to me, inscribed on the fly-leaf, 'George Robertson to Rev. H. M. R. Rupp, B.A.—Proverbs xiii, 12'. The text referred to was, 'Hope deferred maketh the heart sick; but when the desire cometh, it is a tree of life'. G.R. was a grand old man; and I have often wondered why nobody has written the story of his life, for he was the friend and benefactor of so many authors, artists, musicians and public men that the tale should sell like hot cakes.

At the close of my three months' rest, the bishop of Armidale, Dr Moyes,[4] offered me temporary work at Pilliga, on the north-western plains. He was appointing a new vicar there, but wanted someone to hold the fort for four months, as the present occupant of the post was leaving. The bishop said that he only wanted me to conduct services, visit the schools and minister to any sick people. We decided to go. While I was at Weston my wife and my son had presented me with my first (and only) motor car—a Morris Cowley. We drove all the way to Pilliga, about 480 miles from Sydney. There was a little furniture in the Pilliga vicarage, and we resolved to take thither only what we could pack into the hapless car. Our son accompanied us, as he was on

The Beech Orchid, *Dendrobium falcorostrum* FitzG., from Dorrigo. Photographed by Rupp, September 1936. (From the Fordham Album by courtesy of Mrs J. Trotter)

vacation at the time. The Morris responded nobly, and the long journey was accomplished without a single mishap. My story of Pilliga must be reserved for a later chapter; the next one deals with botanical doings at Paterson and Weston.

7 Botany at Paterson and Weston

I had gradually been accumulating a private herbarium of native plants. By the time we settled at Paterson in 1924 it had begun to present serious difficulties; I could not afford to buy expensive cabinets for it, nor was there room for them. Nor did I have the leisure to look after the specimens properly, and to defend them against the attacks of the various 'wogs' which find dried plants to their taste. In 1926 I decided that something must be done if the specimens, numbering more than 2,000, were to be saved. More and more, in recent years, I had been concentrating my attention on the Orchids; and at last I resolved to offer the general collection, with such duplicates of orchids as I could spare, to the Museum of Trinity College, Melbourne. The warden, Dr Behan,[1] gratefully acknowledged the offer; but suggested that it might be better to transfer it to the Botany Department of the university; he had discussed the matter with the head of that department, Professor Ewart,[2] who said that if I were disposed to act on the suggestion, my offer would be warmly welcomed. Very gladly I fell in with this proposal; and when a year or two later I visited Melbourne, Professor Ewart showed me the large cabinet which had been secured to house the collection. Thereafter I practically confined my collecting to Orchids, and began to study them more carefully, though I have never lost my interest in our other plants.

In 1925, during a visit to Sydney, I went out to Turramurra to see Mr Maiden, who had retired from the Botanic Gardens after holding the office of government botanist for thirty years. His health had been failing for some time, and I was shocked to find him so crippled and looking so frail. But he was remarkably cheerful, and in the comfort of his own home was still hard at work on the Eucalypts. I have always been so glad that I did go out to see him, for a few weeks afterwards he died. An admirable tribute to his life and work was paid by the late Professor A. H. S. Lucas in the *Proceedings of the Linnean Society of New South Wales* for 1930.

Lucas was himself a distinguished botanist, specializing in the Algae. He was lecturer in Biology and Chemistry at Trinity College when I was there; and when we met at the Sydney Science Congress he remembered me quite well. I did not remind him how certain of his students used to 'pinch' chloroform from the laboratory for the base purpose of cleaning their pipes; but I daresay he knew all about it.

The Paterson district, though not up to the level of Bulahdelah in the matter of wild flowers, possessed areas of great interest to a botanist. There were remnants of old brush forests in some of the gullies among the hills, and the upper part of Dunn's Creek, behind Hungry Hill, was rich in plant life. On the more open country in the river valley there were many ground orchids. Dr Rogers had named a small orchid which I found at Bulahdelah, and which proved to be 'new', in my honour; and this little *Prasophyllum ruppii* turned up again a mile or two from Paterson. But the outstanding events in my botanical doings of those years were two holidays spent with the late John Hopson of Eccleston, on the Allyn River.[3]

John Hopson was an orchardist and farmer in the

foothills that led up to Barrington Tops. He was also a distinguished entomologist, and a splendid all-round naturalist. A man who had been born in that somewhat remote area, and had spent practically all his life there, might be pardoned for displaying narrow and parochial views of life. But there was nothing narrow or parochial about John Hopson. He was a well-read man, with a singularly broad outlook and a delightful disposition. He was deeply religious, but never paraded the fact, nor tried to thrust his religious convictions down other people's throats. Somehow he heard of my botanical proclivities, and soon let me know that he would not be satisfied until he had me up on Barrington Tops. Needless to say, I was in complete accord with his ambition for me. It was two years before it could be realized.

The proper time for a botanist to visit Barrington Tops is midsummer, and for a long time I could not take a holiday at that period. But he got me up into the foothills for a fortnight in the spring of 1926, and gave me a wonderful time. We rode up the Paterson Valley to Mount Royal, and found orchids galore. There was a dainty little epiphyte which appeared to me to be new, and I sent specimens to Dr Rogers, who confirmed my opinion, but added that it had been found a few days earlier by Mrs H. Curtis (then Miss Hilda Geissmann) on Tambourine Mountain in Queensland. So I was beaten on the post. However, in the Allyn River brushes we found a pretty little *Dendrobium* which also proved to be new.[4] We climbed Bald Knob, a 4,000-foot sentinel at the approaches to the Tops, and there I made my first acquaintance with the magnificent Negrohead Beeches,[5] some of which must have been ancient veterans when Captain James Cook discovered Botany Bay. I confess I cannot see much sense in either of the popular names which have been attached to this grand tree. In southern Queensland it is called 'Antarctic Beech', though there are two species in Australia and several in New Zealand and South America which are much nearer the Antarctic zone. In New South Wales it is the 'Negrohead Beech'— or more commonly 'Niggerhead'. The mountains flanking the upper Paterson and Allyn Rivers form its southern boundary. Thence it extends northward to the Lamington National Park in Queensland.

At last, in January 1928, came the longed-for trip to Barrington Tops. We left Eccleston on a cloudy morning, and drove up the Allyn River (but not in a motor car) for eight miles. The river-oaks here were laden with great clumps of rock lilies,[6] more aptly known in Victoria as King Orchids, since they are not lilies, and are found as often on trees as on rocks.

Then we took to the saddles for the thirteen-mile climb to the plateau about 4,000 feet above us. It was tough going, first through groves of rosewood[7] and other trees of the rainforest, then over a swampy tract which ought to have had an Aboriginal name meaning 'The Abode of Leeches', and then into the fern-tree glens. Three species of fern-trees are found here, the one above 3,000 feet being identical with the common fern-tree of the southern States.[8]

We followed a narrow bridle track along the Allyn Gorge, looking across to a great 12,000-acre beech forest, where occasionally we caught sight of the Beech Orchid, the most beautiful *Dendrobium* possessed by New South Wales.[9] John Hopson was in front leading the packhorse; I lingered for a few minutes to examine some ground-orchids alongside the track, and upon resuming progress my guide was out of sight. Presently I came to a fallen tree, right across the track, and all wet and slippery from the mountain mists. I could not dismount just there, so I put my horse at the obstruction. He got his front

The Buttercup Orchid, *Dendrobium agrostophyllum* F. Muell., a north Queensland species, photographed by Rupp, c.1935. (From the Fordham Album by courtesy of Mrs J. Trotter)

Two Greenhood Orchids from the Barrington Tops, January 1928. Above: *Pterostylis decurva* Rogers. Below: *P. coccina* FitzG. (From glass negatives by H. M. R. Rupp kindly lent by the late Arthur Rupp, Willoughby)

legs over, but could not manage his hind legs! If I attempted to alight on the log, I should risk precipitation into the gorge. Presently a smiling bearded face appeared round a corner in front. 'I thought you'd probably get stuck there,' said John Hopson. Soon I was released, and freed from my weight, the horse scrambled over.

As we approached the plateau, thunder boomed, and we were obviously in for rain. Hopson pointed to a hill on our left, and said 'Carey's Peak'. This is the highest point of the Barringtons, about 5,300 feet above sea level. We hurried on, making for an old hut which then stood about two miles beyond the Peak. (It was subsequently destroyed by fire.) We reached it just in time; rain came pelting down, and although it was midsummer, the air turned bitterly cold. But John Hopson had plenty of dry firewood stacked in the hut, and we soon had a roaring fire going. Grilled steak tasted very good! We camped up there four days, and every day was wet except the last; but we ignored the weather, and botanized in all directions. Midsummer is spring time on the Tops, and the flowers were glorious. I had a theory that we should find orchids up here usually associated with southern areas; and this proved to be correct. One, growing in masses of sphagnum moss, had hitherto only been found on the Blue Mountains; and in the bogs were colonies of a stately Greenhood which I had last seen on the slopes of Mount Barrow in Tasmania.[10] Several new orchids were found, and later I had the pleasure of naming two species of *Prasophyllum*,[11] one after Dr Rogers, and the other in honour of my friend and guide, John Hopson. One of the gems among the orchids, a lilac *Diuris* with dark purple veins,[12] grew in myriads on the moors of the plateau. Despite its abundance here, it has never been found anywhere else, though it has been diligently sought on other highlands.

Our last day broke fine and sunny. John Hopson remarked, 'This is good; I want you to see my property this morning'. I was a little puzzled; but he led me to Carey's Peak, where we dismounted and fastened our horses to saplings. It was no distance to the summit, for the plateau itself is about 5,000 feet high. Hopson took my hand, bade me close my eyes and led me up. The peak is on the very edge of the plateau, and when we reached the top he said, 'Now!' I was really overcome by the grandeur of the panorama; my guide said quietly, 'All this is MINE; it has been all my life'; and I understood. He knew almost every gully and ridge and stream in those beautiful mountains, and loved them as he loved life itself. We looked down the deep gorge of the Allyn to the foothills about Eccleston; then we traced its valley down to the junction with the Paterson at Gresford, and down the Paterson past Vacy and Paterson village to the rich flats of Morpeth; then our eyes followed the Hunter past Raymond Terrace and on to the mangrove swamps that heralded its approach to the ocean at Newcastle. Beyond Newcastle wooded hills, scarcely perceptible

The Rock 'Lily' or King Orchid, *Dendrobium speciosum* var. *hillii*. A plant from the local rainforest 'flowering in Mrs J. D. Tucker's collection, Paterson, September 1926'. (Marjory Loader Collection)

as such in the distance, stretched out to the Hawkesbury and the outskirts of Port Jackson; and far away on the southern horizon stood up the bluff of 'The Gib' at Bowral. To the south-west we could pick out the principal headlands of the Blue Mountains—Mounts Tomah, Wilson, King George, Irvine. To the west the Mount Royal Range broke in gaps which gave glimpses of the Hunter Valley towards Singleton. Eastward we looked along the great scarps of the southern fall from the plateau, and beyond them to Dungog and the ranges about Bulahdelah. Far off, the blue Pacific lay before us, from Port Stephens to the cliffs north of Port Jackson. I said to myself as I gazed, 'This is John Hopson's Garden'; and as such I always think of it. My good friend did not long survive this trip. He lived to express his pleasure in having a new orchid named after him; then one day, while visiting a sick brother in the Paterson valley, he suddenly collapsed, and died of heart failure. His memory is green in the hearts of all who were privileged to enjoy his companionship among his beloved mountains.

When we moved to Weston, Alec Chisholm, the well-known bird man and author of charming books like *Birds and Green Places* wrote expressing his sympathy with me in being transferred from the orchid riches of Paterson to such an unpromising area as the coalfields. But I had already been skirmishing; and I replied offering to eat my hat if within twelve months I had not found more orchids on the coalfields than I had seen in nearly six years at Paterson. Being a journalist, he seized the opportunity to publish this rash offer in a Melbourne paper; but had it been a wager, he would have lost. For the sandy scrubs of Weston and Kurri Kurri carried such a wealth of ground-orchids as I had never seen elsewhere. There were seventeen species of Greenhoods alone, including one new to science.[13] Epiphytes were rare except in the ranges some miles from the town of Weston; but by the end of the twelve months the coalfields had beaten Paterson by two species! Mr M. W. Nichols of Kurri Kurri has been carrying on the search for orchids since I left, and has added several species to those I recorded.

8 From Pilliga to Raymond Terrace

The scattered village of Pilliga lies at the north end of an area of about one and a half million acres of wooded plains usually known as the Pilliga Scrub, extending from the Warrumbungle Range on the south to the Namoi River on the north. Miles away to the east are the Nandewars, about which I have written in connection with Barraba; but in the whole of 'the Scrub' there is no break in the level of the plains except for the banks of the creeks which traverse the area. At Pilliga itself there is an artesian bore, the flow from which has created an artificial creek or long 'billabong'.

We arrived early in the September of a fairly good season—1932. Generally speaking, I am not a lover of the plains. Years before, I used to dislike the long, monotonous stretches of plain along the Macintyre River north-west of Warialda, where the only trees were those in the neighbourhood of the river itself. But the Pilliga Scrub was 'something different'. It was all heavily timbered, with a variety of trees and shrubs which kept my botanical interest constantly on the alert. Eucalypts, angophoras, buddah and belah and bull-oak, brigalows and half a dozen other wattles, cypress pines, quandongs, whitewoods, leopard trees, wilgas, grueys, beefwoods, native pomegranates... so the list runs on.[1]

Among the shrubs there were wattles, too; then there were the semi-shrubby Darling Peas—deep maroon, bright purple, pink, and white; and loveliest of all, the myrtle-bush or native heather,[2] growing in raw sand among the pines, glorious with dense masses of white or pink blossom, the calyces turning

later to deep red, rivalling the Christmas Bush of the coastal belt. Most of the trees were prolific bloomers; some had minute inconspicuous flowers, but buddahs, whitewoods, leopard trees, beefwoods and others made a brave show. The native pomegranate or 'bumble' merits a few words of its own. To see its glory you had to visit it late in the afternoon or before sunrise, for it is a night-bloomer. Large pure white flowers, decked with long streaming stamens and style, and exhaling an exquisite perfume which scented the air for yards around.

Outside the scrub proper grew the Darling lily and the native jonquil, and great patches of bluebells and white everlastings. And the bird life! Emus were abundant. Black swans, ducks and other water birds frequented the Pilliga billabong, and flocks of ibis hunted grasshoppers and other pests nearby. Galahs in their thousands, cockatoos, and lovely parrots of various kinds, were nearly always on view; grey jumpers, spotted bowerbirds, babblers; hosts of beautiful little budgerigars; and among the small fry, finches, wrens and mistletoe-birds. One of the loveliest sights I witnessed was the rising of a great flock of budgerigars between me and the sun in late afternoon; they were like a cloud of gold.

I could not find any ground orchids, though a few are recorded for the area. But a most beautiful form of the Channelled Cymbidium grew on the pines and eucalypts, often in huge clumps; its flowers were bright golden-green with heavy red markings.[3] I had not expected to find an epiphyte on the inland plains, an area of low rainfall; but this Cymbidium only

Rupp's photograph of the Channel-leaved Cymbidium, *C. canaliculatum*, which he distinguished as forma *aureolum* from material collected at Pilliga in October 1932. The flowers were 'golden yellowish-green with red markings'. (Marjory Loader Collection)

grows on trees with hollow trunks or branches, sending its roots down the hollows to feed on the humus of the decaying wood; such hollows usually retain moisture for a long time. One day I came upon a eucalypt which had been blown down in a storm; the hollow trunk was shattered by the fall, exposing the roots of a Cymbidium which had perched about forty feet up. The roots reached right to the base of the tree.

My 'relief work' over, we left Pilliga not without regret, although I do not think we should have cared to stay all through the summer. Bishop Batty of Newcastle was now on a visit to England. There were still no vacancies in the diocese; but Archdeacon Woodd gave me temporary work, and we settled for the time being in East Maitland, where the Rev. K. S. C. Single (now Canon Single) was rector. He had married Annis Kirkland, the only child of my old Colac vicar, the Rev. John Kirkland. The latter had for some time been Sub-dean of Newcastle Cathedral, and had then been appointed rector of Singleton, where he died. His widow, whom I remembered for her unfailing hospitality and kindly disposition, was living near the East Maitland rectory, and it was very pleasant indeed to renew my acquaintance with her and her daughter, whom I recollected as a dainty and affectionate little girl. She grew into a charming and gifted woman, and was greatly loved in the parish. We were deeply grieved a few years later to see the announcement of her sudden death in Maitland hospital. Her mother did not survive her very long.

After we had been some months in Maitland, the Rev. Arthur Holmes of Gresford asked me if I would take charge of that parish for a year, as he had been appointed to organize certain diocesan work. We knew a good deal about Gresford, having lived in the neighbouring parish of Paterson, and I agreed to go. Just then, however, Archdeacon Woodd sent for me, and told me that the parish of Woy Woy, near the border dividing the dioceses of Newcastle and Sydney, was vacant; if I cared to take it, the appointment was mine.

I had heard disconcerting tales of the state of the parish, but we drove down to make inquiries on the spot, and both my wife and I fell in love with the place. True, the result of the inquiries was not very reassuring, but I decided to accept the offer. Mr Holmes released me from our agreement, and to Woy Woy we went. Church buildings were very poor; but there was a very comfortable rectory, on the shores of the beautiful Brisbane Water, a long arm of the Hawkesbury estuary. Church attendance was low, and the parish finances were lower! But when the people got to know us, they rallied splendidly. Woy Woy has a 'floating' population, swelled enormously in holiday seasons, and even every weekend, by swarming crowds from Sydney. This creates difficult problems for clergy and church workers. But we settled in there very happily.

I was now sixty years old[4]; and I told the good folks of Woy Woy that if they were content to keep me, I was ready to stay there till I retired. I had long before made up my mind to retire as soon as I began to feel unequal to the responsibilities of a parish priest. I had seen too many instances of old men hanging on to their parishes long after they had become incapable of the vigour and energy required for spiritual leadership, and I was resolved not to follow their examples. But Woy Woy was not to be my last sphere of parochial activities. We had been there about three years, when the bishop wrote to say that the parish of Raymond Terrace was vacant;

that he intended to nominate me for it, and that it was his personal hope that I would accept it. We did not want to leave. We loved the place and its people, and they seemed to love us. There were other considerations, too, which made us reluctant to go. The bishop, however, pressed his offer. Raymond Terrace, I knew, was a very much 'better' parish; and I sensed that one reason for his choice of me was that he felt I had had rather a 'thin time' since Weston days. At all events, in the end we concluded that I ought to accept.

Woy Woy may have been a poor parish in some respects; but never in all our married life had there been such a demonstration of affectionate goodwill as the people gave us at the farewell gathering in the parish hall. Though we had been asked to keep that evening free, we were not told anything more; the packed hall surprised us, the generous gifts and the tributes paid by the speakers made our hearts very full. Dear and beautiful Woy Woy! We love you still.

And so we came to my last parish, in 1936. Raymond Terrace was a very old settlement at the junction of the Hunter with its largest tributary, the Williams, which rises in the ravines of the southern fall from the Barrington Tops plateau; in some places only a steep ridge separates it from the Allyn. The town lies about sixteen miles north of Newcastle. Its progress has been hampered by the want of a bridge over the Hunter at Hexham, where inevitable delay is caused by the necessity of crossing on a punt. This, I understand, will be remedied in the not-too-distant future by the erection of a bridge[5]; but in these days of shortage of labour and building material, optimism is not very high. At all events, it was the lack of a bridge which caused the great Raymond Terrace butter factory to be erected at Hexham on the other side of the river, some five miles from the town. But 'The Terrace' is gradually coming into its own; shortly before we left, the Masonite Company built its factory there, bridge or no bridge; and other industrial developments are in view. The great arterial road known as the Pacific Highway runs through the town.

St John's Church, Raymond Terrace, is a picturesque stone building, with a beautiful marble sanctuary and altar provided by the Windeyer family, and some good windows.[6] The rectory, over a hundred years old, is a solid stone structure with walls two feet thick, which keep it cool in summer and warm—with the help of some firewood—in winter. At the time of my appointment it was practically uninhabitable. White ants had made

terrible depredations in roof and flooring; there were no 'modern conveniences', and the kitchen was in a separate building. The parish had had its troubles; but it is not for me to suggest where the faults lay. At all events, it had now been decided to cut into building allotments, and sell part of the unnecessarily large grounds on which the rectory and Sunday School stood. The church was across the street. Some parishioners were so aghast at the condition of the rectory that they were disposed to favour the building of a new one in the church grounds, which extended back to the Pacific Highway. I pointed out, however, that the old historic house could be renovated and brought up to date at far less cost than would be involved in a new house of the right kind; this view was also taken by others, and it prevailed. The land sale left ample ground around the house, and was successful enough to provide for a thorough renovation, with something over to spend on a crack in the east end of the church. When at last the bishop came to bless the completed work, he said to me, 'You have the best rectory in the diocese'—and I think he was right.

We were quite happy at Raymond Terrace; and we had the joy of seeing both our daughters happily married there, though both were going to live a long way from us. But when I reached my sixty-sixth milestone, I felt that the time had come for me to relinquish the responsibilities of a parish. It was already obvious that Raymond Terrace was not going to continue to be a rather sleepy little town off the main track of progress; there were great possibilities ahead, and I did not feel equal to the task of dealing with them. It was forty years since I had received deacon's orders in St Paul's Cathedral, Melbourne. Of course many men— notably the evergreen and much-loved archdeacon of Newcastle,[7] who was ordained deacon eleven years before me, and is still going strong in archidiaconal duties, though he has not been a parish priest for many years—can beat this record of an active ministry. However, in 1939 I felt it my duty to retire on pension. Raymond Terrace gave us a kindly farewell, and we left for Sydney. During my three years as rector, the church, the rectory and the Sunday school had all been renovated, at a cost (if I remember rightly) of over £600; but the parish was free of debt. I have still something to say about my life after retirement. As I look back over the past, though there were clouds as well as sunshine, I feel I can say from my heart, 'Surely goodness and mercy have followed me all the days of my life'.

9 And So to the End

A friend intending to call on us asked somebody in our street if he knew where the Rev. Mr Rupp lived. 'Oh, you mean the Orchid Man,' was the reply. Such is the reputation which has come to me. Whether it be a compliment or the reverse, I suppose that in some measure it is deserved, for my enthusiasm for these fascinating plants was known long before I retired from parochial work, and since then most of my time has been occupied with them. But I have often had to disappoint inquirers who have either written or called for my advice on the best methods of cultivation, or how to keep some valuable hybrid in good health. For I am not, and have never professed to be, an 'orchid grower', in the usual sense of that term. I am merely an amateur botanist, who has concentrated on the study of our Australian native orchids, and has written much— perhaps too much—about them. I am no authority at all on the gorgeous exotics to be seen in florists' windows or at the various annual Orchid Shows. I can admire their beauty, but I have no desire to cultivate them myself, and I must confess that I know nothing whatever about any of the endless Cymbidium hybrids that adorn the orchid houses of the ever-increasing company of 'growers'. To me, the humblest little Greenhood or 'Prassie' on the Port Jackson hills is just as fascinating, and more suited to my circumstances; it is an orchid, and shares equally with its flamboyant relatives the remarkable characteristics of that great family of flowering plants.

I did not retire from parish responsibilities from any feeling that I had grown too old to work. When a man admits that he is too old to work, it surely means that for some time he has known that he is 'losing his punch', and cannot tackle the problems of his duties with that sustained energy and 'vim' which they demand. Therefore, in my opinion, he should not wait for the day when he simply cannot struggle on any longer. For in that case he has almost certainly been doing quite inadequately things that ought to be done adequately, and could be better done by a younger man. Few people outside the ranks of the clergy realize, I think, how heavy the responsibilities of a parish priest are; for many of them are never told in Gath, nor published in Ashkelon.[1] But heavy they are; and when I felt no longer able to tackle them as I should, I withdrew. I think my decision was right.

We settled down in the Sydney suburb of Northbridge.[2] The name has no allusion whatever to the Sydney Harbour Bridge, as some folks appear to imagine. It lies on the north side of the old suspension bridge over an arm of Middle Harbour, about four miles from the city. We were fortunate in securing a house with a lovely outlook on to the heights of Castlecrag, across Middle Harbour, and beyond it to Seaforth and Balgowlah. And we can never be 'built in', unless the day should come for a forty-storey skyscraper to arise from Sailor's Bay Gully below us—which is extremely unlikely.

I am still on the roll of the Newcastle clergy; I usually go up for the annual synod, and have several times helped by taking services in my old diocese. I hold no licence in the diocese of Sydney; but no good purpose would be served by entering into the reasons for this. We had not been in Sydney very

Christmas Greeting from HMRRupp

Cymbidium canaliculatum *var.* Sparkesii.
The "Black Orchid" of N. Queensland.
MAREEBA.

SARCOCHILUS FITZGERALDII.

Christmas Greetings from HMRRupp

Christmas greeting cards prepared by the Rev. H. M. R. Rupp from his photographic studies of orchids. (From the Fordham Album by courtesy of Mrs J. Trotter)

long before I was invited by Mr R. H. Anderson,[3] chief botanist of the National Herbarium and now also director of the Botanic Gardens, to become an honorary member of the staff at the Herbarium, with the Orchidaceae as my special responsibility. As may be imagined, I was not slow to accept the invitation. My relations with Mr Anderson, and with every member of his official staff, have been of the happiest nature, and I count it a privilege to call them my friends. Indeed, our 'boss', as he is affectionately called, seems to have a genius for collecting a band of workers who are a very happy family, who can pull together without the slightest friction, and with whose loyal cooperation he has greatly increased the efficiency of the institution. A few years ago there were good prospects of a new building for the Herbarium; but the Second World War has compelled its indefinite postponement.[4]

There had been no orchid specialist for many years, and I found plenty of work to occupy me. The whole orchid section required rearrangement and revision; much of the nomenclature was out of date; and many specimens, being very fragile, had suffered considerable damage in the course of time. I decided to mount them all securely as I went through them, and this took the better part of two years. Among the most valuable of the special collections, as distinguished from the general collection, is that of Ronald Gunn, who collected in Tasmania from 1836 to 1841. Gunn's specimens, all well over a hundred years old, were with few exceptions beautifully pressed, and are in an admirable state of preservation; many, indeed, look as if they had been gathered quite recently. The Deane collection, although not so good, is of interest because Henry Deane[5] was a close associate of R. D. Fitzgerald, whose magnificent work, *Australian Orchids*, published in parts from the 'seventies to the 'nineties of last century, is still a classic of orchid literature. Some hundreds of specimens from New Guinea, Celebes, Sumatra and South Africa, collected by Dr Rudolf Schlechter,[6] are also of special value. Schlechter was rated by the late Dr Rogers as 'a very great botanist, and the greatest of all orchidologists'. He travelled extensively in many parts of the world; and his immense herbarium, or most of it, was deposited in the Berlin Museum, the greater part of which was destroyed in the last war. It is uncertain how much of Schlechter's collection was lost, or where such portion of it as may have been saved is now stored. We are therefore fortunate in having so many duplicates of his, most of them labelled in his own hand. In 1947 I offered Mr Anderson my own fairly comprehensive herbarium of Australian and New Zealand orchids, to become part of the national collection, and the offer was accepted.

Since my retirement I have continued to contribute papers on orchids to the *Proceedings of the Linnean Society of New South Wales* (of which I was elected a corresponding member), the *Victorian Naturalist*, the *Queensland Naturalist*, the *North Queensland Naturalist*, the *Australian Orchid Review* and other journals; and in 1946 I was invited to write an illustrated paper on Australian orchids for the *Journal of the New York Botanical Garden*.

Among the papers contributed by me during these years I may mention the following: (1) prepared in collaboration with Mr E. D. Hatch,[7] of Auckland, N.Z., 'The Relation of the Orchid Flora of Australia to that of New Zealand', and (2) in collaboration with Mr Trevor Hunt,[8] of Ipswich, Queensland, 'A Review of the Genus Dendrobium in Australia'. Both were published by the Linnean Society in Sydney. Though I have never been in New Zealand, the orchids of that country had greatly interested me ever since we lived in Paterson, when a correspondence developed with the late H. B. Matthews of Remuera, Auckland. Mr Matthews was recognized as the leading authority on orchids in the Dominion. We exchanged collections of Australian and New Zealand species, and he sent me a large number of extremely interesting photographs, thereby stimulating my comparative studies of the two floras, which clearly had much in common. Alas! my good friend across the Tasman was gradually going blind, and he did not survive very long after the complete loss of his sight. But some years later his mantle fell upon Edwin Hatch, whose home was also near Auckland, and who has published some

illuminating papers on the orchids of his country in the *Transactions* of the Royal Society of New Zealand. Collaboration by correspondence is a slow business, and Mr Hatch and I took more than twelve months over our 'Relations' paper. The same observation is true concerning the Dendrobium Review, which I regard as perhaps the best contribution towards the knowledge of Australian orchids in which I have shared authorship. Trevor Hunt hopes one day to work out a handbook of the Queensland orchids, which is badly needed, and no one is as well fitted as he to do the job.

The *Victorian Naturalist* is probably one of the best little journals of its kind in the world. During the past twenty-five years or so it has been edited by Charles Barrett,[9] Alec H. Chisholm,[10] and J. H. Willis[11] successively, and all three have kept up a very high standard. I have received generous treatment at their hands in the matter of space for some lengthy papers, which I can only hope have merited the cooperation so freely given. The North Queensland Naturalists' Club, for so long under the capable leadership of Dr H. Flecker of Cairns,[12] has now thoroughly established its own little *Naturalist*, to which I have contributed many articles; for though I have never been able to visit that great northern area of our sister state, I have received scores of valuable specimens from correspondents there, some of which have proved to be new species, and others have been recorded for the first time in Australia. The Brisbane periodical, the *Queensland Naturalist*, has also opened its columns to me freely, chiefly through the medium of the distinguished government botanist, C. T. White, FLS,[13] another good friend.

So it will be gathered that my life since retirement has not been an idle one. And never, surely, was man more blessed than I in the number of his true and lasting friends, old and young. Whether they be colleagues in the ministry of the Church, or men and women botanists and naturalists, or just simply 'friends', I owe them more than I can ever repay, and can but say, God bless them all.

Several whom I have not hitherto mentioned must have a few words here. Alex G. Hamilton was already a veteran retired officer of the New South Wales Education Department when I first knew him some twenty-five years ago.[14] A distinguished naturalist himself, he had in his schoolmaster days been mainly responsible for the introduction of Nature Study into the public schools. He published, through the Linnean Society, 'The Flora of the Mudgee District', 'The Flora of Mount Wilson', and other botanical papers. He corresponded with me for many years, and sent me orchid specimens. On the occasion of a visit to Western Australia he collected orchids for me there, and added materially to my knowledge of western species. After my retirement I found him living with his son, E. A. Hamilton, who was for some years president of the New South Wales Orchid Society; but dear old 'Daddy' Hamilton's active days were nearly over. His eyesight was failing; and later on his mental faculties grew clouded. Much therefore though his loss was felt by his wide circle of friends, it was a relief when at last, quietly and peacefully, he passed away.

The Rev. E. Norman McKie was Presbyterian minister of Guyra, on the New England tableland, for over thirty years.[15] He had once been a bank clerk at Paterson; and when he saw my letter in the Sydney press in connection with the Jennings Smith memorial, he sent a contribution. This opened up a correspondence which soon ripened into friendship, the way for which was paved by the fact that he was as keen a botanist as I was. His interest lay chiefly with the eucalypts; he collaborated with the late W. F. Blakely in a great deal of valuable work on them,[16] and is commemorated by the species *Eucalyptus mckieana*. But he was also interested in grasses and orchids; and nearly all the New England specimens in my herbarium came from him. Ecclesiastically, Norman McKie and I were widely parted, though he was a man of very liberal opinions; but our differences in churchmanship never caused as much as a ripple on the surface of our friendship. Although I knew that his health had been unsatisfactory for several years, I was shocked and grieved to learn that he had died in the Armidale Hospital in June 1948. Guyra people immediately opened a fund to establish a scholarship in his memory at the New England University College.

Mrs Pearl Messmer, of Lindfield, has driven me on many botanical excursions which I should otherwise have missed. Herself a lover of the bush and a most capable botanist who could, and she would, make many valuable contributions to the knowledge of our Australian flora, she gave me much assistance in revising the Australian section of the Orchidaceae at the National Herbarium. For many years her hospitable home has been open to me. She is a lover of birds no less than of orchids; and since the keenness of a bird's eye is proverbial, I think she must have learnt from her feathered friends the secret of acute vision; for once launched into the bush, nothing escapes her observation.

Among the most delightful memories of the past nine years are those of two visits to Taihoa, on Mount Irvine, in the Blue Mountains; the home of the Scrivener family.[17] Years before, in the days of my association with the Australian Board of Missions, I had been in correspondence with Miss Joyce Scrivener, who even in her schooldays had a flair for orchids. This was shared in full measure by her younger sister Gwen; and soon after we came to Sydney they began to enrich my herbarium, and to add to my knowledge of the Blue Mountains orchid flora. Their mother was a sister of the pioneer New Guinea missionary, the Rev. Copland King, who with the Rev. Albert Maclaren started the Anglican Mission there in 1891. Before long I was invited up to Taihoa for an orchid hunt. It was a delightful holiday among the kindest of people. The sisters led me on a five-mile scramble to a curious formation known as 'The Pavements', where the wildflowers were glorious. Five miles off the beaten track in the Blue Mountains is a nice walk for young folks; but when on the return trip I casually mentioned my age (I was then sixty-nine), the girls were horrified. I assured them (and it was true) that I had enjoyed the tramp thoroughly. Only, I needed a spell now and then!

Miss Joyce Scrivener became the wife of Dr Archibald Telfer of Mosman a few years ago; but before that happened there came another invitation to Taihoa; this time my wife and our son (who had recently returned, as related below, after his long and critical illness) were included; and the five members of the family then living at home conspired to give us 'happy days'. Orchids were not so much in evidence this time; if it be suggested that the luscious Mount Irvine blackberries took the edge of one's botanical appetite, I will not deny it! Miss Gwen Scrivener has painted many of the local orchids, combining art and botanical accuracy admirably.

When the Second World War broke out, our son, Arthur, a master at Shore School, enlisted. He was recommended for a commission, but was turned down as 'too old'. The best comment on this fatuous decision is that two years later, after his escape and return from Singapore, he was commissioned as lieutenant. On the day that he left Sydney for Malaya in the *Queen Mary*, I went past Mrs Macquaries Chair in the Domain, and watched the great liner as she made her way down the harbour, with very mixed feelings. Of course his mother and I knew that he had done the right thing, and could not have done otherwise; but we knew, too, that we might never welcome him home again. He wrote interesting letters from various parts of Malaya; and then the Japanese came down. He took part in the battle of

Mersing, after which our forces had to retreat to Singapore Island. On the island he, with a few men of his company, became isolated. The Japanese were now between them and the rest of the battalion. They had some narrow escapes; a bullet shattered a tin of bully-beef in Arthur's haversack, but he was not touched. Feeling sure that the Australians were by this time safe in Singapore city, they decided to try and get round by water. They secured a small motor boat, but the engine would not function, and they hid all next day in a steaming mangrove swamp trying to put it right. At last they succeeded, and made their way round the west and south of the island.

At the entrance of Singapore harbour, they were dismayed to learn from an English officer in another boat that the whole place was in the hands of the Japanese. He advised them to get away if they could. Luckily they found a tin of petrol; and it was resolved to attempt to cross to Sumatra, hoping that they might be able to get on from there to Java, which they supposed to be still in Allied hands.

The journey across the Straits of Malacca (normally a matter of twenty-four hours) in their frail and leaking boat took four days, and they had practically no food beyond a small ration of rice. At last, the petrol exhausted, they landed on a small island off the Sumatran coast. Here dwelt an old Dutchman, who not only gave them a chart, but also supplied them with two tins of petrol! They set off again; for a day or two they could see nothing of Sumatra but vast mangrove swamps. At last they came to the mouth of a river, and reached a village. The villagers had very little in the way of food, but sent them to another settlement fifty miles upstream, on a copra boat. Food was scarce there too. Thence they tramped twelve miles through jungle to a Dutch factory. The Dutch fed and clothed them, gave them money, and sent them by car to the railhead in the north of the island, whence they were taken by train to Padang on the west coast. It was from here that they hoped to ship to Java; but news came that Java, too, had fallen to the Japanese. To their great joy an Australian cruiser turned up; but the commander stated that his orders were to take any *British* troops he might find; nothing was said about Australians. It sounds incredible, but they were left behind to fend for themselves. Some other Australians came in; and after some days they commandeered a crazy little tub to take them over to Colombo. Had the notorious Bay of Bengal been in a bad mood, they would probably never have been heard of again. But they reached Colombo, exhausted and weary; the

Australian General Hospital took them in hand, and later on sent them home.

Meanwhile we had heard nothing since before the fall of Singapore. At last a cable came, 'Still going strong'; but the authorities refused to give us any clue to its origin, or to say when it was sent. For all we knew, it might have been sent before Singapore fell. (Actually, it came from Colombo.)

One day I was writing in my 'den', when I heard a queer noise from the kitchen, and ran out to see if anything had happened to my wife. She and her son were locked in embrace at the door! For us, the dark clouds were lifted. Naturally, Arthur thought we had known of his safety at Colombo; the troopship on which he travelled home went direct to Melbourne, and the New South Wales men were rushed straight to the train for Sydney, so that he could not let us know of his impending arrival. But little we cared about that; he was home! After some weeks' leave, he was sent to train troops in various camps, and was gazetted lieutenant. Then he was sent to New Guinea.

We had a few cheerful letters; but one day came a message from the Minister for the Army, telling us that Lieutenant A. R. Rupp was seriously ill in Moresby hospital. Before long 'seriously' became 'dangerously'. Scrub typhus had claimed him. The next six weeks I do not care to recall in detail, for he was literally at death's door. Some folks would say it was merely a coincidence; I prefer to see Providence, in the fact that a cousin of his, Sergeant E. F. Dowe, was dispenser at Moresby hospital. He saw Arthur every day, and wrote or cabled to us continually. We can never forget his goodness. Arthur had been desperately ill for nearly a month when, one night, I had a strange psychic experience. I do not think I can publish it; it must suffice to say that from that night I lost my fear that he would die. A week or two later his condition was reported improved. But he was nine months in various military hospitals before he was discharged—with his memory very badly impaired.

The manner of his discharge was typical of the callousness of those who were responsible; but happily the Red Cross folks intervened and brought him home. The Repatriation Department did nothing to help him; he was sent for by their medical staff two or three times, and merely told to be patient. But a refugee German psychiatrist, who had escaped from the Nazis and settled in Sydney, somehow heard of his case, and said he would like to see him. An appointment was made. After going through his medical papers and questioning him, Dr F———

offered to treat him; but refused any fee.[18] In a month we could see improvement; in twelve months he was a different man. Soon afterwards he was able to drive his car again; and later on he was invited to rejoin the staff at Shore School. Now he is happily married to a wife who captured his father's and mother's hearts as she did his. So we can sing our *Laus Deo*!

In 1942 I suffered a bad attack of arthritis. My doctor, whom we had known on the coalfields years before, ordered my removal to Royal North Shore Hospital, where I received the kindest of treatment; but I was in a bad way, and it looked as though this might be 'the beginning of the end'. But the specialist visiting the hospital tried me with certain injections, warning me, however, that most people could not take them without ill effects. As none appeared, and I began to move my limbs without pain, he increased the dose. In three months neighbours at Northbridge were surprised to find me walking about that hilly suburb quite normally, insomuch that one dubbed me 'the miracle-man'. That was more than six years ago, so I think we can call it a real 'cure', and I never cease to be grateful to Dr A—— for his treatment.[19]

So here I am, past my seventy-sixth milepost, a bit slower at climbing hills than I used to be, and not so fond of long walks, but still fairly active. Not long ago, rather than wait an hour and a half for a bus on a hot day, I walked from Chatswood station to Hunters Hill—four and one half miles with a climb at the end—and was none the worse for it.

In November 1948 I was surprised to receive a letter from the Royal Society of New South Wales, informing me that I had been awarded the Clarke Memorial Medal for 1949, in recognition of what the council of the society was pleased to call my 'distinguished contributions to the knowledge of Australian Orchidaceae'. This is an honour which I never dreamed of as likely to come my way; but it would be idle to pretend I am not gratified to learn that my work in connection with the flowers I love so well is considered worthy of recognition. The Clarke Medal was established in 1878, to honour the memory of the Rev. William Branwhite Clarke, first rector of St Thomas's Church, North Sydney, and according to the *Australian Encyclopaedia* (1927 edition) 'the acknowledged leader of scientific thought in Australia'. He was primarily a geologist; and it was chiefly due to his investigations that the great Victorian goldfields were discovered nearly a century ago. But he was a good all-round naturalist, and wrote papers on such diverse subjects as 'Deep Sea Soundings' and 'The Effect of Forest Vegetation on Climate'. The great east window of St Thomas's Church is a tribute to his memory on the ecclesiastical side.

I suppose I should not omit some reference here to the publication of *The Orchids of New South Wales* at the end of 1943. Mr Anderson, chief botanist at the National Herbarium, had been most anxious for some years past to bring out a new 'Flora of New South Wales'. The only existing official handbook, that of Moore and Betche, published in 1893, although an admirable work, had long been both out of date and out of print. Mr Anderson therefore asked me whether I could prepare a volume on the Orchids, as the first part of the projected new 'Flora'. I need scarcely say that I regarded this request as a great compliment, and gladly undertook to do my best. In the Introduction to the text I have acknowledged the generous help given by Mr Anderson himself and other members of the Herbarium staff; and have also recorded the valuable assistance of some members of the outside public, in enabling us to meet the heavy cost of the plates. All but two of the latter are the work of Mr George Scammell, who generously offered us as many of his exquisite line-drawings as we cared to include. The book was admirably produced by the Australasian Medical Publishing Co. of Sydney. It has been favourably received, both in Australia and overseas; but though only five years have elapsed since its publication, already it is in need of certain corrections and supplementary data[20]; for, as Baldwin Spencer remarked to me so many long years ago, 'Science never stands still'. It is hoped that before long this volume will have the companionship of similar textbooks on other families of New South Wales plants, now being prepared by various members of the Herbarium staff. In 1947 the Department of Agriculture published an extremely valuable addition to New South Wales botanical literature in Mr Anderson's *Trees of New South Wales*, a smaller edition of which had been issued in 1932.

'I remember.' Mysterious words, those, aren't they? Who and what is this indefinable 'I', that can look and listen back by means of the strange faculty we call 'memory', to things and sounds and persons of the long ago? Yes, I see the reeds waving in the wind on the edge of a broad brown stream. I hear them rustle as they sway. I hear the music of the warblers as they watch their pretty nests hidden in those long reeds and the laughing voices of schoolboys splashing about under the willows opposite. Yet sixty years have come and gone since I was there. An old man ponders over these mysteries; but cannot fathom them, nor can he probe the Future. A new world is being born, but he will not be here to share its achievements.

There are prophets who foretell nothing but disaster and darkness ahead. I do not prophesy. I,

too, have known doubts and misgivings, but life has taught me that my faith is based on a Rock that no storms can break. So in life's eventide I am content. As Barbauld wrote:

Life! I know not what thou art,
But I know that thou and I must part;
And when, or how, or where we met,
I own to me's a secret yet.

Life! we've been long together,
Through pleasant and through cloudy weather;
'Tis hard to part when friends are dear;
Perhaps 't will cost a sigh, a tear;
Then steal away, give little warning,
Choose thine own time;
Say not Good Night, but in some brighter clime
Bid me Good Morning.[21]

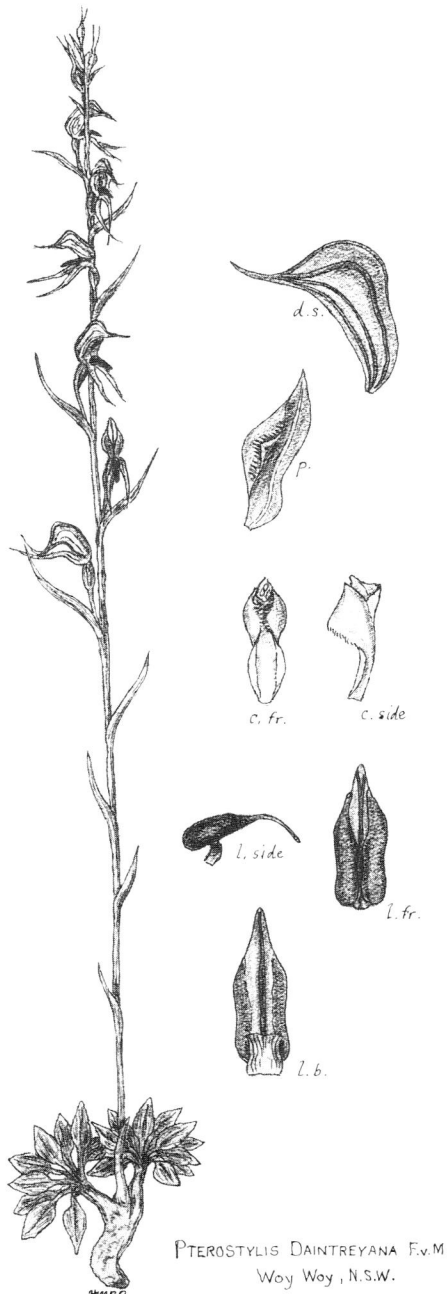

PTEROSTYLIS DAINTREYANA F.v.M.
Woy Woy, N.S.W.

In 1948 the staff of the National Herbarium prevailed upon Rupp to have a formal photographic portrait taken. In a letter to Trevor Hunt, he observed, 'I'd better go and brush up with a view to my execution at 11:45'. The 'execution' occurred at the studio of Howard Harris, Sydney, on 25 November. (Photograph by courtesy of the late Mrs Eileen Cox)

The Botanist as Artist

A Supplement of Sketches

In the course of his botanical work, Rupp sketched profusely as he recorded observations, communicated them to others, and illustrated publications. His letters were frequently illustrated with drawings to clarify aspects of plant habit or floral structure. Other sketches, such as those reproduced here, were carefully prepared on separate sheets which could be enclosed with letters or specimens, sent to publishers, or simply kept on hand for reference. Most were Indian ink line drawings, some were lightly coloured with crayon, and many included enlargements of distinctive features.

Although Rupp's sketches, like his handwriting and printing, were generally bold, clear and accurate, he tended to decry his artistic ability. As he declared to J. H. Willis in 1945, 'it is well understood by now (or ought to be) that I am no artist—I never had any training whatever'. Perhaps 'training' was hardly required for his purpose. In any case, it will be appreciated from the accompanying examples selected from the numerous sketches presented to Joyce and Gwen Scrivener and to Marjory Loader, that however untutored, Rupp could have justifiably felt rather more pleased with his efforts as a botanical illustrator.

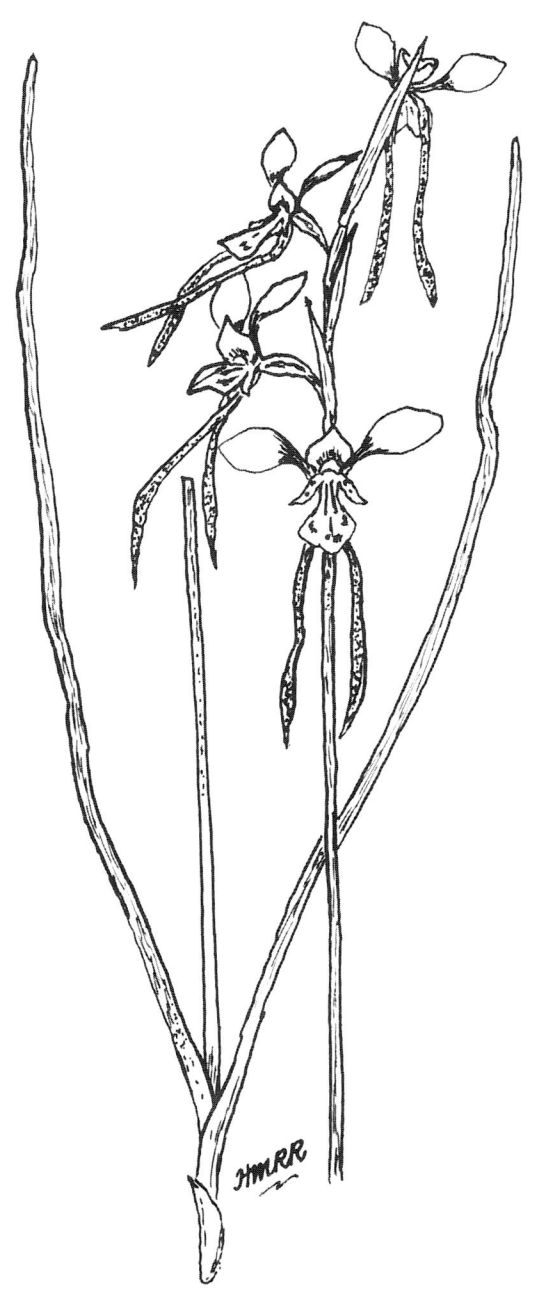

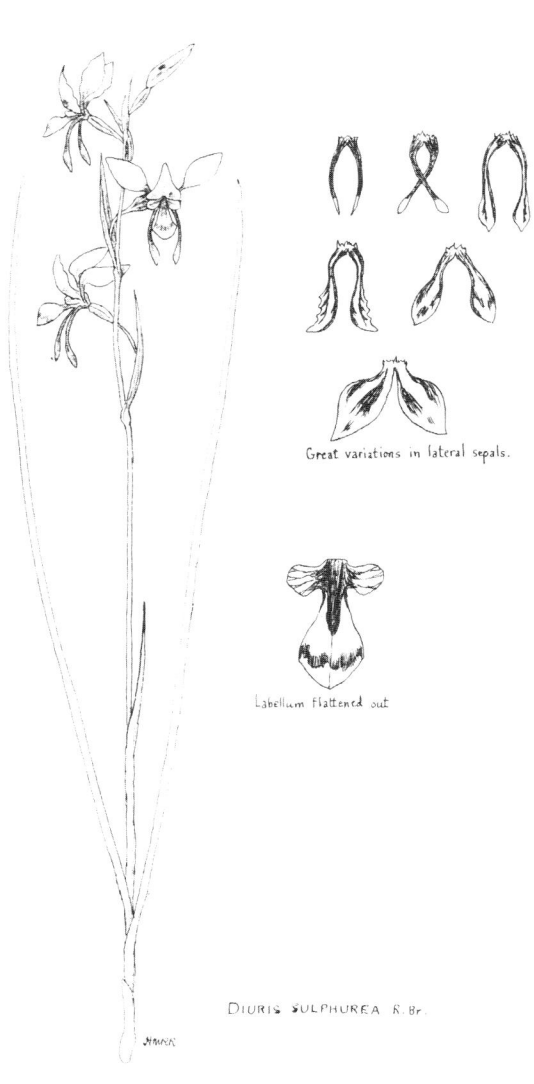

In 1940, Rupp described this Double-tail Orchid as *Diuris colemanae* in honour of Edith Coleman. It is now included in R. D. FitzGerald's variable species, *D. tricolor.* (Scrivener Collection, RBGS)

Diuris sulphurea R.Br., the Double-tail Orchid which is also known as the Tiger or Hornet Orchid, which Rupp found at Copmanhurst, Bulahdelah, Paterson and Weston between 1911 and 1930. (Marjory Loader Collection)

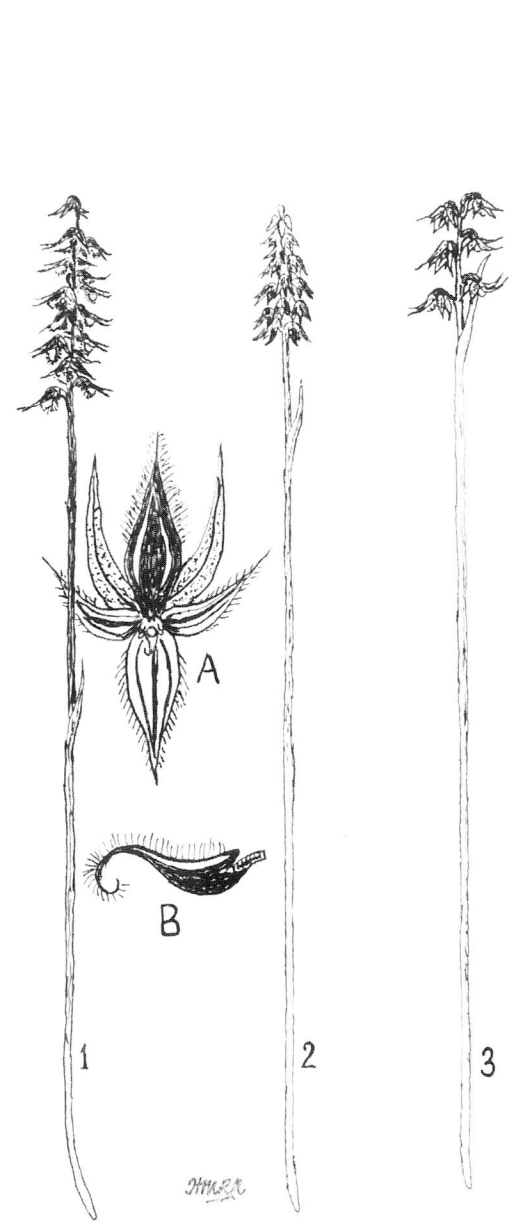

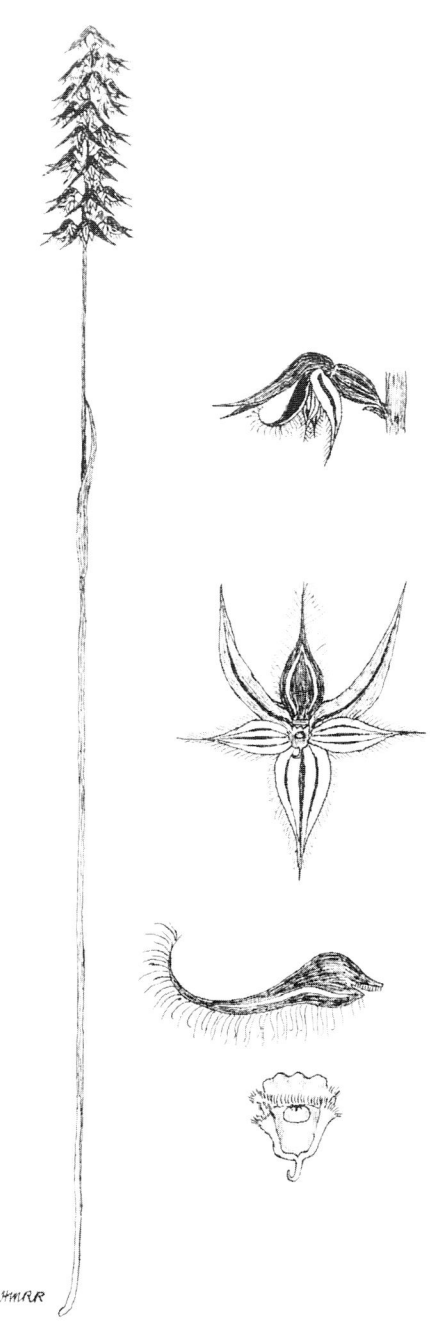

PRASOPHYLLUM ACUMINATUM Rogers.
Paterson, N.S.W.

Three species of Midge Orchid, or 'the little Prassies' as Rupp termed them: 1. *Prasophyllum acuminatum* Rogers; 2. *P. trifidum* Rupp; 3. *P. archeri* Hook.f.; A. flower of *P. acuminatum* flattened and enlarged; B. labellum of *P. acuminatum* from the side. Sketches prepared for an article in the *Journal of the New York Botanical Garden,* 1946. (Scrivener Collection, RBGS)

A Midge Orchid, *Prasophyllum acuminatum* Rogers, drawn in 1927 from a specimen collected at Paterson. (Scrivener Collection, RBGS)

PRASOPHYLLUM ROGERSII, Rupp.

Barrington Tops , N.S.W.

The Leek Orchid Rupp named in honour of his friend
Dr R. S. Rogers in 1928. (Marjory Loader Collection)

PRASOPHYLLUM ROGERSII Rupp.

Barrington Tops.

The Leek Orchid which Rupp found on the Barrington
Tops in January 1928 and named for his friend Dr R.
S. Rogers of Adelaide. Rupp considered that in this
sketch the 'lateral sepals were shewn too acute'.
(Scrivener Collection, RBGS)

(Right) The Midge Orchid which Rupp found on the Barrington Tops, January 1928, and named to honour his companion and guide, John Hopson of Eccleston. This is now referred to *P. nudum* Hook.f. (Scrivener Collection, RBGS)

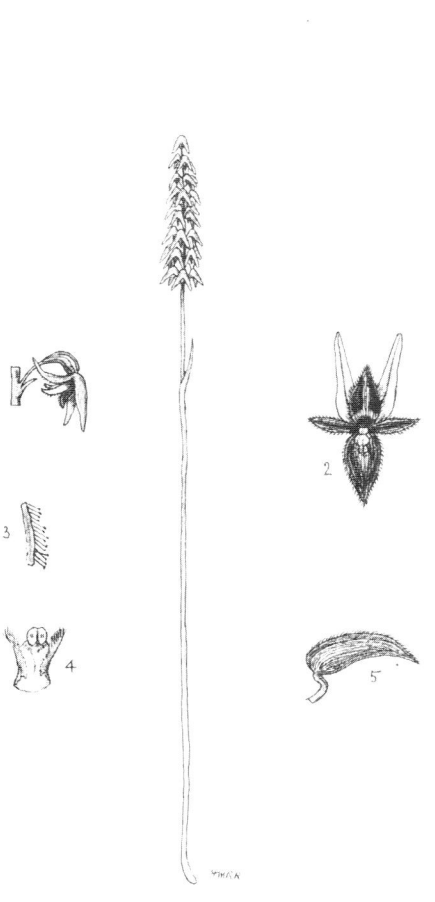

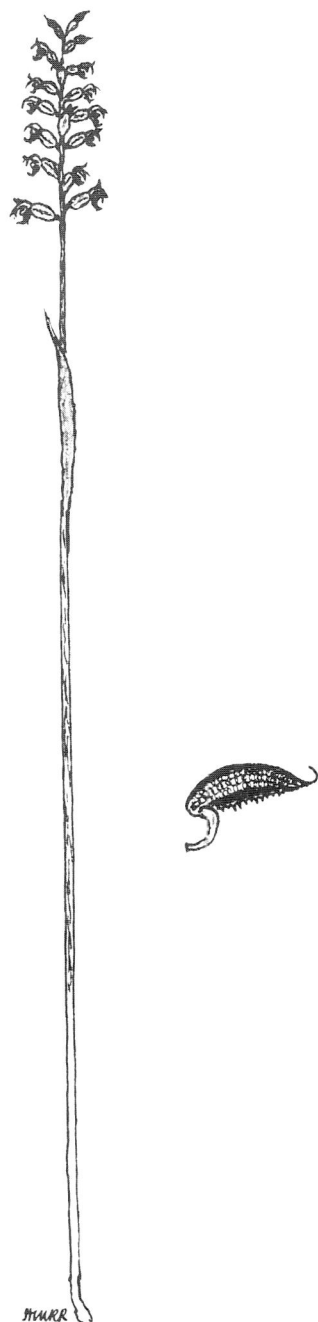

PRASOPHYLLUM RUPPII Rogers.
Paterson.

Enlargements: 1, Flower from side. 2, Flower, front. 3, Hairs. 4, Column, front. 5, Labellum, side.

The Midge Orchid which Dr R. S. Rogers named in Rupp's honour in 1927. (Scrivener Collection, RBGS)

Prasophyllum Hopsonii Rupp, with enlargement of labellum.

CHILOGLOTTIS GUNNII Lindl.

A. Barrington Tops, N.S.W.
B.   "   "   shewing stem-elongation after flowering.
C. Two-flowered form, Launceston, Tas.

The Bird Orchid, *Chiloglottis gunnii* Lindley. (Marjory Loader Collection)

Labellum and base of column, enlarged

CALOCHILUS CAMPESTRIS R.Br.
Weston, N.S.W.

The Copper Bearded Orchid, *Calochilus campestris* R.Br., sketched from a specimen found at Weston in the spring of 1930. (Marjory Loader Collection)

Diuris venosa Rupp.
Barrington Tops.

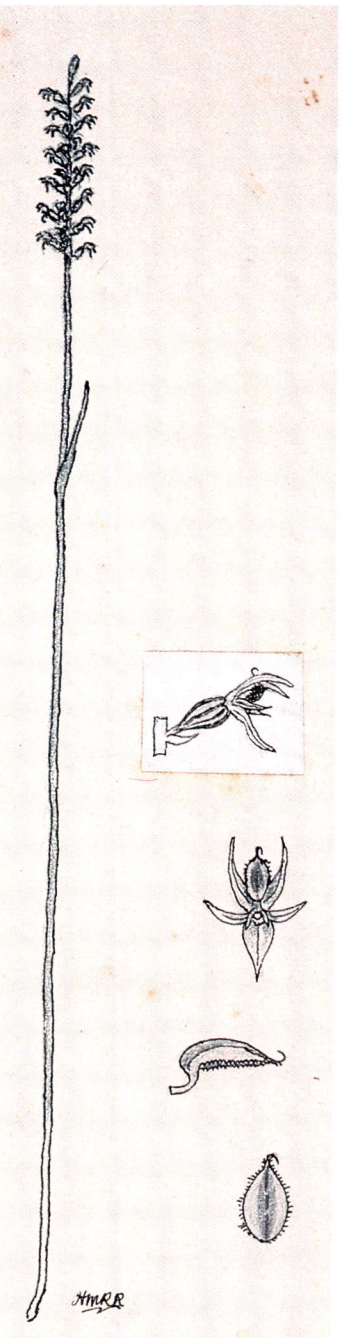

PRASOPHYLLUM HOPSONII Rupp.

Barrington Tops.

DIURIS CUNEATA Fitzg.
Bowen Island, Jervis Bay.

AMRR

DENDROBIUM JOHANNIS, Reichb.f.

Cairns.

X 2

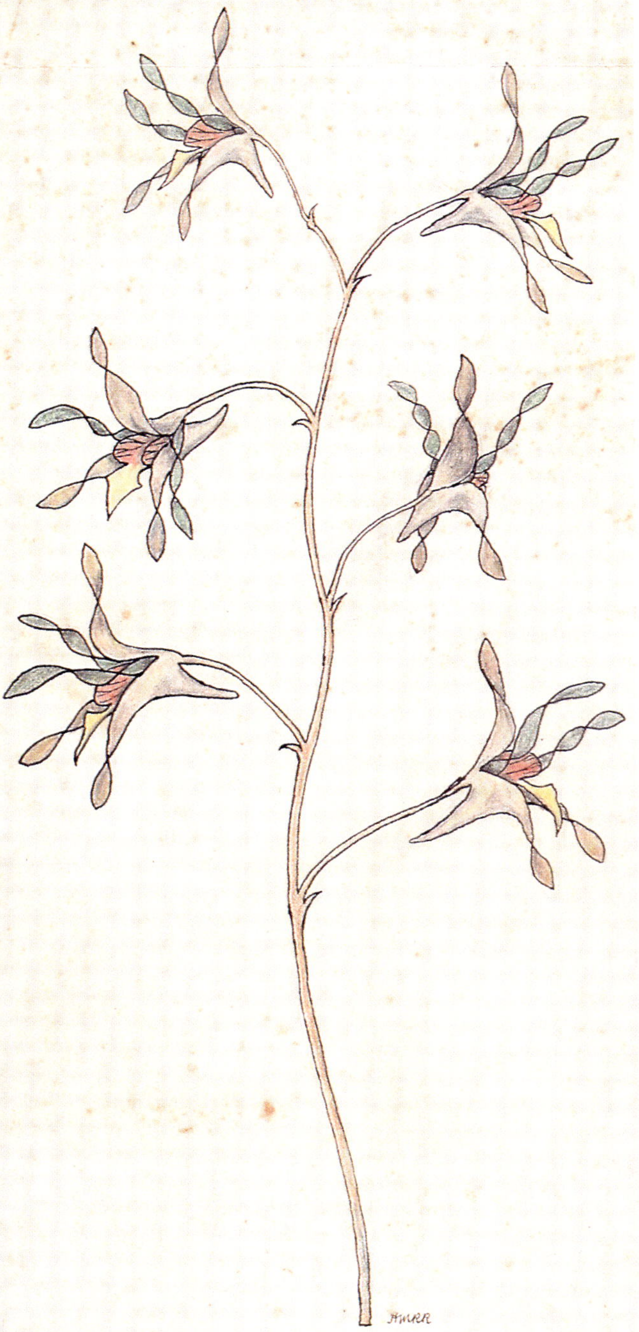

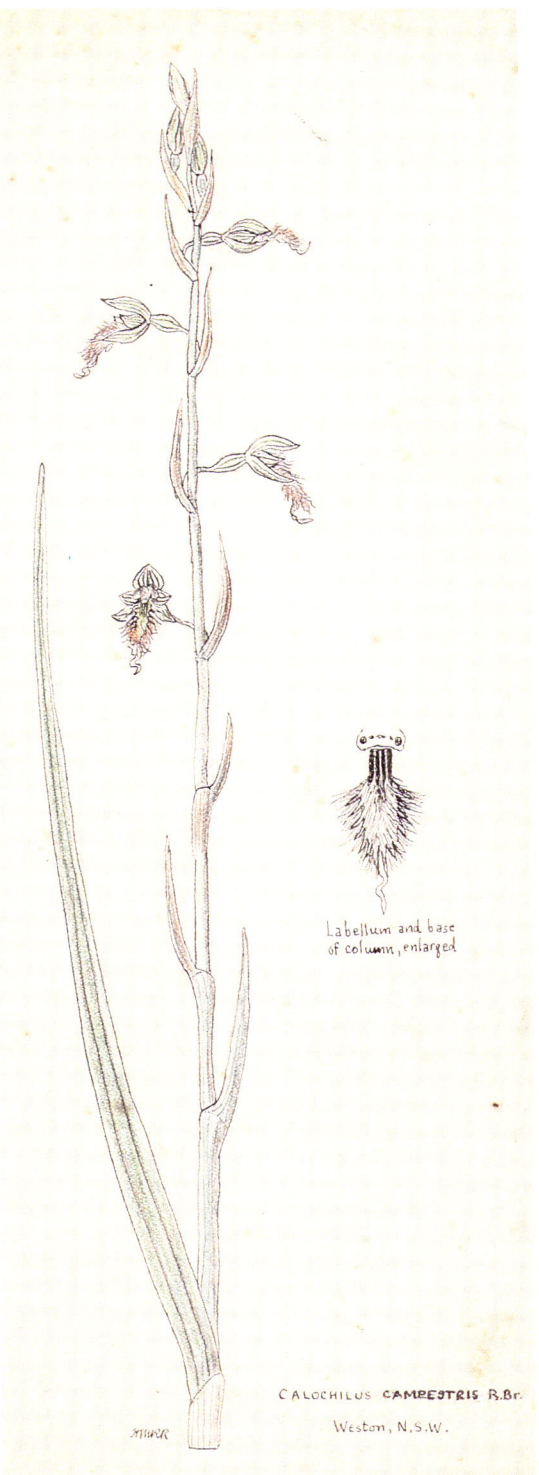

Labellum and base
of column, enlarged
AMRR

CALOCHILUS CAMPESTRIS R.Br.

Weston, N.S.W.

PTEROSTYLIS GRANDIFLORA R.Br.

CHILOGLOTTIS GUNNII Lindl.

A. Barrington Tops, N.S.W.
B.   "   "   shewing stem-elongation after flowering.
C. Two-flowered form, Launceston, Tas.

PRASOPHYLLUM ROGERSII, Rupp.

Barrington Tops, N.S.W.

PTEROSTYLIS BAPTISTII Fitzg.
Belmont, N.S.W.

CALOCHILUS ROBERTSONII Benth.
Paterson.

Labellum and base
of column, enlarged

DENDROBIUM TOFFTII, Bailey.
Cairns.

Sarcochilus Weinthalii Bailey.
TOOWOOMBA, Q.
KYOGLE, N.S.W.

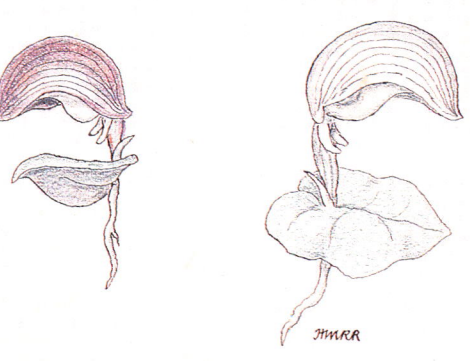

CORYSANTHES ACONITIFLORES Salisb.
Bullahdelah.
(C. BICALCARATA R. Br.)
The white-flowering form is not uncommon.

CYMBIDIUM CANALICULATUM, R.Br.

Forma aureolum, Rupp.

Pilliga Scrub, N.S.W.

HWRR

with abnormal flower at base.

CALOCHILUS PALUDOSUS R.Br.
Weston.

CALOCHILUS ROBERTSONII Benth.
Paterson.

The Red Bearded Orchid, *Calochilus paludosus* R.Br., of which Rupp found specimens nearly a metre tall growing near Weston in the spring of 1930. (Marjory Loader Collection)

Robertson's Bearded Orchid, *Calochilus robertsonii* Benth., sketched from specimens Rupp collected at Paterson in 1926. (Marjory Loader Collection)

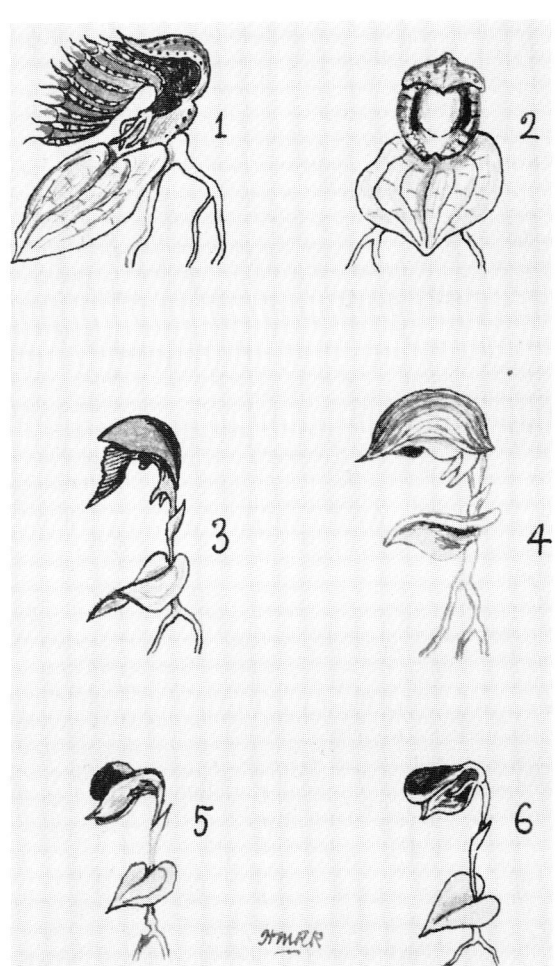

The Nodding Greenhood or Parrot's Beak Orchid, *Pterostylis nutans* R.Br. Rupp noted that the second specimen shows 'disintegration of basal rosette' of leaves 'in dense undergrowth at Paterson'. The smallest sketches depict R. D. FitzGerald's variety *hispidula* from Bulahdelah, now raised to specific rank. (Marjory Loader Collection)

Helmet Orchids, sketched in 1946: 1. *Corybas fimbriatus* (R.Br.) Reichb.f.; 2. *C. diemenicus* (Lindley) Reichb.f.; 3. *C. undulatus* (Cunn.) Rupp; 4. *C. aconitiflorus* Salisb.; 5. *C. unguiculatus* (R.Br.) Reichb.f.; 6. *C. fordhamii* (Rupp) Rupp. (Scrivener Collection, RBGS)

Aberrant forms of the Snake Tongue Greenhood, *Pterostylis ophioglossa* R.Br., from Weston, 1930. (Marjory Loader Collection)

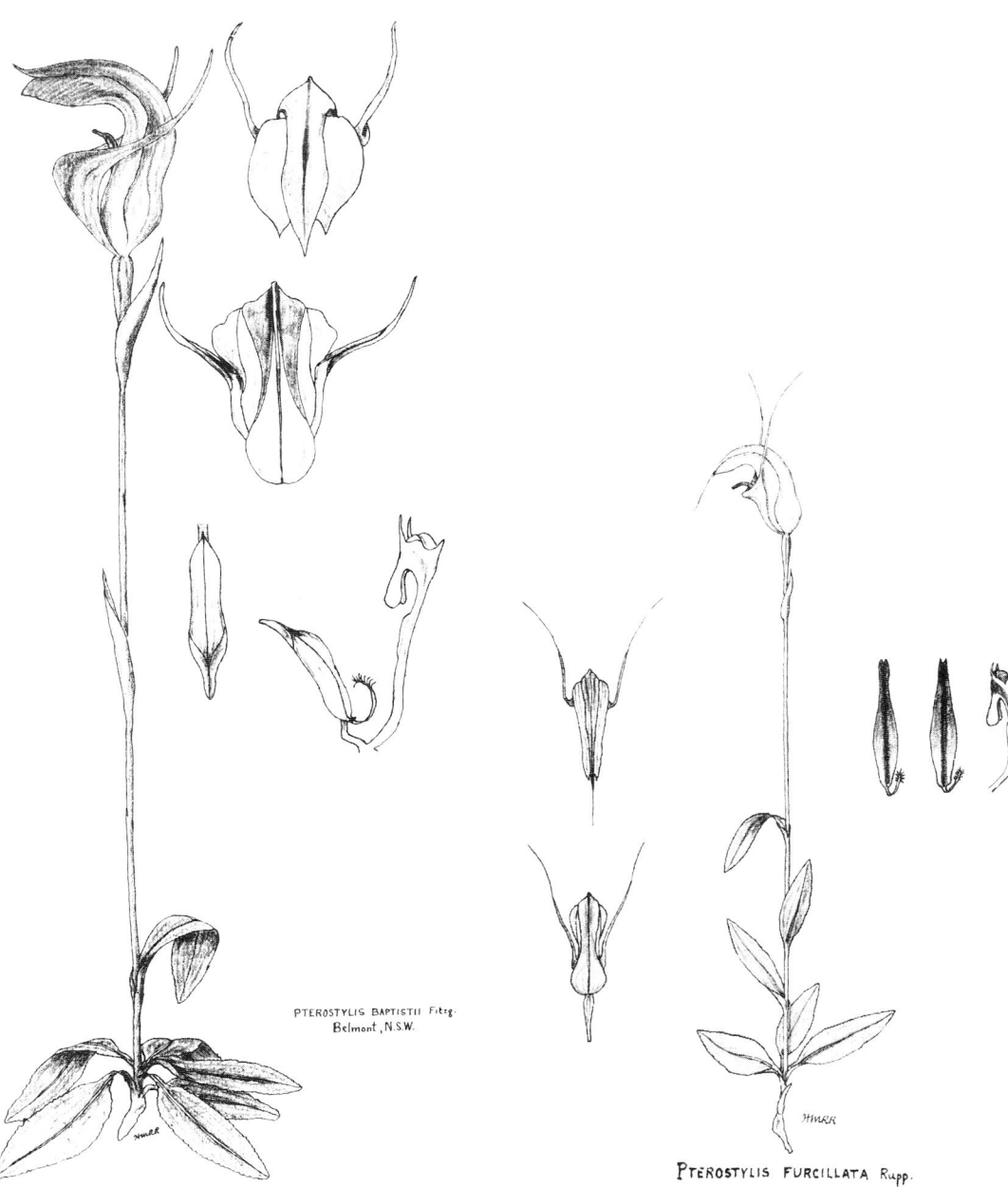

PTEROSTYLIS BAPTISTII Fitzg.
Belmont, N.S.W.

PTEROSTYLIS FURCILLATA Rupp.
South Maitland Coalfields. 1930.

Rupp considered *Pterostylis baptistii* FitzG. to be 'King of the Greenhoods'. This sketch, 'shewing top & back of flower' and 'labellum & column enlarged', was prepared from a specimen collected by Florence Rupp at Belmont in September 1926. (Marjory Loader Collection)

Described from specimens Rupp found near Kurri Kurri Hospital and at Abermain, *Pterostylis furcillata* Rupp is now considered to be a natural hybrid between *P. obtusa* R.Br. and *P. ophioglossa* R.Br. (Scrivener Collection, RBGS)

Pterostylis nana R.Br.
and conspecific forms.

1, 2, *P. pyramidalis* Lindl. (Western Australia.)
3, 4, *P. puberula* Hook.f. (New Zealand.)
5—8, *P. nana* R.Br. 5, South Australia. 6, Tasmania. 7, Victoria. 8, Western Australia.

The Dwarf or Snail Greenhood, *Pterostylis nana* R.Br., is now held to be quite distinct from the West Australian Leafy Snail Greenhood, *P. pyramidalis* Lindley, and not 'conspecific'. (Marjory Loader Collection)

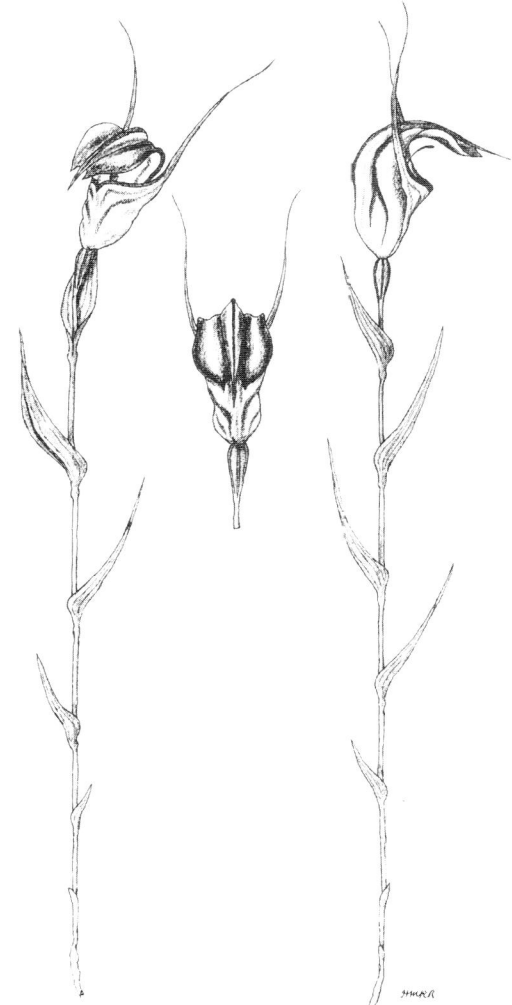

PTEROSTYLIS GRANDIFLORA R.Br.

Rupp considered that although 'by no means the largest-flowered species' of Greenhood, *Pterostylis grandiflora* R.Br. 'may be acknowledged the most beautiful'. Sketched from a Bulahdelah specimen, probably in May 1924. (Marjory Loader Collection)

DENDROBIUM JOHANNIS, Reichb.f.

Cairns.

The Brown Antelope Orchid, *Dendrobium johannis*
Reichb.f., of north Queensland, sketched from a
specimen collected by W. F. Tierney in October 1932.
(Marjory Loader Collection)

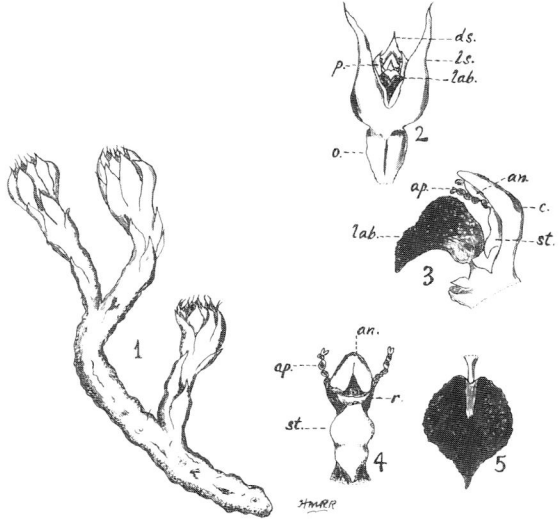

The subterranean orchid Rupp named *Cryptanthemis
slateri* (now known as *Rhizanthella slateri*): 1. Plant
(natural size in original); 2. Flower, front view, greatly
enlarged (lab = labellum, ds = dorsal sepal, ls = lateral
sepal, o = ovary, p = petal); 3. Column and labellum
from the side, greatly enlarged (c = column, ap =
appendage, an = anther, st = stigma); 4. Column from
the front, greatly enlarged (r = rostellum); 5. Labellum
from above, greatly enlarged. (Marjory Loader
Collection)

DENDROBIUM TOFFTII, Bailey.
Cairns.

Drawn from a specimen collected by W. F. Tierney in March 1933, the Blue Orchid of North Queensland and New Guinea is now known as *D. nindii* W. Hill. (Marjory Loader Collection)

Rupp's *Sarcochilus harriganae* described in 1938 from Dorrigo specimens, is now referred to *Sarcochilus spathulatus* Rogers. (Scrivener Collection, RBGS)

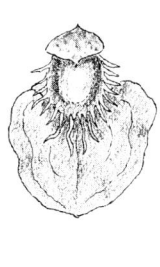
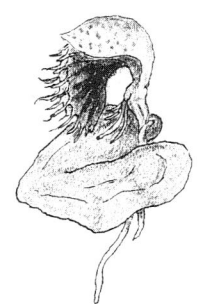

CORYSANTHES PRUINOSA Cunn.

Paterson. 1924

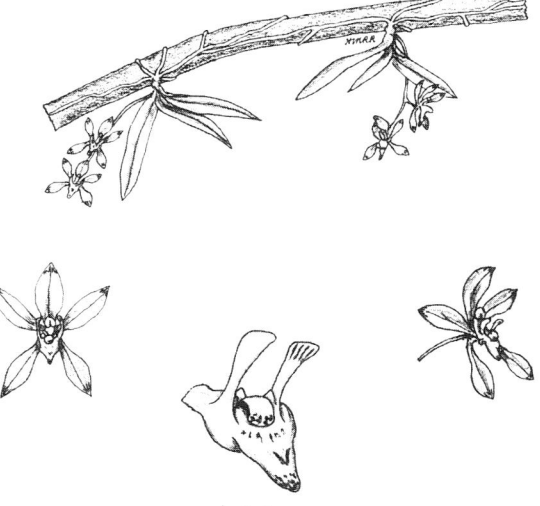

Labellum.

BULBOPHYLLUM WEINTHALII Rogers.

X ½

Dorrigo.

SARCOCHILUS SPATHULATUS Rogers.

Tambourine Mtn., Q. Foothills of Barrington Tops, N.S.Wales.

The Blotched Bulbophyllum which Dr R. S. Rogers named in 1933 to honour Ferdinand August Weinthal. (Marjory Loader Collection)

Rupp's friend Alick Dockrill proposed the genus *Parasarcochilus* in 1967 to accommodate this and other species. However, this epiphyte is again known as *Sarcochilus spathulatus* Rogers. This specimen was sketched 'about 1929'. (Scrivener Collection, RBGS)

Sarcochilus Weinthalii *Bailey*
TOOWOOMBA, Q.
KYOGLE, N.S.W.

The Blotched Sarcochilus was named in 1903 by F. M. Bailey from specimens discovered by F. A. Weinthal near Toowoomba. Rupp's specimen was found by Weinthal, near Kyogle, in August 1936. Enlargements show details of a flower including the column and labellum. (Marjorie Loader Collection)

CYMBIDIUM CANALICULATUM, R.Br.
Forma *aureolum* Rupp.
Tillipa Scrub, N.S.W.

SARCOCHILUS OLIVACEUS *Lindl.*

Rupp described this form of the widely distributed and variable Channel-leaved Cymbidium, *Cymbidium canaliculatum* R.Br., in 1934. He drew attention to the 'abnormal flower at the base' of the raceme sketched. (Marjory Loader Collection)

Often found growing on Lawyer Cane in or near rainforests, this specimen of the Lawyer Orchid was sketched about 1925 when Rupp was stationed at Paterson. (Marjory Loader Collection)

Bibliography of an Orchidologist

Early in 1947, Rupp advised Trevor Hunt that in response to the urging of several friends, he had attempted 'to tabulate' his 'orchid publications'. He estimated that 'the horrible total' then stood at 139 papers containing descriptions of about forty new species—and of varieties, 'who knows?'[1] Six years later, Dr J. H. Willis, then botanist at the National Herbarium of Victoria, was still concerned that a complete list of Rupp's botanical writings had not been compiled. Accordingly, Rupp obliged Willis with a list of some 120 papers, excluding about seventy which had appeared in the *Victorian Naturalist*, for Willis, as a former editor, was well acquainted with those contributions.

The quest for Rupp's writings continued after his death, with lists and revised lists passing between Dr Willis and Mr Knowles Mair, two of Rupp's keenest admirers. By the end of October 1956, Mair had listed 202 papers published 'up to early 1954'. Meanwhile, Willis had listed 'a few' more papers 'up to 1953' and the two botanists combined their efforts considering, in Mair's words, that it was 'obligatory that we have a complete reference' to all of Rupp's papers.[2]

By the end of 1957 Dr Willis had compiled a bibliography comprising two books and 223 papers. Now published at last, this Bibliography has been rearranged in chronological order at the compiler's request, and with his permission, 112 additional entries have been inserted, so that 'popular' and scientific items, ecclesiastical and botanical, are recorded. The books are listed separately, with articles and papers listed under their respective years of publication in alphabetical order of the titles of

the journals in which they appeared. The arrangement under journal titles is again chronological.

Of special significance and interest among the additional items are the early contributions to the *Geelong Grammar School Quarterly* (kindly located by Mr Michael D. de B. Collins Persse, keeper of the School's Archives), those to the *Riverine Grazier*, and the series which appeared in the *Sydney Mail* and the *Australian Woman's Mirror* during the 1920s and 1930s. The *Mirror* was a *Bulletin* publication which appeared each Tuesday and boasted a circulation of over 167,000. Here, Rupp's articles were in good company for other contributors included A. H. Chisholm, Edith Coleman, Florence Irby, Isobel Bowden, Keith McKeown, Douglas Stewart, Roderic Quinn, Jessie Urquhart, Mary Gilmore, Steele Rudd, J. H. M. Abbott, Katherine Susannah Pritchard and other naturalists and littérateurs. In these articles, Rupp revealed the kind of enthusiasm he hoped to generate in his readers, emphasizing the beauty of native orchids, the excitement of hunting them, and their mysterious qualities:

There is a vein of inquisitiveness in the average human being, even though it be considered 'bad form' to reveal it, so most people are attracted secretly, if not avowedly, by anything which is unconventional, out of the ordinary. Orchids are the most unconventional family of flowering plants in the world. You never know what they are going to do next.[3]

Hence it was perfectly natural for people to study orchids rather than more 'conventional' forms of plant life. Rupp even took the opportunity to seek information about a group of animals to which he had something of an aversion: 'I should be very pleased to hear from any *Mirror* readers who have observed insects investigating orchid flowers'.[4] Edith Coleman and Keith McKeown must have enjoyed reading this request.

Notwithstanding fairly intensive recent searches, and the earlier efforts of Dr Willis and Mr Mair, it is still likely that some publications have been omitted, although over 340 items are listed. To offset any such loss and to render the Bibliography somewhat more complete, about fifty of Rupp's more significant unpublished writings (excluding letters) have also been included.

Abbreviations

ABMR   ABM Review (Australian Board of Missions)

AD Arm.   Archives of the Diocese of Armidale (Armidale Diocesan Registry)

ADB   Australian Dictionary of Biography

AD Newc.   Archives of the Diocese of Newcastle (Auchmuty Library, University of Newcastle)

AO   Archives Office of New South Wales

AOR   Australian Orchid Review

Aust.   The Australasian (Melbourne)

Aust. Nat.   Australian Naturalist

AWM   Australian Woman's Mirror

Biog. Reg.   A Biographical Register 1788–1939 (by H.J. Gibbney and A.G. Smith)

Contrib. N.S.W. Nat. Herb.   Contributions from the New South Wales National Herbarium

Corian   The Corian (Geelong Church of England Grammar School)

GGSQ   Geelong Grammar School Quarterly

HRA   Historical Records of Australia

J. Roy.Soc. W.A.   Journal of the Royal Society of Western Australia

ML   Mitchell Library, State Library of New South Wales

NDC   Newcastle Diocesan Churchman

N.Q. Nat.   North Queensland Naturalist

Proc. Linn. Soc. N.S.W.   Proceedings of the Linnean Society of New South Wales

Proc. Roy. Soc. N.S.W.   Proceedings of the Royal Society of New South Wales

Proc. Roy. Soc. Q.   Proceedings of the Royal Society of Queensland

Proc. Roy. Soc. S.A.   Transactions and Proceedings of the Royal Society of South Australia

Proc. Roy. Soc. Vic.   Proceedings of the Royal Society of Victoria.

Q. Nat.   Queensland Naturalist

RBGM   Library of the Royal Botanic Gardens, Melbourne

RBGS   Library of the Royal Botanic Gardens, Sydney

Riv. Graz.   The Riverine Grazier (Hay, N.S.W.)

SM   The Sydney Mail

SMH   The Sydney Morning Herald

Vic. Nat.   The Victorian Naturalist

WWA   Who's Who in Australia

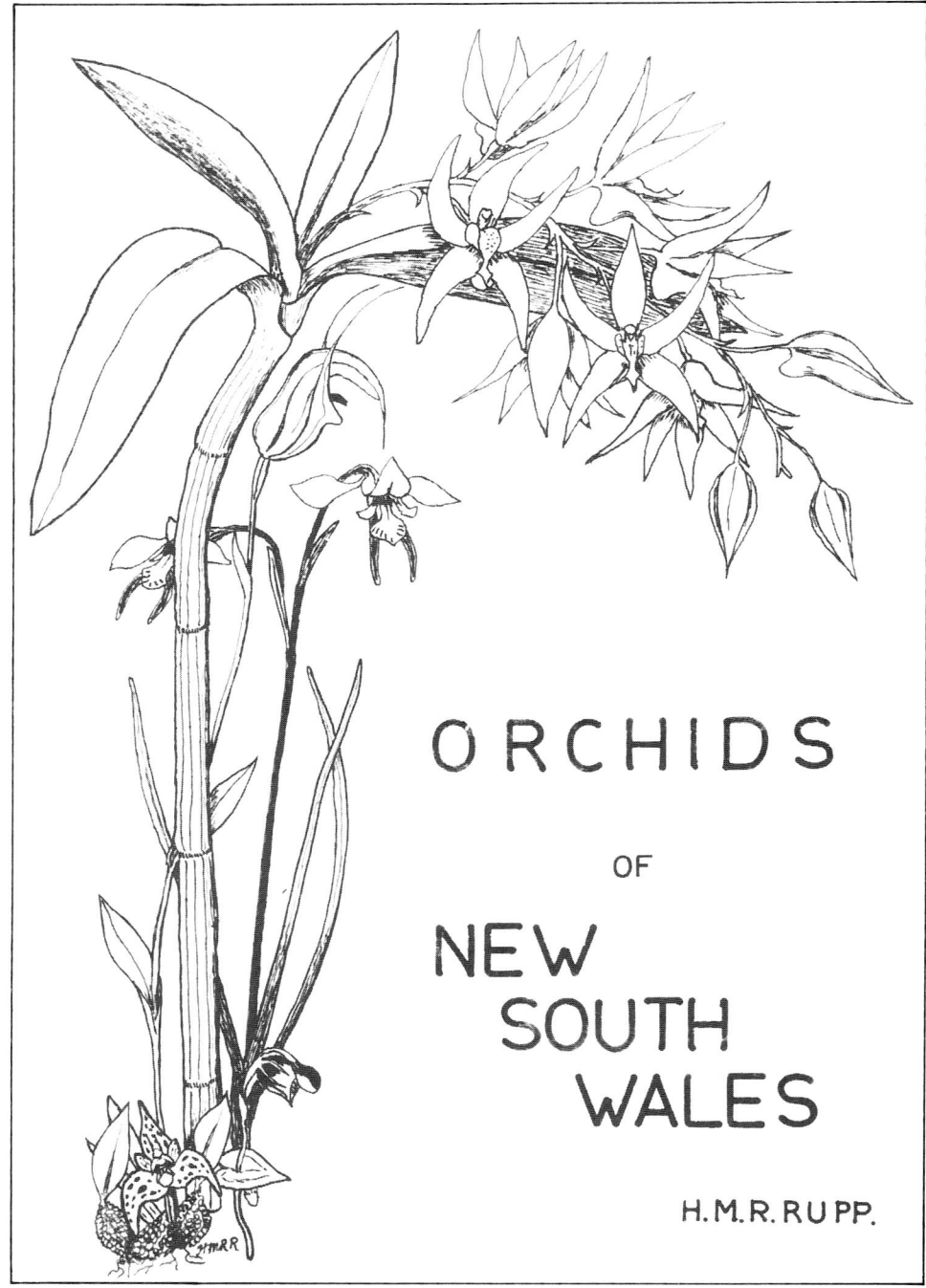

Rupp's design for the dustwrapper of *The Orchids of New South Wales,* depicting the Beech Orchid, *Dendrobium falcorostrum* FitzG.; a hybrid Greenhood, *Pterostylis furcillata* Rupp; Veined Doubletail Orchid, *Diuris venosa* Rupp; Banded Helmet Orchid, *Corybas fordhamii* Rupp; and Blotched Bulbophyllum, *Bulbophyllum weinthalii* Rogers. (From the original drawing which Rupp presented to Miss Gwen Scrivener in December 1943 and is now in the Library of the Royal Botanic Gardens, Sydney)

GUIDE TO THE ORCHIDS
OF NEW SOUTH WALES

By
H. M. R. RUPP, B.A.
Rector of St Mary's, Weston, N.S.W.

WITH FOREWORD BY
EDWIN CHEEL
President Linnean Society of N.S.W.

AUSTRALIA:
ANGUS & ROBERTSON LIMITED
89 CASTLEREAGH STREET, SYDNEY
1930

THE
ORCHIDS
OF
NEW SOUTH WALES

BY
The Rev. H. M. R. RUPP
B.A. (Melb.)

Issued from the National Herbarium, Sydney, as a part of the
FLORA OF NEW SOUTH WALES.
December, 1943.

Title page of Rupp's first book on New South Wales orchids.

Title page of Rupp's second book on New South Wales orchids.

A. Published Works

I. Books

S. Paul's Church, Paterson. . .a Souvenir of the Eighty-fourth Anniversary, Wednesday, November 27th, 1929. Paterson (?) 1929. Illust. (With A. P. Elkin.)

*Guide to the Orchids of New South Wales. . .*with a Foreword by Edwin Cheel. Angus and Robertson, Sydney 1930. Illust.

The Orchids of New South Wales, National Herbarium, Sydney 1943. Illust.

The Orchids of New South Wales, Facsimile edition, with a Supplement by D. J. McGillivray and Foreword by K. Mair. National Herbarium, Sydney 1969. Illust.

II. Articles and Scientific Papers

1889
'The Camp', *GGSQ*, December, pp.13–16.
1890
'The Cadet Camp', *GGSQ*, October, pp.33–4.
'The Wannon District', *GGSQ*, December, pp.28–30.
1891
'At Spring Creek' (*re* wreck of *Joseph H. Scammell*, 7 May 1891), *GGSQ*, July, pp.19–20.
'Orchids', *GGSQ*, December, pp.66–7.

1892
'A Trip among the Grampians', *GGSQ*, April, pp.5–9.
1894
'At Mungadel' (poem), *GGSQ*, December, p.5.
'Notes on Some Indigenous Plants of the Hay District', *Riv. Graz.*, 31 December.
1895
Letter to Editor (*re* misprints in a previous article, with reference to *Euphorbia eremophila*), *Riv. Graz.*, 4 January.
'Some Problems and Theories in Science', *Riv. Graz.*, 11 January.
Letter to Editor (*re* Moa in N.Z.), *Riv. Graz.*, 18 January.
'Western Victoria', *Riv. Graz.*, 1 February.
Letter to Editor (*re* collection of seeds for Baron von Mueller), *Riv. Graz.*, 19 February.
1896
'The late Baron von Mueller', *GGSQ*, October, pp.7–8.
'Life's Way' (poem) *GGSQ*, December, p.12.
1897
'Evening' (poem), *GGSQ*, July, p.7. (Another poem, 'A Winter Trip to the Erskine Falls'[Lorne] in the October issue [p.22] may also have been Rupp's,

although it was signed simply 'M' instead of the
usual 'M.R.')

1903
'Impressions of Tamworth' (quoted from a letter), *GGSQ*,
October, pp.17–18.

1912
'The Clarence River', *GGSQ*, July, pp.19–21.

1915
'The Missionary Duty in Country Parishes', *ABMR*, 1
August, pp.116–17.

1916
'Wake Up!', *ABMR*, 1 December, pp.178–9.

1923
'Bulahdelah—Parish Notes', *NDC*, 2 July, p.32.
'Bulahdelah—Parish Notes', *NDC*, 1 October, p.29.
'Bulahdelah—Parish Notes', *NDC*, 1 November,
pp.32–3.

1924
'Notes on the Habits of Certain Orchids' (illust.), *Aust.
Nat.*, **5**: 166–71.
'Bulahdelah—Parish Notes', *NDC*, 1 September, p.27.
'Bulahdelah—Parish Notes', *NDC*, 1 November, p.32.

1925
'On the Orchids of the Bullahdelah District of New
South Wales', *Aust. Nat.*, **5**: 217–28.
'Paterson—Parish Notes', *NDC*, 1 January, p.33.
'Paterson—Parish Notes', *NDC*, 2 March, p.30.
'Paterson—Parish Notes', *NDC*, 1 May, pp.35–6.
'Paterson—Parish Notes', *NDC*, 1 July, p.33.
'Paterson—Parish Notes', *NDC*, 1 September, p.35.
'Notes on Species of *Pterostylis*' (illust), *Proc. Linn. Soc.
N.S.W.*, **50**: 299–310.
'Cult of the Orchid', *SMH*, 1 December.

1926
'The Lake of Flowers', *Aust.*, 3 April.
'Ten Days on the Upper Allyn River (N.S.W.)', *Aust.
Nat.*, **6**: 4–7, 17–20.
'Habits of Certain Orchids', *Aust. Nat.*, **6**: 16.
'Orchids: Part I', *Corian*, May, pp.16–18.
'Orchids: Part II', *Corian*, August, pp.91–4.
'Paterson—Parish Notes', *NDC*, 1 March, p.31.
'Paterson—Parish Notes', *NDC*, 1 May, p.34.
'Paterson—Parish Notes', *NDC*, 1 July, p.30.
'Paterson—Parish Notes', *NDC*, 2 August, p.31.
'Paterson—Parish Notes', *NDC*, 1 November, p.24.
'Further Notes on Species of *Pterostylis*', *Proc. Linn. Soc.
N.S.W.*, **51**: 184–6.
'Description of a New Species of *Diuris* (*D. venosa*) from
Barrington Tops, N.S.W.' (illust.), *Proc. Linn. Soc.
N.S.W.*, **51**: 313–4.
'Australian Orchids: Many Species of Great Beauty'
(*Dendrobium* and *Diuris*; illust.), *SM*, 27 October,
pp.12–13.
'Notes on the Genus *Corysanthes*', *Vic. Nat.*, **43**: 25–8.
'A New Species of *Diuris*' (*D. venosa* sp. nov.; illust.),
Vic. Nat., **43**: 153–4. (September).

1927
'Greenhood Orchids of the Paterson District (N.S.W.)'
(illust.), *Aust. Nat.*, **6**: 57–62.
'Orchids of the Paterson District (N.S.W.)', *Aust. Nat.*,
7: 7–12.
'Paterson—Parish Notes', *NDC*, 1 March, p.33.
'Paterson—Parish Notes', *NDC*, 1 July, p.38.
'Paterson—Parish Notes', *NDC*, 1 September, p.32.
'Paterson—Parish Notes', *NDC*, 1 November, pp.30–1.

'Paterson—Parish Notes', *NDC*, 1 December, p.30.
'A New *Dendrobium* (*D. tenuissimum* sp. nov.) for New
South Wales and Queensland' (illust.), *Proc. Linn.
Soc. N.S.W.*, **52**: 570–1.
'Beauties of the Bush: Australian Orchids' (*Sarcochilus* and
some terrestrials, e.g., *Phaius* and *Calanthe*; illust.),
SM, 23 February, pp.13, 50.
'Beauties of the Bush: Our Orchids' (the Greenhoods;
illust.), *SM*, 1 June, p.14.
'Beauties of the Bush: Our Orchids' (*Prasophyllum* and
Caladenia; illust.), *SM*, 21 September, p.15.
'Beauties of the Bush: Our Orchids' (*Corysanthes* or
Corybas; Acianthus and *Dendrobium*; illust.), *SM*, 16
November, p.19.

1928
'Paterson—Parish Notes', *NDC*, 1 February, p.30.
'Paterson—Parish Notes', *NDC*, 2 April, pp.29–30.
'Paterson—Parish Notes', *NDC*, 1 May, p.31.
'Paterson—Parish Notes', *NDC*, 1 June, p.46.
'Paterson—Parish Notes', *NDC*, 2 July, pp.36–7.
'Paterson—Parish Notes', *NDC*, 1 September, p.30.
'A Review of the Australian Species of *Corysanthes*
(Orchidaceae), including description of a new species,
C. dilatata', (in collaboration with W. H. Nicholls;
illust.), *Proc. Linn. Soc. N.S.W.*, **53**: 80–9.
'Terrestrial Orchids of Barrington Tops, N.S.W.'
(including *Prasophyllum rogersii* and *P. hopsonii* spp.
nov.; illus.), *Proc. Linn. Soc. N.S.W.*, **53**: 336–42.
'Notes on *Corysanthes* and Some Species of *Pterostylis* and
Caladenia (illust.), *Proc. Linn. Soc. N.S.W.*, **53**: 551–4.
'Our Beautiful Orchids: Where and How they Grow'
(illust.), *SM* 4 January, pp.19, 50.
'Australian Orchids: Searching for New and Rare Plants'
(illust.), *SM*, 14 March, pp.24, 58.
'Wanderings on the Upper Clarence' (illust.), *SM*, 2
May, pp.2–8.
'More About Our Orchids', *SM*, 11 July, p.25.
'Orchids of the Early Spring' (illust.), *SM*, 5 September,
p.17.
'The Glory of the Orchids' (illust.), *SM*, 26 September,
p.17.

1929
'Paterson—Parish Notes', *NDC*, 1 October, p.32.
'Paterson—Parish Notes', *NDC*, 1 November, pp.31–2.
'Variations in Certain Orchids' (*Dendrobium speciosum* var.
gracillimum var. nov., *Prasophyllum intricatum* and
Pterostylis ophioglossa var. *collina* var. nov.; illust.), *Proc.
Linn. Soc. N.S.W.*, **54**: 550–2.
'Forms and Habits of Certain Orchids', *Vic. Nat.*, **46**:
3–8. (May).

1930
'Paterson—Parish Notes', *NDC*, 1 January, pp.29–30.
'Paterson—Parish Notes', *NDC*, 1 February, p.31.
'Paterson—Parish Notes', *NDC*, 1 May, pp.33–4.
'Notes on the Autumn Orchids of the South Maitland
Coalfields, N.S.W., with Description of a New
Species of *Pterostylis*' (*P. furcillata* sp. nov.; illust.),
Proc. Linn. Soc. N.S.W., **55**: 413–16.
'Notes and Comments' (on *Diuris dendrobioides* and *D.
punctata*), *Vic. Nat.*, **46**: 184–5. (January).
'The Orchid *Prasophyllum nigricans* R.Br.', *Vic Nat.*, **47**:
29–30. (June).

1931
'Further Notes on the Orchids of the South Maitland
Coalfields, with Descriptions of a New *Dendrobium*

Plate XIV.

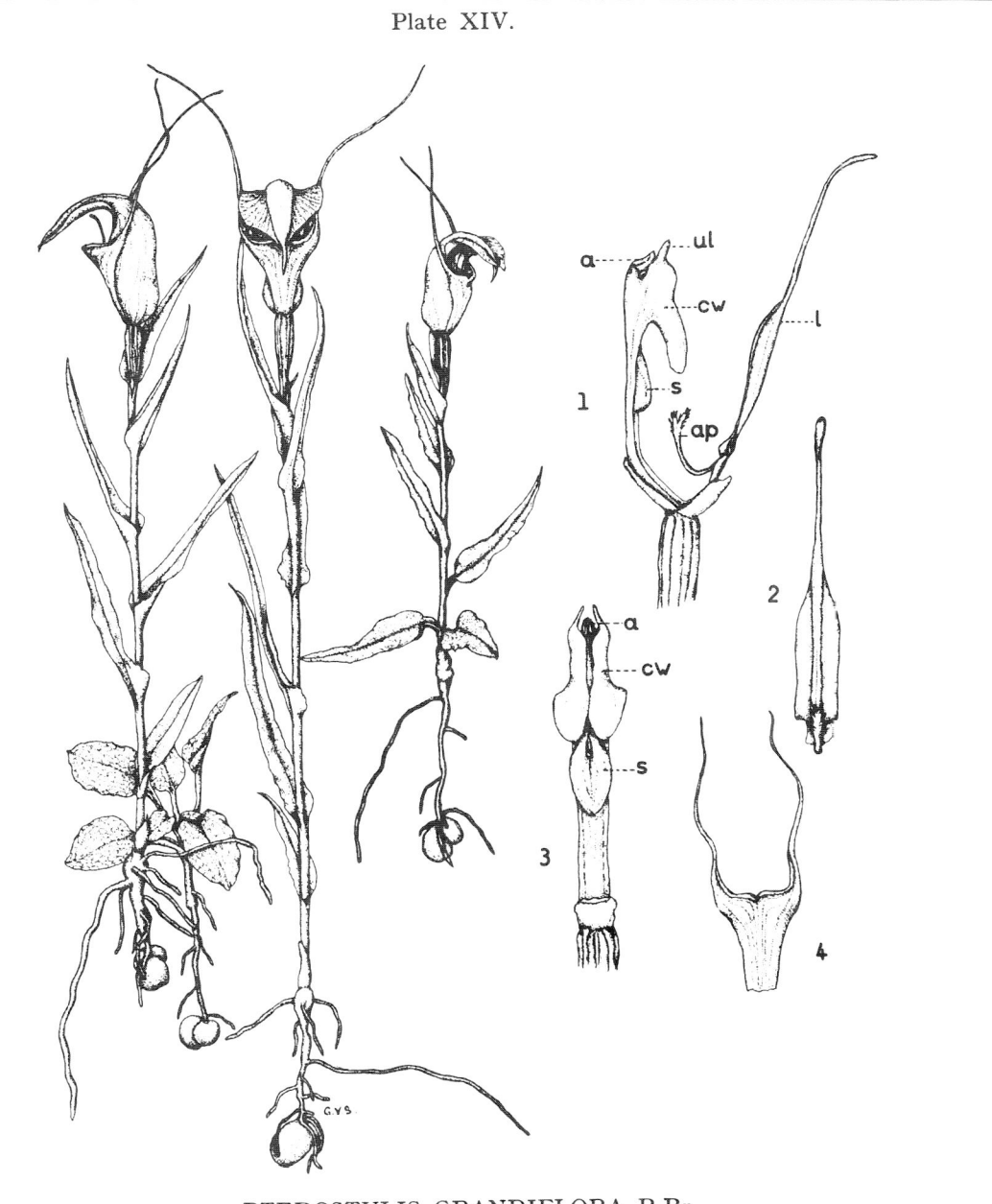

PTEROSTYLIS GRANDIFLORA R.Br.

Plants about two-thirds natural size. 1, Column and labellum, from the side. Note the filiform clavate tip of the labellum in this species. 2, Labellum, from above. 3, Column, from the front. The top of the ovary is shown at the base of this figure. 4, Lateral sepals, from the front. 1 to 3 enlarged. *a*, anther. *ap*, appendage. *cw*, column wings. *l*, labellum. *s*, stigma. *ul*, upper lobe of the column wings.

One of George Vance Scammell's finely executed line drawings in Rupp's *The Orchids of New South Wales*, 1943. Rupp declared that 'Scammell's plates could not have been bettered by anyone'.

from Bullahdelah' (*D. kestevenii* sp. nov.; illust.), *Proc. Linn. Soc. N.S.W.*, **56**: 133–8.

'Notes on New South Wales Orchids: I' (*Corysanthes undulata, Caladenia angustata, C. testacea, Dendrobium teretifolium* var. *fairfaxii* etc.; illust.), *Proc. Linn. Soc. N.S.W.*, **56**: 458–60.

1932

'Cinderella of the Garden: Some of the Beauties of Australia's Orchid Family' (illust.), *AWM*, **8**(18): 48–9, (29 March).

'Australian Native Orchids: How to Find Them and How to Grow Them' (illust.), *AWM*, **8**(34): 10, 53. (19 July).

'Australian Orchids—Identification and Culture' (illust.), *AWM*, **8**(38): 48. (16 August).

'Growing Ground Orchids' (illust.), *AWM*, **8**(46): 11. (11 October).

'Libels on our Orchids' (illust.), *AWM*, **9**(2): 47, 57. (6 December).

'Weston—Parish Notes', *NDC*, 1 June, pp.41–2.

'Notes on New South Wales Orchids: II. Discovery of a Remarkable New Genus and Species at Bullahdelah' (*Cryptanthemis slateri* gen. et sp. nov.; illust.), *Proc. Linn. Soc. N.S.W.*, **57**: 57–63.

'Off the Beaten Trail: On the Upper Clarence River', *SM*, 27 July, pp.36–7.

'Early Days: The Paterson Valleys: Part I' (with A. P. Elkin), *SMH*, 27 August.

'Early Days: The Paterson Valleys: Part II' (with A. P. Elkin), *SMH*, 3 September.

'A New Australian Subterranean Orchid' (*Cryptanthemis slateri*; illust.), *Vic. Nat.*, **49**: 102–4. (August).

'Australian and New Zealand Orchids', *Vic. Nat.*, **49**: 151–2. (October).

'Plant Life in the Pilliga Scrub' (illust.), *Vic. Nat.*, **49**: 187–90. (December).

1933

'A Botanist's Story: Part I', *Aust.*, 8 July.

'A Botanist's Story: Part II', *Aust.*, 15 July.

'A Botanist's Story: Part III', *Aust.*, 22 July.

'Birds and Flowers on Inland Plains' (illust.), *AWM*, **9**(8): 10, 39. (17 January).

'Notes on New South Wales and Queensland Orchids' (forms of *Dendrobium, Cryptanthemis slateri*, etc.; illust.), *Proc. Linn. Soc. N.S.W.*, **58**: 221–8.

'The Genus *Pterostylis* R.Br.: A New Scheme of Classification, with Notes on the Distribution of the Australian Species' (illust.), *Proc. Linn. Soc. N.S.W.*, **58**: 421–8.

'Notes on Certain Species of *Caladenia*' (illust.), *Vic. Nat.*, **49**: 259–61. (March).

'An Interesting Tasmanian Orchid' (*Townsonia viridis*; illust.), *Vic. Nat.*, **50**: 18–21. (May).

'A Rejoinder to Dr Lindinger' (re *Cryptanthemis slateri*), *Vic. Nat.*, **50**: 46–7. (June).

'Three Rare Orchids' (*Dendrobium tofftii, D. moorei* and *D. gracilicaule* var. *howeanum*; illust.), *Vic. Nat.*, **50**: 168–9. (November).

1934

'More of our Lovely Orchids' (illust.), *AWM*, **10**(15): 11. (6 March).

'Orchid Mysteries' (illust.), *AWM*, **10**(30): 10, 39. (19 June).

'Orchid Mysteries: II' (illust.), *AWM*, **10**(33): 9, 47. (10 July).

'Orchid Mysteries: III' (illust.), *AWM*, **10**(36): 10. (31 July).

'Spring Amid the Orchids' (illust.), *AWM*, **10**(49): 9. (30 October).

'Did You Know This About Orchids?' (illust.), *AWM*, **11**(1): 47. (27 November).

'A New Orchid from Proserpine, North Queensland' (*Cleisostoma orbiculare* sp. nov.; illust.), *N.Q. Nat.*, **2**: 13–14.

'Some Orchids of the Proserpine District, North Queensland', *N.Q. Nat.*, **2**: 19–20.

'*Bulbophyllum macphersonii* nom. nov.', *N.Q. Nat.*, **2**: 21.

'Notes on Some North Queensland Orchids', *N.Q. Nat.*, **3**: 10–11.

'Notes on Australian Orchids: I. A Review of the Genus *Cymbidium* in Australia I' (including *C. canaliculatum* var. *marginatum* var. nov. and var. *sparkesii* (Rendle) comb. nov.; illust.), *Proc. Linn. Soc. N.S.W.*, **59**: 93–100.

'The Habitat, Character, and Floral Structure of *Cryptanthemis slateri* Rupp.' (illust.), *Proc. Linn. Soc. N.S.W.*, **59**: 118–22.

'Notes on Certain Species of *Dendrobium*' (*D. bowmanii* and *D. mortii*; illust.), *Q. Nat.*, **9**: 51–2.

'A New Species of *Calochilus*' (*C. grandiflorus*; illust.), *Vic. Nat.*, **50**: 239–40. (February).

'A Thrice-named Orchid' (*Bulbophyllum macphersonii*; illust.), *Vic. Nat.*, **51**: 81–2. (July).

'The Scented Sun-orchid' (*Thelymitra aristata*; illust.), *Vic. Nat.*, **51**: 170–1. (November).

'Orchid Hunting in Northern New South Wales', *Vic. Nat.*, **51**: 187–90. (December).

1935

'Among the Dendrobs' (illust.), *AWM*, **11**(8): 9, 47. (15 January).

'Gems Among our Orchids' (illust.), *AWM*, **11**(16): 10, 43. (12 March).

'Orchid Pygmies' (illust.), *AWM*, **11**(26): 11. (21 May).

'Where to Look for Orchids: Regions Where their Fairy Beauty may be Found' (illust.), *AWM*, **11**(35): 19. (23 July).

'Where to Look for Orchids: II. Types of Terrain Haunted by these Shy Native Beauties' (illust.), *AWM*, **11**(52): 10, 27. (19 November).

'Woy Woy—Parish Notes', *NDC*, 1 August, pp.33–4.

'Woy Woy—Parish Notes', *NDC*, 1 October, pp.34–5.

'Further Notes on North Queensland Orchids' (*Cheirostylis ovata* and *Nervilia dallachyana*; illust.), *N.Q. Nat.*, **3**: 18–19.

'The Pollination of *Cymbidium iridifolium* A. Cunn.' (in collaboration with Kenneth Macpherson), *N.Q. Nat.*, **3**: 26.

'North Queensland Orchid Records', *N.Q. Nat.*, **3**: 42–3.

'Further Notes on North Queensland Orchids' (including *Cleisostoma cornutum* sp. nov.; illust.), *N.Q. Nat.*, **4**: 9–11.

'Notes on Australian Orchids: II. A Review of the Species *Dendrobium teretifolium* R. Br.' (including description of var. *fasciculatum* var. nov.), *Proc. Linn. Soc. N.S.W.*, **60**: 155–8.

'Notes on Two Closely Allied Dendrobs' (*D. delicatum* and *D. kestevenii*; illust.), *Q. Nat.*, **9**: 61–4.

'*Liparis habenarina* F.v.M. in New South Wales' (illust.), *Vic. Nat.*, **52**: 12–13. (May).

'Discovery of a New Zealand Orchid on Lord Howe

Island' (*Bulbophyllum tuberculatum*; illust.), *Vic. Nat.*, **52**: 73. (August).

'Orchid Notes from New South Wales', *Vic. Nat.*, **52**: 145-6. (November).

1936

'A Problem in Orchids' (relations between *Sarcochilus fitzgeraldii*, *S. rubicentrum* and *S. hartmannii*; illust.), *AOR*, **1**(1): 9-10. (January).

'*Cymbidium canaliculatum*' (illust.), *AOR*, **1**(3): 11-12. (July-September).

'Two Beautiful Immigrants: Orchids that May be Grown in New Surroundings' (*Dendrobium densiflorum*—India; *D. cunninghamii*—N.Z.), *AWM*, **12**(14): 13. (25 February).

'The Orchids of Spring: Botanist H.M.R. Rupp's Bush House' (illust.), *AWM*, **12**(42): 48. (8 September).

'Woy Woy—Parish Notes', *NDC*, 1 February, p.32.

'Woy Woy—Parish Notes', *NDC*, 2 March, p.34.

'Woy Woy—Parish Notes', *NDC*, 1 June, p.42.

'Raymond Terrace—Parish Notes', *NDC*, 1 July, p.30.

'Further Notes on Orchid Pollination' (in collaboration with Kenneth MacPherson), *N.Q. Nat.*, **4**: 25-6.

'A Northern Form of *Sarcochilus hillii* F. Muell. (var. *tricalliatus* var. nov.)', *N.Q. Nat.*, **4**: 31.

'Australian Species of *Cymbidium*'. *Orchidologia Zeylanica*, **3**: 130.

'A Mushroom Note' (*Lepiota rhacodes*), *Vic. Nat.*, **52**: 175. (February).

'Notes on Some Unusual Colourings in Orchid Flowers' (*Diuris*, *Dendrobium*, etc.), *Vic. Nat.*, **53**: 141. (December).

1937

'*Liparis simmondsii*: A Recent Addition to the N.S.W. Orchid Flora', (illust.), *AOR*, **2**(1): 14. (March).

'Notes on a Rare North Queensland Orchid' (*Anoectochilus yatesae*; illust.), *AOR*, **2**(2): 19. (June).

'Some Notes on Nomenclature, with an Index to the Meaning of Names of Australian Orchid Genera: I', *AOR*, **2**(4): 27-8. (December).

'Raymond Terrace—Parish Notes', *NDC*, 1 October, p.20.

'Raymond Terrace—Parish Notes', *NDC*, 1 November, p.44.

'*Cleisostoma tridentatum* Lindl. (syn. *C. cornutum* Rupp)', *N.Q. Nat.*, **5**(49): 3.

'New North Queensland Records of Orchids' (*Pterostylis curta* R.Br. and *P. ophioglossa* R.Br. var. *collina* Rupp), *N.Q. Nat.*, **5**(49): 3-4.

'A Diminutive North Queensland Orchid' (*Oberonia pusilla* Bailey; illust.), *N.Q. Nat.*, **5**(51): 2.

'*Phreatia baileyana* (Bail.) Schltr.', *N.Q. Nat.*, **6**(52): 3.

'*Pterostylis curta* R.Br.' (illust.), *N.Q. Nat.*, **6**(52): 3.

'*Dendrobium fleckeri* Rupp', *N.Q. Nat.*, **6**(52): 3-4.

'The Greenhood Orchids of Australia', *Orchidologia Zeylanica*, **4**: 136.

'A Giant Australian Orchid' (*Galeola ledgeriana*), *Orchidologia Zeylanica* **4**: 155.

'A Census of the Orchids of New South Wales, 1937', *Proc. Linn. Soc. N.S.W.*, **62**: 27-31.

'Notes on Australian Orchids: III. A Review of the Genus *Cymbidium* in Australia II', (illust.), *Proc. Linn. Soc. N.S.W.*, **62**: 299-302.

'Two New Dendrobs for North Queensland' (*D. fleckeri* sp. nov. and *D. carrii* sp. nov.; illust.; in collaboration with C. T. White), *Q. Nat.*, **10**: 25-6.

'Notes on the Distribution of Certain Orchids', *Vic. Nat.*, **53**: 189-91. (March).

'"Freakishness" in Orchids' (*Dendrobium* spp.; illust.), *Vic. Nat.*, **54**: 6-7. (May).

'A New Epiphytic Orchid from Dorrigo N.S.W.' (*Cleisostoma gemmatum* sp. nov.; illust.), *Vic. Nat.*, **54**: 112-13. (December).

1938

'Some Notes on Nomenclature: II' *AOR*, **3**: 25-6. (March).

'Recent Orchid Discoveries in New South Wales and Queensland, with some Reflections' (including *Cryptanthemis slateri*) (illust.), *AOR*, **3**: 37-40. (June).

'The Genus *Cleisostoma* in Australia' (illust.), *AOR*, **3**: 82. (September).

'Notes from my Bush House', *AOR*, **3**: 111, 121. (December).

'Raymond Terrace—Parish Notes', *NDC*, 1 July, p.213.

'The Genus *Corysanthes* in Australia and New Zealand', *J. Roy. Soc. W.A.*, **24**: 63-4.

'Note on *Bulbophyllum prenticei* F. Muell.', *N.Q. Nat.*, **7**(56): 2-3.

'A New *Sarcochilus* (Orchidaceae) from the Dorrigo' (*S. harriganae* sp. nov.; illust.), *Proc. Linn. Soc. N.S.W.*, **63**: 128.

'A New Orchid for South Queensland' (*Acianthus ledwardii* sp. nov.; illust.), *Q. Nat.*, **10**: 113-14.

'*Cleisostoma gemmatum* Rupp (= *C. purpuratum*, nom. nov.)', *Vic. Nat.*, **54**: 190. (April).

1939

'Fitzgerald's "Australian Orchids": Unpublished Plates brought to light: The Mysterious *Cymbidium gomphocarpum*', *AOR*, **4**: 65-6. (September).

'A Beautiful N.S.W. Ground Orchid' (*Thelymitra venosa* var. *magnifica* var. nov.; illust.), *AOR*, **4**: 81. (September).

'*Cymbidium canaliculatum*' (illust.), *AOR*, **4**: 83-4. (September).

'The Genus *Cryptostylis*' (illust.), *AOR*, **4**: 103-4. (December).

'A New *Pterostylis* (*P. alveata* Garnet)', *AOR*, **4**: 112. (December).

'*Dendrobium delicatum* and *D. kestevenii*' (illust.), *AOR*, **4**: 124-5. (December).

'*Dipodium stenocheilum* Schwarz', *N.Q. Nat.*, **8**(60): 2-3.

'Fitzgerald's *Australian Orchids*', *N.Q. Nat.*, **8**(60): No. 60: 3-4.

'*Dendrobium schneiderae* Bailey—a New Northern Form (var. *major* var. nov.)', *Q. Nat.*, **11**: 3-4.

'Winter Ramblings behind Barrington Tops', *Vic. Nat.*, **56**: 71-2. (September).

1940

'Which Australian Orchid has the Most Extensive Range of Habitat?', *AOR*, **5**: 13. (March).

'*Dendrobium linguiforme* var. *nugentii* Bail.', *AOR*, **5**: 17. (March).

'Two New Orchid Records in Queensland (*Pterostylis woollsii* and *Chiloglottis trapeziformis*)', *AOR*, **5**: 19. (March).

'Treasures of an Orchid Herbarium', *AOR*, **5**: 24-5. (March).

'The Eucaladenias of New South Wales' (illust.), *AOR*, **5**: 83-7. (September).

'A Note on *Dendrobium taylori* (F. Muell.) FitzG. and *D. hispidum* A. Rich.', *AOR*, **5**: 96. (December).

'A New Australian Species of *Pterostylis* (*P. allantoidea* Rogers)', *AOR*, **5**: 96. (December).

'Two New *Caladenia* Records for N.S.W. (*C. cardiochila* and *C. praecox*)', *AOR*, **5**: 112. (December).

'*Thrixspermum album* (Ridl.) Schltr., the North Queensland Orchid hitherto known as *Cleisostoma congestum* Bailey', *N.Q. Nat.*, **9**(64): 1–3.

'A Distinctive Form of *Caladenia carnea* R.Br. (var. *gracillima* var. nov.)' (illust.), *Q. Nat.*, **11**: 86–7.

'A Note on Two Tasmanian Caladenias' (*C. pallida* and *C. patersonii* var. *rosea*; illust.), *Vic. Nat.*, **56**: 142–3. (January).

'A Puzzling New South Wales Orchid' (*Liparis habenarina*; illust.), *Vic. Nat.*, **57**: 41–2. (June).

'A New *Diuris* of the Inland' (*D. colemanae* sp. nov.; illust.), *Vic. Nat.*, **57**: 63–5. (July).

1941

'Memories of an Orchid Lover: I', *AOR*, **6**: 41–2. (June).

'Memories of an Orchid Lover: II', *AOR*, **6**: 64–5. (September).

'*Epipogum roseum*', *AOR*, **6**: 66. (September).

'Contributions to the New South Wales *Orchidaceae*' (with descriptions of two new species, *Pterostylis longicurva* and *Diuris rhomboidalis*; illust.), *Contrib. N.S.W. Nat. Herb.*, **1**: 125–8.

'Fifty Years Ago', *Corian*, May, pp.96–8.

'The Orange-blossom Orchid (*Sarcochilus falcatus*). *N.Q. Nat.*, **10**(66): 1.

'*Dendrobium wilkianum* sp. nov.' (illust.), *N.Q. Nat.*, **10**(67): 4.

'Some New South Wales Orchid Records' (*re* list compiled by N. A. Wakefield), *Vic. Nat.*, **57**: 169–70. (January).

'An Orchid Novelty' (*Prasophyllum rogersii*), *Vic. Nat.*, **57**: 206. (April).

'The Breaking-up of the Genus *Cleisostoma* (Orchidaceae) in Australia', *Vic. Nat.*, **57**: 216–20. (April).

'Two New Species of *Prasophyllum*' (*P. trifidum* sp. nov., and *P. aureoviride* sp. nov.), *Vic. Nat.*, **58**: 21–2. (June).

'An Early Victorian Botanist' (John George Robertson), *Vic. Nat.*, **58**: 30–1. (June).

'The Breaking-up of the Genus *Cleisostoma* in Australia' (*Sarcanthus purpuratus* comb. nov), *Vic. Nat.*, **58**: 41. (July).

'A New Species of *Corysanthes*' (*C. fordhamii* sp. nov.; illust.), *Vic. Nat.*, **58**: 83–4. (October).

1942

'Notes on New South Wales Orchids: Additions and Alterations to Previously Published Lists', *Aust. Nat.*, **11**: 45–8.

'Brief Notes on Some Native Orchids', *AOR*, **7**: 19. (March).

'In Memoriam: Richard Sanders Rogers, M.A., M.D., D.Sc. (1862–1942)' (portrait), *AOR*, **7**: 22–4. (June).

'The Occurrence of Giants and Dwarfs in Certain Species of Australian Orchids', *AOR*, **7**: 50. (September).

'*Caladenia fitzgeraldii* nom. nov.', *AOR*, **7**: 64. (December).

'*Dendrobium quadrilobium* Rolfe?', *N.Q. Nat.*, **10**(68): 1–2.

'A Supplementary Note on *Dendrobium wilkianum*', *N.Q. Nat.*, **10**(68): 4.

'A Previously Undescribed Orchid from North

Queensland' (*Saccolabium tierneyanum* sp. nov.; illust.), *Q. Nat.*, **12**: 18–19.

'Note on *Dendrobium carrii* Rupp et White', *Q. Nat.*, **12**: 19–20.

'Note on *Pterostylis daintreyana* and *Pterostylis parviflora* in Queensland', *Q. Nat.*, **12**: 35.

'*Prasophyllum woollsii* F.v.M.', *Vic. Nat.*, **58**: 134. (January).

'Robert Brown's *Lyperanthus ellipticus*' (with description of *Rimacola* gen. nov.), *Vic. Nat.*, **58**: 187–8. (April).

'Notes on Certain Species of *Caladenia*', *Vic. Nat.*, **58**: 197–9. (April).

'*Corybas* or *Corysanthes*?', *Vic. Nat.*, **59**: 60–1. (August).

'The Section *Genoplesium* in the Genus *Prasophyllum*: Parts I and II' (*P. horburyanum*, *P. elmae*, *P. nichollsianum*, *P. unicum*, *P. sagittiferum*, *P. wilsoniense* and *P. plumosum* (nov. spp.; illust.), *Vic. Nat.*, **59**: 121–8, 137–40. (November).

1943

'*Epidendrum obrienianum*', *AOR*, **8**: 14. (March).

'The Determination of Species', *AOR*, **8**: 41. 47. (September).

'*Dendrobium dicuphum* F. Muell.' (illust.), *N.Q. Nat.*, **11**(69): 2–3.

'Additions to the *Orchidaceae* of New South Wales', (*Pterostylis longipetala* sp. nov. and *Thelymitra rubra* var. *magnanthera* var. nov.), *Proc. Linn. Soc. N.S.W.*, **68**: 9–10.

'A New Terrestrial for South Queensland' (*Prasophyllum parvicallum* sp. nov.; illust.), *Q. Nat.*, **12**: 52–3.

'The Section *Genoplesium* in the Genus *Prasophyllum*: Part III. Notes and New Records, etc.', *Vic. Nat.*, **59**: 160–3. (January).

'A New Species of *Calochilus*' (*C. gracillimus* sp. nov.), *Vic. Nat.*, **60**: 28. (June).

1944

'The Native Orchids of the County of Cumberland, New South Wales', *AOR*, **9**: 57–60. (December).

'*Dipodium punctatum*, var. *stenocheilum* (Schwarz) stat. nov.', *N.Q. Nat.*, **11**(71): 2–3.

'*Dendrobium adae*' (illust.), *N.Q. Nat.*, **11**(72): 3.

'Notes on Australian Orchids: IV' (*Diuris punctata* var. *sulfurea* var. nov., *Sarcochilus fitzgeraldii* var. *aemulus* var. nov.), *Proc. Linn. Soc. N.S.W.*, **69**: 73–5.

'A Critical Revision of R. D. Fitzgerald's *Australian Orchids*', *Proc. Linn. Soc. N.S.W.*, **69**: 274–8.

'What Constitutes a New Botanical Species?', *Vic. Nat.*, **60**: 135–8. (January).

'A New Species of *Thelymitra*' (*T. retecta* sp. nov.; illust.), *Vic. Nat.*, **60**: 176. (March).

'A New Species of *Pterostylis* from Portland', (*P. celans* sp. nov.; illust.), *Vic. Nat.*, **61**: 106–7. (October).

'The Orchid *Acianthus fornicatus*', *Vic. Nat.*, **61**: 125–6. (November).

1945

'A Sydney Surburban *Pterostylis*' (*P. mitchellii* Lindl.; illust.), *AOR*, **10**: 19–20. (March).

'Memories of an Orchid Lover: III', *AOR*, **10**: 31–2. (June).

'What is an Orchid?' (illust.), *AOR*, **10**: 52–4. (September).

'*Dendrobium bifalce* Lindl.', *N.Q. Nat.*, **12**(76): 1–2.

'New Species of *Dendrobium* from Babinda' (*D. luteocilium* sp. nov.; illust.), *N.Q. Nat.*, **13**(77): 1–2.

'Relation of the Orchid Flora of Australia to that of New

Zealand, with Description of a New Orchid Genus for New Zealand', (*Aporostylis* gen. nov.; in collaboration with E. D. Hatch), *Proc. Linn. Soc. N.S.W.*, **70**: 53–61.

'Notes on New South Wales Orchids: A New Species and Some New Records' (*Thelymitra purpurata* sp. nov.; illust.), *Proc. Linn. Soc. N.S.W.*, **70**: 288–90.

'A Papuan Orchid—*Dendrobium coplandii* (Bailey) comb. nov.' (illust.), *Proc. Roy. Soc. Q.*, **56**: 125.

'*Dendrobium adae* F. M. Bailey' (illust.), *Q. Nat.*, **12**: 114.

'A New Species of *Dendrobium* from North Queensland', (*D. ancorarium* sp. nov.; illust.), *Q. Nat.*, **12**: 115–16.

'*Dendrobium gracillimum*, stat. nov.', *Vic. Nat.*, **61**: 200–1. (March).

'Australian Orchids, 1900–1945 (Additions to R. D. FitzGerald's "Australian Orchids", 1875–1895)', *Vic. Nat.*, **62**: 65–73, 92. (August–September).

1946

'The Study of Orchidology in Australia', *AOR*, **11**: 10–11. (March).

'Some Orchid Questions: Answers to Queries from Major R. S. Davis, U.S. Army, Manila', *AOR*, **11**: 18. (March).

'Notes on the Australian Hyacinth Orchid, *Dipodium punctatum*' (illust.), *AOR*, **11**: 32–4. (June).

'Two New Orchids from Western Australia' (*Monodenia australiensis* sp. nov. and *Thelymitra cucullata* sp. nov.; illust.), *AOR*, **11**: 70–2. (September).

'An Orchid Garden in Launceston, Tasmania' (Neil Burrows), *AOR*, **11**: 73–4. (September).

'Some Recollections of an Old Boy', *Corian*, December, pp.199–200.

'Something about Australian Orchids' (illust.), *Journal of the New York Botanical Garden*, **47**: 172–80. (July).

'A New Orchid Genus and Species from North Queensland' (*Monilabium hamatum* gen. et sp. nov.; illust.), *N.Q. Nat.*, **13**(78): 2–4.

'A Review of the Species *Caladenia carnea* R.Br.' (including description of *C. aurantiaca* stat. nov.), *Proc. Linn. Soc. N.S.W.*, **71**: 278–81.

'Notes on Australian Orchids: V. A Review of the Genus *Calochilus* R.Br., *Acianthus caudatus* var. *pallidus* var. nov., and *Caladenia carnea* var. *minor et exigua*' (N.Z. forms in Aust.; illust.), *Proc. Linn. Soc. N.S.W.*, **71**: 287–91.

'*Dendrobium adae* F.M. Bailey and *D. ancorarium* Rupp: A Dimorphic Species?' *Q. Nat.*, **13**: 12–13.

'Orchid Lore' (letter to editor), *SMH*, 19 March.

'*Pterostylis boormanii* Rupp', (illust.), *Vic. Nat.*, **62**: 159–62. (January).

'An Enterprising Orchid' (*Epidendrum obrienianum*), *Vic. Nat.*, **62**: 188. (February).

'Introduction to F. Fordham's paper "Pollination of *Calochilus campestris*"', *Vic. Nat.*, **62**: 199. (March).

'Australian and New Zealand Orchids', *Vic. Nat.*, **63**: 45–7. (June).

'Notes on the Purple *Diuris*', *Vic. Nat.*, **63**: 178–80. (December).

1947

'A Review of the Species *Pterostylis parviflora* R.Br.', *Aust. Nat.*, **11**: 160–2.

'The Orchid Flora of the Blue Mountains of New South Wales', *AOR*, **12**: 32–3. (June).

'A Review of the Genus *Dendrobium* in Australia' (in collaboration with T. E. Hunt), *Proc. Linn. Soc. N.S.W.*, **72**: 233–51.

'A Missing *Pterostylis*' (illust.), *Q. Nat.*, **13**: 56.

'A Variable *Diuris*' (*D. aurea*; illust.), *Vic. Nat.*, **63**: 193–5. (Jan.).

'A New Species of *Prasophyllum*' (*P. uroglossum* sp. nov.), *Vic. Nat.*, **64**: 3. (March).

'The Epiphytic Orchids of Victoria and Tasmania', *Vic. Nat.*, **64**: 117–19. (October).

1948

'Nomenclature of *Calanthe veratrifolia* R Br.', *N.Q. Nat.*, **15**(86): 18–19.

'Some Notes on the Distribution of Orchids—Species Common to North Queensland and S.E. Australia', *N.Q. Nat.*, **16**(89): 2–5.

'The Orchid Flora of the Central Western Slopes of New South Wales' (with descriptions of new species of *Diuris*, viz. *D. althoferi*, *D. cucullata*, *D. cuneilabris*, and *Caladenia caerulea* and var. *heliotropica* var. nov.; illust.), *Proc. Linn. Soc. N.S.W.*, **73**: 130–6.

'*Oberonia muelleriana* Schltr.', *Q. Nat.*, **13**: 116–18.

'A Natural Hybrid between Two Orchid Genera' (*Glossodia* and *Caladenia*), *Vic. Nat.*, **64**: 186. (February).

'Taxonomic Difficulties in the Genus *Diuris*', *Vic. Nat.*, **64**: 241–3. (April).

'The Monotypic Orchids of Australia and Tasmania', *Vic. Nat.*, **65**: 35–7. (June).

'The Section *Genoplesium* in the Genus *Prasophyllum*' (with *P. anomalum*, *P. bowdenae*, *P. mucronatum*, *P. mollissimum* and *P. obovatum* spp. nov.; illust.), *Vic. Nat.*, **65**: 141–54. (October).

'What is *Caladenia gracilis* R.Br.?' (illust.), *Vic. Nat.*, **65**: 173–6. (November).

'Some Orchids of Barrington Tops' (illust.), *Wild Life*, **10**: 251–2.

1949

'(*Dendrobium*) *bigibbum* or *phalaenopsis*?', *AOR*, **14**: 34–5. (June).

'A Review of the Genus *Bulbophyllum* (Orchidaceae) in Australia' (with descriptions of new species; illust.; in collaboration with Trevor E. Hunt). *Proc. Roy. Soc. q.*, **60**: 55–68.

'Robert Brown's *Genoplesium baueri* (Orchidaceae)', *Vic. Nat.*, **66**: 75–9. (August).

'Giant Rocks (Wallabadah Rock near Blandford, N.S.W.)' *Vic. Nat.*, **66**: 89. (September).

'"By Their Fruits" by Margaret Willis' (book review), *Vic. Nat.*, **66**: 96–7. (September).

1950

'Schlechter's Orchid Specimens in the National Herbarium of N.S.W.', *AOR*, **15**: 45. (June).

'In Memoriam: Oakes Ames', *AOR*, **15**: 65. (September).

'A Memoir of Robert Brown', *AOR*, **15**: 110. (December).

'The Schlechter Collection of *Orchidaceae* in the National Herbarium of New South Wales', *Contrib. N.S.W. Nat. Herb.*, **1**: 304–11.

'Notes on Australian Orchids, with Descriptions of Three New Species' (*Bulbophyllum waughense*, *Diuris latifolia* and *Prasophyllum albiglans*; illust.), *Contrib. N.S.W. Nat. Herb.*, **1**: 317–21.

'Note on *Bulbophyllum evasum*' (illust.), *N.Q. Nat.*, **17**(94): n.p.

1951

'Australian Orchids', *Australian Junior Encyclopaedia*, **2**: 705–9.

'In Memoriam: William Henry Nicholls (1885–1951)'
(portrait), *AOR*, **16**: 84–6. (September).
'In Memoriam: Edith Coleman', *AOR*, **16**: 122.
(December).
'A Review of the Australian Species of *Sarcochilus*
(Orchidaceae) including. . .a new species, *S.
tricalliatus*', (illust.), *Proc. Linn. Soc. N.S.W.*, **76**:
49–56.
'Proposed New Genus of Orchids' (*Rhinerrhiza* gen.
nov.), *Vic. Nat.*, **67**: 206–10. (February).
'In the Althofer Country', *Vic. Nat.*, **68**: 67–8. (August).
'Resupination in Orchids', *Vic. Nat.*, **68**: 75–6.
(September).
1952
'Robert David Fitzgerald', *AOR*, **17**: 22–3. (March).
' ''Wild Orchids of Britain'' by V.S. Summerhayes'
(book review), *AOR*, **17**: 52. (June).
'The Genus *Bulbophyllum* in Australia', *AOR*, **17**: 97–8.
(September).
'*Saccolabium loaderanum* sp. nov.' (illust.), *N.Q. Nat.*,
20(101): 17–18.
1953
'A New Australian *Dendrobium*' (*D. masonii* sp. nov;
illust.) *AOR*, **18**: 18 (March).
'A Beautiful *Dendrobium* from New Guinea' (*D.
ostrinoglossum* sp. nov.), *AOR*, **18**: 58. (June).
'A New Species of *Eria*' (*E. linariiflora*; illust.), *AOR*, **18**:
63, 67. (June).
'Another New *Saccolabium* (Orchidaceae) from North
Queensland' (*S. subluteum* sp. nov.; illust.), *N.Q. Nat.*,
21(105): n.p. (June).
'A Note on the Orchid *Saccolabium brevilabre* (F. Muell.)

Rupp.' (illust.), *N.Q. Nat.*, **21**(105): n.p. (June).
'Notes on Australian Orchids' (including *Dendrobium
elobatum* sp. nov.; illust.), *Vic. Nat.*, **69**: 116–20.
(January).
'Memories of Victorian Orchids', *Vic. Nat.*, **69**: 145–6.
(March).
'*Drymoanthus minutus* (Orchidaceae)' (illust.), *Vic. Nat.*,
69: 155. (April).
'A New Species of *Chiloglottis*' (*C. dockrillii* sp. nov.;
illust.), *Vic. Nat.*, **70**: 54–5. (July).
'The Winter Spider Orchid, a New Species from
Wongan Hills' (*Caladenia glossodiphylla* sp. nov.; illust.;
with Rica Erickson), *Western Australian Naturalist*, **4**:
64–5. (December).
1954
'Orchid Notes from North of the Murray' (including
Caladenia holmesii sp. nov.; illust.), *Vic. Nat.*, **70**:
179–80. (February).
1955
'The Naming of Orchid Species', *AOR*, **20**: 80. (June).
'Orchid Notes from New South Wales and Queensland'
(including *Cryptostylis hunteriana* Nicholls, *Dendrobium
glabrum* J.J. Smith and *Cymbidium gomphocarpum*
FitzG.), *Vic. Nat.*, **72**: 14. (May).
'Additional Species of the Genus *Diuris* (Orchidaceae) in
New South Wales' (*D. maculosissima*, *D. goonooensis*
and *D. curtifolia* spp. nov.), *Vic. Nat.*, **72**: 110–11.
(November).
1958
'Orchids' (by R. S. Rogers for 1st edn, 1926; revised by
H. M. R. Rupp, 1952), *Australian Encyclopaedia*, **6**:
408–13.

B. Unpublished Material

1892
Catalogue of Genera & Species of Native Plants
obtained in the District of Coleraine (exclusive of
the Wando Valley). J. W. Willis Collection,
Melbourne.
Catalogue of Indigenous Genera & Species obtainable
near Wando Dale (122 species). Date may be 1894.
J. H. Willis Collection, Melbourne.
Orchid Records: I. Coleraine, Victoria. MS exercise
book. RBGS.
1892–194–
Records of Orchids, 1892–194–. MS exercise book.
RBGS.
1894–95
A Catalogue of Indigenous Species of Plants from Hay
District, Murrumbidgee River, Riverina, N.S.W.
Species collected during November and December
1894, and January and February 1895. RBGS.
1896
A Census of Native Plants to be found in or about the

Neighbourhood of Buninyong. J. H. Willis Collection,
Melbourne.
Catalogue of Plants Merri Merri Creek (North
Coburg & District). J. H. Willis Collection,
Melbourne.
Orchid Records: II. Merri Merri Creek, Victoria. MS
exercise book. RBGS.
Orchid Records: III. Buninyong, Victoria. MS exercise
book. RBGS.
1897
Descriptive Catalogue of Native Victorian Plants in the
Rusden Museum, Trinity College, Melbourne (148
species). J. H. Willis Collection, Melbourne.
Kingston and Surrounding Localities (188 species). J. H.
Willis Collection, Melbourne.
1898
A Census of Native Victorian Plants obtained in the
Neighbourhood of BEEAC; and in the adjoining
localities. J. H. Willis Collection, Melbourne.

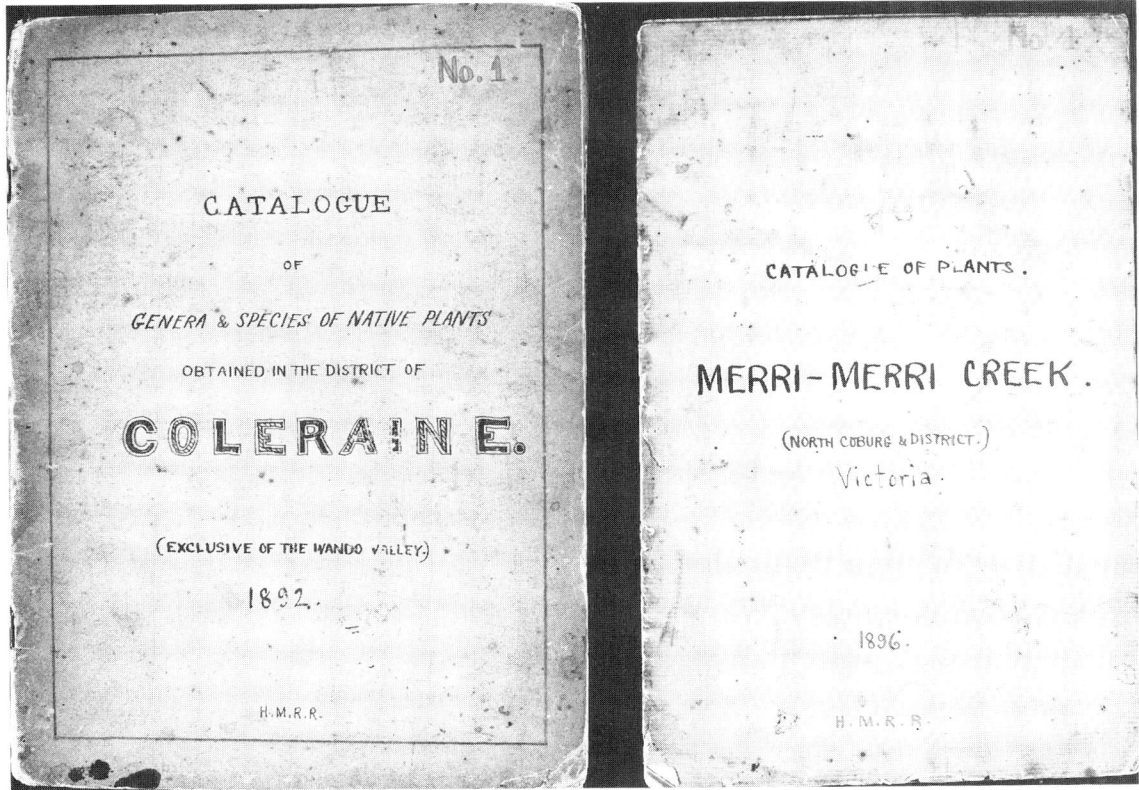

Rupp's botanical notebooks numbers 1 and 4, listing plants of the Coleraine and Merri Creek districts, 1892 and 1896. (By courtesy of J. H. Willis and I. C. Clarke)

1898–1902
  Orchid Records: IV. Beeac, Victoria. MS exercise book. RBGS.
1904
  Flora of Tamworth (with map). MS exercise book. RBGS.
  Orchid Records: V. Tamworth, N.S.W. MS exercise book. RBGS.
1905
  Orchid Records: VI. Warialda, N.S.W. MS exercise book. RBGS.
  Plants of Burnett & Stapylton Counties (Warialda) and Lane Cove, N.S.W. RBGS.
  Warialda—Native Plants found in the Counties of Burnett & Stapylton and part of Arrawatta. MS notebook. RBGS.
1911
  Native Plants collected in the Copmanhurst District, Upper Clarence River District, N.S.W. MS notebook. RBGS.
  Orchid Records: VII. Copmanhurst, N.S.W. MS exercise book. RBGS.
1914
  Native Plants collected at Glen Innes & Bolivia. January. MS exercise book. RBGS.

  Orchid Records: VIII. Barraba, N.S.W. MS exercise book. RBGS.
1914–19
  Orchid Records: IX. Lane Cove River, NSW. MS exercise book. RBGS.
1919–22
  Orchid Records: X. Tasmania. MS exercise book. RBGS.
1921–2
  Flora of Launceston, Mt Barrow, Turner's Marsh and the Lower Tamar, Northern Tasmania. RBGS.
1923–4
  Census of Plants collected in the Parish of Bulahdelah, N.S.W. April 1923–September 1924 (with map and flowering times). RBGS.
  Orchid Records: XI. Bulahdelah, N.S.W. MS exercise book. RBGS.
  Orchids of Bulahdelah District. Single leaf MS. RBGS.
1924–5
  Alphabetical Index of the Herbarium of Australian Plants of the Rev. H. M. R. Rupp (with 18 pp. of personal, botanical and geographical notes). Large foolscap MS. Loader Collection, copy in RBGS.
  Orchids of the Paterson District to 10 July 1925. 4 pp. MS. RBGS.

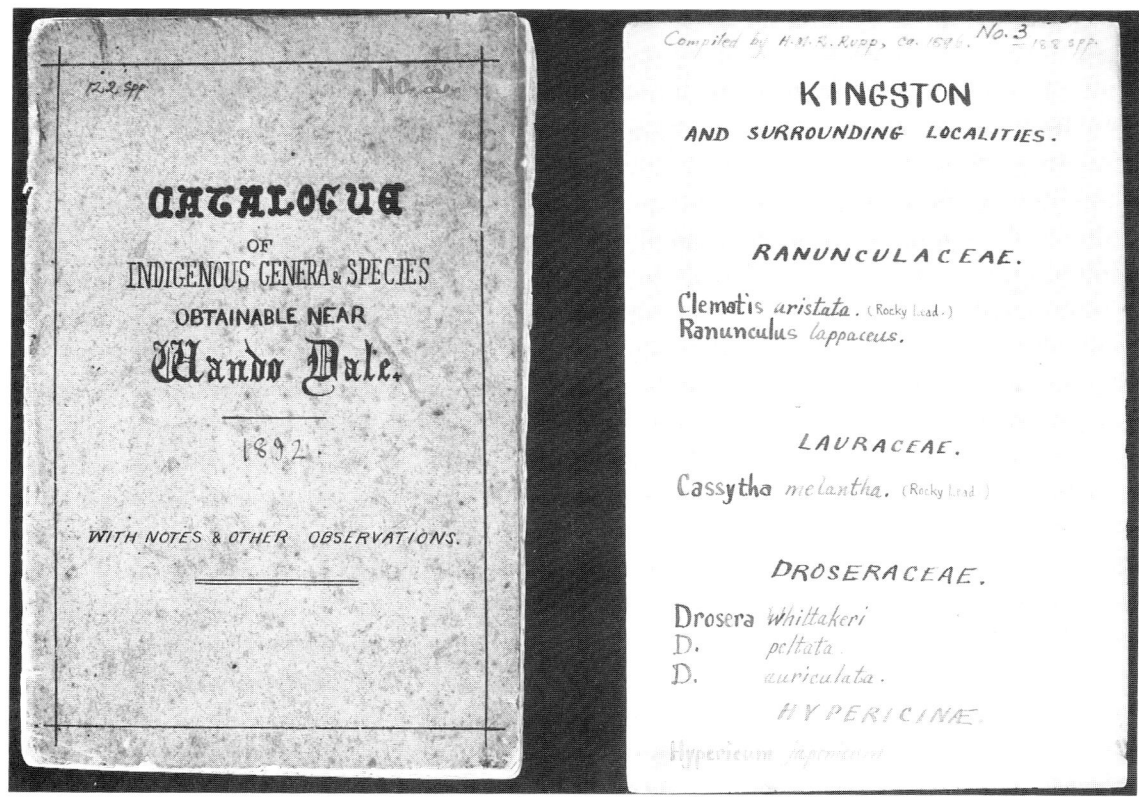

Rupp's botanical notebooks numbers 2 and 3, listing plants of the Wando Dale and Kingston districts, 1892 (or 1894) and 1897. (By courtesy of J. H. Willis and I. C. Clarke)

1924-9
Flora of Paterson District, N.S.W. MS exercise book.
  RBGS.
Orchid Records: XII. Paterson, N.S.W. MS exercise
  book. RBGS.
1925
Alphabetical Index to the Herbarium of Rev. H. M. R.
  Rupp (Orchids excluded). MS notebook. Botany
  Department, University of Melbourne.
1925, 1928
Orchid Records: XIII. Eccleston & Barrington Tops,
  N.S.W. (Oct. 1925, Jan. 1928). MS exercise book.
  RBGS.
1926
Notes on Various Plants Represented in the Herbarium
  of the Rev. H. M. R. Rupp, B.A. (Paterson, Jan.).
  MS notebook. Botany Department, University of
  Melbourne.
1930-2
Orchid Records: XIV. South Maitland Coalfields. MS
  exercise book. RBGS.
1932
Indigenous Flora of the Pilliga Scrub (N.S.W.) observed
  by H.M.R. Rupp Sept.-Dec. RBGS.

1932-3
My Australian Hobby or The Recollections of an
  Amateur Botanist. Manuscript in possession of Mrs
  Rachel Cox, Armidale, N.S.W. Written December
  1932 to January 1933.
1933
The Genus *Pterostylis* . . . in Australia. Notes on
  Classification and Distribution. RBGS.
1933-6
The Orchid Flora of Brisbane Water. List of Genera and
  Species. Typescript. Loader Collection. Copy in
  RBGS.
Orchids of Brisbane Water. MS exercise book. RBGS.
Orchid Records: XVII. Woy Woy, N.S.W. MS exercise
  book. RBGS.
1934
Orchid Records: XV. Bellinger Valley & Dorrigo
  Highlands, N.S.W. Oct. MS exercise book. RBGS.
1936
Orchid Records: XVI. Brunswick River, N.S.W. Aug.-
  Sept. MS exercise book. RBGS.
1936-9
Orchid Records: XVIII. Raymond Terrace, N.S.W. MS
  exercise book. RBGS.

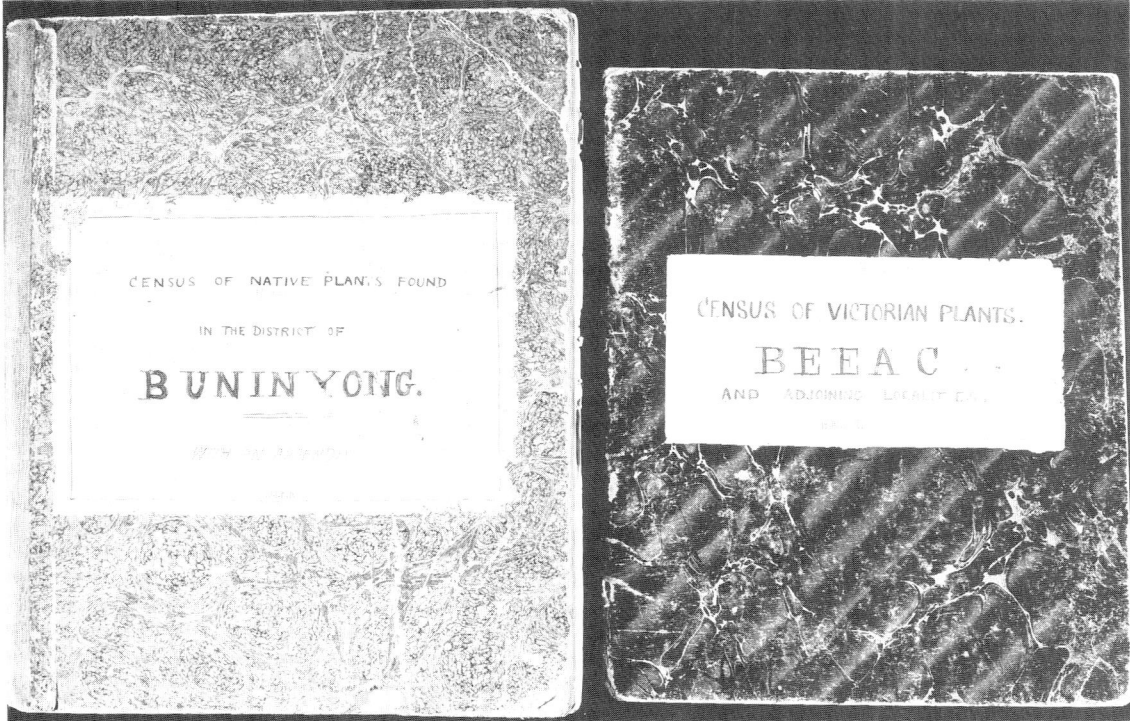

Rupp's botanical notebooks listing plants of the Buninyong and Beeac districts, 1896 and 1898. (By courtesy of J. H. Willis and I. C. Clarke)

1939–194–

Orchid Records: XIX. Middle Harbour–French's Forest, N.S.W. MS exercise book. RBGS.

1946

The Unpublished Plates of R. D. Fitzgerald's 'Australian Orchids' in the Mitchell Library, Sydney. Typescript November. Filed with D 248, Mitchell Library, State Library of New South Wales.

1948

The Orchid Flora of New England. Single leaf, typescript. With Fordham correspondence. RBGS.

1948–9

Restrospect. Typescript—copies in possession of Mrs Rachel Cox, Armidale, N.S.W. and of the late Mrs Eileen Cox, Noosa, Qld. Written November 1948 to January 1949.

1950?–1951

Rev. H. M. R. Rupp...Personal History, by self. Single leaf, typescript. Brief biography, two copies, almost identical but one typed after the move to Willoughby in October 1951. One copy in possession of Mrs Rachel Cox, Armidale; the other in J. H. Willis Collection, Melbourne. Photocopies of both lodged in RBGS.

Undated

H. M. R. Rupp: Herbarium of Australian and New Zealand Orchids. n.d. List of Species. MS exercise book. (Cover inscribed 'The Orchid Herbarium of H. M. R. Rupp'.) RBGS.

Notes

Part I: The Orchid Man

Chapter 1: Despair and Deliverance

1. More fully, the *G. H. Wappaus,* apparently named for a Hamburg merchant and shipowner, George Heinrich Wappaus (from Mr Robert Wuchatsch, Melbourne).
2. Contemporary newspaper reports vary.
3. *Melbourne Morning Herald,* 9 March 1849.
4. *Melbourne Daily News,* 14 March 1849.
5. *Melbourne Morning Herald,* 16 March 1849; *Argus,* 16 March 1849.
6. *Argus,* 16 March 1849. C. H. Ebden (1811–67), pastoralist and businessman, aided the cause of German immigration and was elected three times to represent Port Phillip in the N.S.W. Legislative Council. See *ADB,* 1, pp. 349–51.
7. See *Melbourne Morning Herald, Melbourne Daily News* and *Argus,* March 1849, *passim.*
8. *Melbourne Morning Herald, Melbourne Daily News,* 7 March 1849.
9. G. M. Trevelyan, *English Social History,* London 1946, p. 557.
10. R. N. Wuchatsch, *Westgarthtown: The German Settlement at Thomastown,* Melbourne 1985, p. 5.
11. *Op. cit.,* p. 3.
12. See *ADB,* 2, pp. 33–4.
13. Wuchatsch, *op. cit.,* pp. 3, 5.
14. Westgarth, *Personal Recollections of Early Melbourne and Victoria,* Melbourne 1888, p. 119.
15. *Op. cit.,* p. 121.
16. *Ibid.*
17. It arrived at Melbourne on 11 February 1849. *Argus,* 13 February 1849.
18. Westgarth, letter of 17 October 1848. *HRA,* I, Vol. XXVI, p. 709.
19. *Melbourne Daily News, Melbourne Daily Herald,* 7 March 1849. The *Port Phillip Gazette* recorded the date of departure as 28 October.
20. Westgarth, letter of 17 October 1848, *loc. cit.* For evidence of the differences between Westgarth, Delius, Earl Grey and the Colonial Land and Emigration Office, see *HRA,* I, Vol. XXVI, pp. 662–6; 706–13. See also Lieutenant-Governor C.J. La Trobe to Earl Grey, 17 February 1852, in Despatches from Governor of Victoria, 1852, p. 709. ML, A2341.

21. Westgarth, *Personal Recollections,* p. 121.
22. Family details from official sources, e.g., Emigration Papers, in N.S.W. State Archives, AO 9/2662 and AO COD114; Marriage Certificates of C. L. H. Rupp; Naturalization Papers, Australian Archives, Canberra, CRS A712, Item 97/E5944.
23. Family tradition is that the deaths occurred during the voyage, although Edmund Finn ('Garryowen') in his *Chronicles of Early Melbourne 1835 to 1852,* Melbourne 1888, recorded that the children 'lost their parents embarking at Hamburg' (II, p. 660).
24. See the second introductory paragraph of H. M. R. Rupp's 'Retrospect'.
25. Certificate of Baptism (copy of 20 October 1862) in possession of Mrs R. M. Cox, Armidale (kindly translated by Mr R. K. O. Eckermann, Armidale). Mr David Thiele of St Paul's College, Walla Walla, N.S.W. has not as yet discovered documentary evidence of any special links between the Thiele and Rupp families on the *Wappaus.* However, Gottlieb Thiele, a tailor, was 'the unofficial leader of the "Old Lutheran" party on the ship', and his subsequent activity in visiting other German settlers around Melbourne, and greeting new German arrivals on the docks, was consistent with making representations on behalf of the orphaned children.
26. *Review of Reviews* (Australian edn), 20 August 1895, p. 147.
27. Isaac Selby, *The Old Pioneers' Memorial History of Melbourne,* Melbourne 1924, p. 128.

Chapter 2: New Kith and Kin

1. Shipping Lists, MB2/39/3, Archives Office, Hobart; Naturalization Certificate, CRS A725, Vol. 1, p. 4, Australian Archives, Canberra; *Hobart Town Courier,* 6 November 1835; Death Certificate of 3 March 1882 (3545); *Biog. Reg.,* II, p. 276.
2. Advertisement dated 26 November 1835, published in *Hobart Town Courier,* 4 and 11 December 1835.
3. Shipping Lists, MB2/43, Archives Office, Hobart; Naturalization Certificate, *loc. cit.*
4. F. P. Labilliere, *Early History of the Colony of Victoria,* London 1878, II, pp. 172–4; see *HRA* I, vol.XIX, pp.87–8.

5. H. G. Turner, *A History of the Colony of Victoria*, Melbourne (1904) 1973, I, p. 394; R. R. McNicholl, *The Early Days of the Melbourne Club*, Melbourne 1976, p. 10.

6. *Melbourne Advertiser* (No. 1), 1 January 1838. See also Nos 2 and 3, 8 and 15 January 1838.

7. *Op. cit.*, No. 8, 19 February 1838. Rucker's advertisement was drafted 6 February 1838 — some issues are missing. Issue No. 8 also advised that Rucker was currently landing an extended range of goods.

8. Charles Swanston (1789-1850), managing director of Derwent Bank, 1831-49. For this story it is interesting to note that he was 'largely responsible for the endowment of Geelong Grammar School'. *ADB*, 2, pp. 500-1.

9. See W. F. A. Rucker, Business Papers, La Trobe Library, State Library of Victoria, Boxes 38, 95, 96, 109. These records were presented by Rucker's son, W. S. Rucker, after his father's death in 1882. The initial capital of the Melbourne Agency of the Derwent Bank comprised £900 in notes, £100 in silver and £5 in copper.

10. *Melbourne Advertiser*, 1 January 1838, *et seq.* See also P. L. Brown (ed), *Clyde Company Papers*, Vol. II, *1836-1840*. London 1952, p. 146.

11. Ian Wynd, *Geelong: The Pivot*, Melbourne 1971, p. 12; R. D. Boys, *First Years in Port Phillip 1834-1842*, Melbourne 1959, pp. 58-9; *Investigator* (Geelong), November 1967, pp. 245-6. Champion shortly ran the store on his own account. He became a magistrate and an early trustee of Geelong Grammar School. P. L. Brown, *Geelong Grammar School: The First Historical Phase*, Geelong 1970, p. 21.

12. McNicholl, *op. cit.*, p. 2.

13. See *Port Phillip Gazette* between 27 October 1838 and 19 January 1839.

14. Edmund Finn, *Chronicles of Early Melbourne 1835 to 1852*, Melbourne 1888, II, p. 542.

15. Paul de Serville, *Port Phillip Gentlemen*, Melbourne 1980, p. 65; McNicholl, *op. cit.*, p. 97.

16. C. P. Billot, *The Life and Times of John Pascoe Fawkner*, Melbourne 1985, p. 213; Finn, *op. cit.*, I, p. 88.

17. *Port Phillip Gazette*, 6 July 1839.

18. De Serville, *op. cit.*, p. 212.

19. Finn, *op. cit., passim*; De Serville, *op. cit.*, p. 212.

20. McNicholl, *op. cit.*, pp. 16-17; De Serville, *op. cit.*, p. 111, where the year is given as 1840.

21. Andrew Lemon, *The Northcote Side of the River*, Melbourne 1983, pp. 9, 10.

22. The building was destroyed in 1925 to make way for the Anglican Church of the Epiphany (later Eastern Orthodox). Lemon, *op. cit.*, pp. 9-13; W. G. Swift, *The History of Northcote*, Northcote 1928, pp. 5, 12, 13.

23. J. B. Were & Son, *The House of Were*, Melbourne 1954, p. 30.

24. Finn, *op. cit.*, II, p. 991.

25. Finn, *op. cit.*, II, pp. 707-10, 991; *House of Were*, pp. 30-1. See also L. A. Schumer, *Henry Dendy and his Emigrants*, East Malvern 1975, pp. 39, 41; Billot, *op. cit.*, p. 231.

26. *Port Phillip Gazette*, 18 February 1845.

27. *Port Phillip Gazette*, 25 February 1843.

28. Lemon, *op. cit.*, p. 11.

29. Letter from John Gillon & Co, Leith, 23 May 1842, in Governor Gipps's Despatches, Vol. 45, p. 54. ML, A1234.

30. Lemon, *op. cit.*, p. 11: Rucker Papers, Box 38/1 La Trobe Library, State Library of Victoria.

31 Naturalization Certificate, CRS A725, Vol. 1, p. 4, Australian Archives, Canberra. N.S.W. Legislative Council XI Vic. 39.

32. Finn, *op. cit.*, I, p. 495.

33. Rucker Papers, *loc. cit.*

34. *Argus, Melbourne Morning Herald, Melbourne Daily News* and *Port Phillip Gazette*, 9 May 1850.

35. Finn, *op. cit.*, I. p. 448; *Argus*, 20 August 1850.

36. *Op. cit.*, II, p. 514.

37. St Peter's Marriage Register, 1851 (No. 351); *Argus*, 5 August 1851.

38. St Peter's Baptismal Register, 1852 (No. 1493); *Argus*, 3 July 1852. He was baptized William Sigismund (*sic*) on 26 November 1852.

39 Finn, *op. cit.*, I, p. 109.

Chapter 3: Country and Calling

1. H. M. R. Rupp, Biographical Notes. in possession of Mrs Rachel Cox, Armidale. In the absence (so far) of evidence of Mr Brookfield's school, one wonders whether this may have been the 'tolerably efficient establishment in South Swanston Street under the mastership of Mr Edward Butterfield ... an able though not personally popular individual'. (Edmund Finn, *Chronicles of Early Melbourne 1835 to 1852*, Melbourne 1888, II, p. 638.) This school moved to Stephen Street in mid-1850; it offered 'the Greek, Latin, French, German, and English languages, with the Mathematics, Drawing, and the usual branches of an English Education'. *Argus*, 10 May 1850; *Melbourne Morning Herald*, 9 May 1850.

2. *Glen Innes Examiner*, 4 October 1917; *The Church Chronicle* (Ballarat), 10 November 1917, p. 185.

3. *Argus*, 24 December 1850.

4. *Argus*, 26 December 1850.

5. Finn, *op. cit.*, II, p. 660.

6. *Glen Innes Examiner*, 4 October 1917; *Church Chronicle*, 10 November 1917.

7. Information from K. J. Cable, 8 December 1986.

8. For the examiners, see *ADB*, 4, and *Argus*, 5 August 1895. St James's was rebuilt on the corner of King and Batman Streets in 1914.

9. Record of Examinations for Holy Orders, Diocesan Archives, Melbourne.

10. Subscription Book, 1847-65, Diocesan Archives, Melbourne.

11. *The Church Gazette* (Melbourne) 1 January 1863, p. 203. The text is II Corinthians iv.1.

12. *ADB*, 3, pp. 217-19. It has been recorded that C. L. H. Rupp had been 'an old pupil of Dr Braim' (J. W. Powling, *Centenary 1856-1956: St John's Church, Port Fairy*, Ballarat, 1957, p. 26), but it is not known where or when.

13. Apparently named Mary Anne Catherine, after her mother, she signed the marriage register accordingly, but was known as Marie Ann. Marriage Certificate; *Geelong Advertiser*, 23 January 1865.

14. *ADB*, 2, p. 402; E. Morris Miller, *Australian Literature: A Bibliography to 1938* (extended to 1950 by F. T. Macartney), Sydney 1956, pp. 411-12.

15. *Geelong Advertiser*, 29 August 1876; 29 January 1878; 4 April 1960.

16. H. M. R. Rupp, Biographical Notes, from Mrs R. Cox, Armidale.

17. *Belfast Gazette*, 5 August 1864.

18. *Melbourne Church News*, 2 January 1868, p. 9; Niel Gunson: *The Good Country: Cranbourne Shire*, Melbourne 1968, pp. 66-7.

19. For clerical career, see *Yearbook of the Church of England in the Diocese of Ballarat for 1904*, p. 14; Crockford's *Clerical Directory*, 1921-4; *Church of England Messenger* (Melbourne), 19 October

1917. In September 1988 it was found that, despite population movements and changes in parish boundaries, handsome stone churches (not all built by Rupp's time) are still in use at Port Fairy (St John's), Learmonth (All Saints), Koroit (St Paul's), Coleraine (Holy Trinity) and Buninyong (Holy Trinity). The last service at Holy Trinity, Kingston, was held in 1969. Since 1970 the old building has been modified for domestic use after it was apparently perfunctorily closed and sold by the diocesan authorities. Its parsonage has gone.

20. Naturalization Papers, CRS A712 Item 97/E5994. Australian Archives, Canberra. Original Certificate of Naturalization (in possession of Miss Elizabeth Rupp, Willoughby, N.S.W.) is dated 6 July 1897.

21. *Church Chronicle,* 10 November 1917. Rupp was inducted to Learmonth (his last parish) for the second time on 12 October 1898. *Ballarat Church Chronicle,* 2 November 1898, p. 185.

22. *Church Chronicle for the Diocese of Ballarat,* 2 January 1905, p. 13; 1 February 1905, p. 26; 1 March 1905, p. 45; 1 April 1905, pp. 52–3. His address was 163 Lydiard Street North, Ballarat.

23. *Church Standard,* 5 October 1917, p. 2; *Church Chronicle,* 10 November 1917, p. 185. He had taken a special interest in the Lake Condah Aboriginal Mission, the Indian Mission, the New Guinea Mission and the Chinese Mission in Australia as represented by Cheok Hong Cheong (1853?–1928). (See *ADB,* 3, pp. 385–6.) The monument in Glen Innes Cemetery records his name as 'C. L. Herman Rupp'.

24. See *Argus,* 9 November 1921; 25 March 1935; 9 January 1947. The Pettets were buried in Brighton Cemetery.

Chapter 4: 'Tiresome Monkey' but Promising Naturalist

1. *Belfast Gazette,* 17 January 1873; Death Certificate of 13 January 1873 (No. 363).

2. *Belfast Gazette, loc. cit.* A marble monument (by Chambers & Clutten of Melbourne), in part inscribed 'Marie A. C. Rupp, the beloved wife of C. L. Herman Rupp', remains in good condition within a wrought-iron enclosure in Port Fairy Cemetery.

3. *Belfast Gazette,* 24 January 1873.

4. Baptismal Register: St John's, Belfast, 20 January 1873 (No. 1821).

5. H. M. R. Rupp to E. M. Miller, 26 August 1949. Moir Collection, Box 4, La Trobe Library, State Library of Victoria.

6. *ADB,* 1, pp, 329–30.

7. H. M. R. Rupp to E. M. Miller, 16 and 26 August 1949. Moir Collection, *loc. cit.* The vicarage at Coleraine was similar — an iron house, very cold in winter but 'a crematory in summer'. See *Church News* (Hamilton), June 1890, p. 312. The old Coleraine vicarage has long since been replaced by a handsome building, while the old Koroit vicarage (no longer a church residence) has been greatly modified.

8. The grave has a cross monument, erected by the widow, and a desk monument erected for the Bread and Cheese Club, Melbourne, by Markwell Bros in 1954–5, when the grave was refurbished. See Moir Collection *loc. cit.* In 1949 Edmund Morris Miller (1881–1964), bibliographer and one-time vice-chancellor of the University of Tasmania (*ADB,* 10, pp. 507–9) had considerable correspondence with H. M. R. Rupp relating to H.H.R. and Koroit. He sent Rupp a copy of his article 'Richard Mahony's Grave' before it was published in *Meanjin,* Spring 1949, pp. 177–80. See also *Bohemia,* Bread and Cheese Club, Melbourne, Vol. 12, No. 5, 1 November 1956.

9. Baptismal Register, Koroit, 12 November 1879, p. 41, No. 1146.

10. H. H. Richardson, *The Fortunes of Richard Mahony,* London 1917–29. For the Ruckers, see Book III, *Ultima Thule,* Part III, Chapter VII.

11. She married J. G. Robertson at the end of 1895, and in 1912 returned to Australia on 'a flying six-weeks' visit to test my memories'. E. Morris Miller, *Australian Literature* (extended to 1950 by F. T. Macartney), Sydney 1956, p. 401.

12. H. H. Richardson, *Myself When Young,* Melbourne 1948. See review by Uther Barker in *Southerly,* Vol. 10, No. 2, 1949, pp. 89–92, where various peculiarities of this work are noted.

13. H. H. Richardson, *Myself When Young,* p. 28.

14. *Op. cit.,* p. 29. Florence turned fourteen in 1880. See also Jean Uhl, *Still Stands the Schoolhouse by the Road,* Koroit 1987.

15. H. M. R. Rupp recalled that the policeman was Sergeant Gray, who had a daughter named Tilly. Letter to E. M. Miller, 16 August 1949. Moir Collection, *loc. cit.* The post office, police station, state school and courthouse still stand in their adjacent positions in the township.

16. H. M. R. Rupp to E. M. Miller, 26 August 1949. Moir Collection, *loc. cit.*

17. H. M. R Rupp to E. M. Miller, 8 August 1949. Moir Collection, *loc. cit.*

18. Florence Monypenny (née Rupp) in Nettie Palmer, *Henry Handel Richardson: A Study,* Sydney 1950, p. 14. Note also the photograph of the Richardsons (opposite p. 4) lent by Mrs Monypenny.

19. Rupp to Gwen Scrivener, 26 December 1948. RBGS.

20. H. M. R. Rupp to E. M. Miller, 8 August 1949. Moir Collection, *loc. cit.*

21. E. M. Miller to H. M. R. Rupp, 12 August 1949. Original in possession of Mrs R. Cox, Armidale.

22. H. M. R. Rupp to E. M. Miller, 26 August 1949. Moir Collection, *loc. cit.*

23. H. M. R. Rupp, 'Memories of an Orchid Lover', *AOR,* June 1941, p. 41. Rupp was not the only orchid enthusiast to be so motivated. See *AOR,* December 1945, p. 83.

24. H. M. R. Rupp to E. M. Miller, 8 August 1949. Moir Collection, *loc. cit.*

25. *Ibid.* Rupp identified these species as 'a large and richly-coloured form of *Caladenia patersonii* ... and the lovely *Diuris longifolia,* with its perfect blend of yellow, brown, and purple. These were my first loves among the Orchids, nor have they ever been despised among the multitude of their relatives which have come my way since.' *AOR,* June 1941, p. 41. Rupp later included *Caladenia dilatata.* See *Vic. Nat.,* March 1953, p. 145.

26. H. M. R. Rupp to E. M. Miller, 16 August 1949. Moir Collection, *loc. cit.*

27. H. M. R. Rupp to E. M. Miller, 17 August 1949. Moir Collection, *loc. cit.*

28. E.g., Florence Monypenny (née Rupp) and Charles Dod, 'the young postmaster who had initiated Mrs Richardson into her work'. Nettie Palmer, *op. cit.,* pp. 14–15.

29. H. M. R. Rupp to E. M. Miller, 26 August 1949. Moir Collection, *loc. cit.*

30. H. M. R. Rupp, Biographical Notes, from Mrs R. Cox, Armidale.

31. Death Certificate of 3 March 1882 (No. 3545). His home was at 31 Darling Street. See also *Argus,* 6 March 1882.

32. Matthew XXV.40.

33. Will of W. F. A. Rucker, 4 February 1882. Registry of Probates, Melbourne.

34. Probate documents of 5 April 1882. Registry of Probates, Melbourne.

35. J. R. Hopkins (1827–97) served on the Winchelsea Council for thirty-two years, and in the Victorian Legislative Assembly. He was previously (1850) married to Eliza Ann Armytage (d. 1885) by whom he had thirteen children, and later (1892) to Alice Roberta Purkiss who survived him. *ADB*, 4, pp. 420–1.

36. *Geelong Advertiser*, 14 November 1889; Death Certificate of 13 November 1889 (No. 16739). The gravestone in the Eastern Cemetery, Geelong, situated not far from those of H. M. R. Rupp's grandparents, the Rowcrofts, and of his aunt and uncle, the Wilsons, states her age as fifty-four.

37. After service with several banks, W. S. Rucker (1852–1901) took over his father's accountancy business in 1882. In 1889 he was joined by Robert William Berry Mackenzie to form the firm of Rucker and Mackenzie, Public Accountants, 59 Queen Street, Melbourne. See James Smith (ed.), *The Cyclopedia of Victoria*, Melbourne 1903, I, p. 398.

38. *Argus*, 10 June, 1901.

39. *The Hamilton and Western District Church News,* 26 March 1890, p. 289.

40. H. M. R. Rupp to E. M. Miller, 26 August 1949. Moir Collection, *loc. cit.*

41. C. L. H. Rupp, 'Episodes in a Holiday Excursion', *Church News* (Hamilton), August 1891–December 1892, drawn from The Diary of the Reverend C. L. H. Rupp on a Tour of England, Scotland, Europe, and the Holy Land, 1890–1891, 2 vols. Typescript, in possession of Miss E. Rupp, Willoughby, N.S.W.

Chapter 5: Geelong Days

1. A. P. Rowcroft, son of Horatio, was a brother of Oriana Maria Wilson and of Mary Anne (Marie Ann) Rupp, Montague's mother. After eight years as commercial master as the senior school, he opened the Junior Grammar School in 1878 and remained in charge until the end of 1890. *Church of England Grammar School, Geelong: History and Register, Jubilee, 1907* (unpaged). The old Junior School bluestone building still stands as The Source restaurant at the corner of Moorabool and Maud Streets, Geelong. Once the home of a Dr Crook, it was bought, with some adjoining property, by J. B. Wilson early in 1878 for £1,820. *Geelong Advertiser*, 15 March 1878. Close by, the Maud Street facade of the old Senior School remains as a handsome, if somewhat forlorn, fragment of the original magnificent bluestone building. See also Weston Bate, *Light Blue Down Under*, Melbourne 1990.

2. J. B. Wilson (1828–95) joined the staff of Geelong Grammar in 1858, and was headmaster, 1863–95. On 7 April 1863, in Christ Church, Geelong, he married Oriana Maria, Horatio Rowcroft's third daughter. (*Geelong Advertiser*, 10 April 1863.) A keen and competent marine biologist, Wilson was well known to Baron von Mueller, Professor Baldwin Spencer, Sir Frederick McCoy and A. H. S. Lucas, all of whom were also known to H. M. R. Rupp. See *ADB*, 6, pp. 417–8; J. H. Maiden in *Vic. Nat.*, November 1908, pp. 116–7; *Corian*, September 1971, pp. 169–79.

3. The fees were then £7 10s. 0d. per quarter.

4. Some of these book prizes, with their informative bookplates, are in the archives of Geelong Grammar School, others remain in family possession. *Church of England Grammar School, Geelong, 1907*, Chapter X; *GGSQ*, 1885–92, *passim*.

5. See Bibliography.

6. H. M. R. Rupp, School Recollections, MS enclosed with letter of 22 January 1956 to former School Archivist, John Ponder. Geelong Grammar School Archives.

7. Rupp in *Corian*, December 1946, p. 199.

8. Phillip L. Brown in *ADB*, 6, p. 418.

9. It has been said that the wreck of this American ship 'sparked off a wave of smuggling and pilfering without parallel in the Geelong district'. J. K. Loney, *Wrecks Along the Great Ocean Road*, n.p., 1976, p. 21.

10. Rupp, School Recollections, *loc. cit.* For a detailed and rather delightful account of the sites of 'the recognized School camps' see James Lister Cuthbertson, 'The School Camps', in *Church of England Grammar School, Geelong, 1907*, Chapter XI.

11. Rupp, 'Memories of Victorian Orchids', *Vic. Nat.* March 1953, p. 145.

12. Rupp, School Recollections, *loc. cit.* It is interesting to note that, in October 1886, the Rev. Charles Rupp established a Young Men's Friendly Society at Coleraine for those aged fourteen or more. One activity was working through the Rev. R. Appleton's *God and Nature,* involving studies of light, air, rocks, plants, human anatomy and other aspects of natural science. It was hoped 'to obtain a microscope' to aid these studies. See *Church News* (Hamilton), 1 October 1886, p. 38; 1 November 1886, pp. 41, 42.

13. *GGSQ*, April 1890, p. 30.

14. Rupp in *Corian*, December 1946, p. 200.

15. *Ibid.*

16. *GGSQ*, October 1891, pp. 33–4. 39.

17. *Op. cit.*, December 1890, pp. 4–6, 20.

18. *Op. cit.*, December 1891, pp 7–12.

19. *Op. cit.*, July 1890, p. 19.

20. C. L. H. Rupp, Diary of ... a Tour to England, Scotland, Europe and the Holy Land, 1890–1891, pp. 6, 7. Original in possession of Miss E Rupp, Willoughby, N.S.W.

21. H. M. R. Rupp, 'Fifty Years Ago', *Corian*, May 1941, p. 98.

22. *Church of England Grammar School, Geelong, 1907* (unpaged).

23. H. M. R. Rupp, 'Fifty Years Ago', *loc. cit.*, p. 96. The internal quotation is from J. L. Cuthbertson's poem, 'Stet Fortuna Domus'.

24. *GGSQ*, April 1892, p. 47.

25. Rupp visited 'Mungadal' (*sic*) to the west of Hay during the university vacation, November 1894 to February 1895, spent with his sister and brother-in-law. See Part I, Chap. 6.

26. See Bibliography.

27. H. M. R. Rupp, 'Orchids', *Corian*, May 1926, p. 16

28. *Ibid.*

29. *Op. cit.*, August 1926, p. 91.

30. *Corian*, May 1941, pp. 96–8.

31. H. M. R. Rupp, 'Some Recollections of an Old Boy', *Corian*, December 1946, pp. 199–200.

32. See *Corian*, August 1949, p. 139, and May 1950, p 76; and Rupp to J. Ponder, 22 January 1956. Geelong Grammar School Archives.

33. Rupp to J. L. Cuthbertson, 21 October 1905. Geelong Grammar School Archives.

34. The scholarship was established in 1883 through a gift of £1,000 from Frederick William Armytage (1838–1912), one of the school's first pupils, in memory of his wife. See *ADB*, 3, p. 52.

Chapter 6: The Season of Trinity

1. H. M. R. Rupp, School Recollections, MS enclosed with letter of 22 January 1956. Geelong Grammar Archives.

2. This was a theological scholarship commemorating the pioneer 'Apostle of the Western District', the Rev. Francis

Thomas Cusack Russell (1823–76); see *ADB,* 2, pp. 405–6. It was intended to assist students proposing to serve in the diocese of Ballarat. The Rev. Charles Rupp expressed his pleasure at this award in *Church News* (Hamilton), 23 December 1892, p. 531.

3. Rupp to Dr Alexander Leeper, 26 August 1893, and associated papers in Trinity College Archives. At one stage Charles Rupp was obliged 'to incur a liability to the Bank' in order to meet 'terminal charges punctually'.

4. J. B. Wilson to Rupp, 23 September 1893. Original in possession of Mrs Marjory Loader, Dural, N.S.W. The Orchid mentioned is known as Red Beaks, specimens of which turn black on drying. The two works used by Wilson would have been Otto Wilhelm Thomé, *Text-book of Structural and Physiological Botany,* London 1877 (7th ed, 1891), and Arthur Dendy and A. H. S. Lucas, *An Introduction to the Study of Botany,* Melbourne 1892. The latter book, dedicated to Baron von Mueller and introduced by Professor Walter Baldwin Spencer, was written by two Melbourne University lecturers, both well-known to Rupp.

5. Examination Records, Registrar's Office, University of Melbourne.

6. For John Dickson Wyselaskie (1818–83), who endowed scholarships in six disciplines at the University of Melbourne, see *ADB,* 6, pp. 446–7.

7. Rupp to Trinity College Council, 13 March 1895, and associated papers in Trinity College Archives.

8. Rupp to T. E. Hunt, 24 July 1946 and 23 April 1947, RBGS.

9. *Alma Mater,* September 1896.

10. James Grant, *Perspective of a Century: A Volume for the Centenary of Trinity College, Melbourne,* Melbourne 1972, p. 144. Rupp mentioned some of these rowers in his memoirs.

11. He later had difficulty in recalling his associate editors, who were (in 1897) C. F. Belcher ('Lord Chief Justice von Belscher'), A. E. Peacock ('His Eminence Cardinal O'Peacock') and T. K. Pitt ('The Most Rev. Thomas Kerosene Fitz-Pitt'). Rupp himself was dubbed 'The Right Rev. Hech Mon Rabid Mac Rupp'.

12. Copy in Leeper Library, Trinity College. The last issue, which was printed in a more sophisticated manner, appeared in 1902.

13. *Alma Mater,* August 1896; May 1897.

14. *Op. cit.,* August 1897. Unfortunately the topic was not recorded.

15. Grant, *op. cit.,* p. 158. See also *Alma Mater,* June 1896.

16. *Alma Mater,* June, 1896.

17. Grant, *op. cit.,* pp. 23, 55.

18. H. M. R. Rupp, Catalogue of Plants, Merri Merri Creek, 1896. By courtesy of Dr J. H. Willis. See also Ian Clarke, 'A Species List for the Merri Creek area (Melbourne, Vic.) compiled in 1896', *Vic. Nat.,* February 1990, pp. 28–34.

19. H. M. R. Rupp, Catalogue of Genera & Species of Native Plants obtained in the District of Coleraine (exclusive of the Wando Valley) (1892); Catalogue of Indigenous Genera & Species obtainable near Wando Vale (122 species) (1892); A Census of Native Plants to be found in or about the Neighbourhood of Buninyong (1896); Kingston and Surrounding Localities (188 species) (*c.* 1897). Manuscript notebooks by courtesy of Dr J. H. Willis, Brighton.

20. H. M. R. Rupp, A Catalogue of Indigenous Species of Plants from Hay District … Species collected during November and December 1894, and January and February 1895. Manuscript RBGS.

21. Rupp Papers in possession of Mrs Marjory Loader, Dural, N.S.W.

22. *Riv. Graz.,* 4 January 1895. The plant is known as Desert Spurge or Caustic Bush.

23. Mueller to Rupp, 13 February 1895, Marjory Loader Collection.

24. Mueller to Rupp, undated fragment of letter, January–February 1895, Marjory Loader Collection.

25. Mueller to Rupp, 1 February 1895, Marjory Loader Collection.

26. Mueller to Rupp, 21 January 1895, Marjory Loader Collection.

27. *Riv. Graz.,* 19 February 1895.

28. Peacock placed his fellow student in exalted company: Charles Darwin (1809–82), naturalist and evolutionist; John Tyndall (1820–93), physicist; Professor Walter Baldwin Spencer (1860–1929), biologist and anthropologist; and Thomas Henry Huxley (1825–95), biologist. Charles Eaton is rather more obscure, but possibly refers to an American actor of that name (1813–42), famed for his portrayal of certain Shakespearean characters, including Shylock, whom Rupp had also portrayed.

29. For G. W. Rusden (1819–1903), see *ADB,* 6, pp. 72–3.

30. Trinity College Calendar, 1897. Donors of exhibits included the Ballarat School of Mines, Dr Alexander Leeper, Rupp and G. W. Rusden himself.

31. *Ibid.;* see also *Alma Mater,* September 1896.

32. The H. M. R. Rupp Botanical Collection, Trinity College, Melbourne — notes accompanying letter, R. L. Sharwood to J. H. Willis, 26 June 1969, Trinity College Archives.

33. R. L. Sharwood to J. H. Willis, 2 May 1969, and reply of 6 May 1969. Trinity College Archives.

34. *Alma Mater,* September 1896. Rupp compiled a Descriptive Catalogue of Native Victorian Plants in the Rusden Museum, Trinity College, Melbourne, which listed 146 species, plus two additional species. Manuscript notebook, presently held in the Department of Botany, University of Melbourne.

35. Frans A. Stafleu and Richard S. Cowan, *Taxonomic Literature,* Utrecht 1983, Vol. IV, p. 991.

Chapter 7: Cures of Souls, Habitats of Plants

1. *Church News* (Hamilton), 30 November 1892, p. 583.

2. Reader's Licence, original document in possession of Miss E. Rupp, Willoughby, N.S.W.

3. Unfortunately, the earlier records of ordinands' examinations in the archives of the diocese of Ballarat are not currently accessible.

4. Document in Diocesan Archives, Melbourne.

5. *Ibid.*

6. Letters Dimissory, 26 May 1899. Registrar of diocese of Ballarat to the bishop of Melbourne. Diocesan Archives, Melbourne.

7. *Ibid.*

8. Oaths and Declaration Book, 1892–1900. Diocesan Archives, Melbourne.

9. *Church of England Messenger* (Melbourne), 1 June 1899.

10. *Alma Mater,* September 1899, p. 63.

11. *Argus,* 2 October 1902. The old brick parsonage is no longer a church residence.

12. *Ballarat Star,* 3 October 1902.

13. A lichen-covered pedestal monument, with marble ledger by F. W. Commons, still stands within the cast-iron railed enclosure in Learmonth Cemetery. C. L. H. Rupp's name

and date of death ('on Michaelmas Eve 1917') were added later, giving the impression that he, too, was buried there.

14. Based upon *Year Book of the Diocese of Newcastle, 1938,* West Maitland 1938, p. 22.

15. Rupp to A. H. Chisholm, 21 April 1931. RBGS.

16. Rupp to T. E. Hunt, 25 March 1945. RBGS.

17. Rupp to T. E. Hunt, 17 August 1946. RBGS.

18. Rupp to H. G. Curtis (née Geissmann) 11 March 1930. RBGS.

19. Rupp to L. A. Gilbert, 13 November 1947; 3 December 1949. RBGS.

20. Rupp to T. E. Hunt, 17 March 1947. RBGS. The reference is to St Matthew xxiii. 24, regarding the hypocrisy of the Scribes and Pharisees.

21. Rupp to L. A. Gilbert, 13 November 1947. RBGS.

22. Rupp to L. A. Gilbert, 9 December 1949. RBGS.

23. Rupp to T. E. Hunt, 9 March 1947. RBGS.

24. Rupp to A. H. Chisholm, 15 May 1926. RBGS.

25. Rupp to H. G. Geissmann (later Curtis) 14 May 1926. RBGS.

26. Rupp to G. W. Althofer, 16 October 1945; Rupp to T. E. Hunt, 26 January 1947. RBGS.

27. See L. A. Gilbert, 'Plants and Parsons in Nineteenth Century New South Wales', *Historical Records of Australian Science,* Vol. 5, No. 3, November 1982, pp. 17–32.

28. Eldest daughter of Richard A. Dowe (1848–1915), solicitor, and Mary, née Bloomfield (1849–1930). Further details are given in notes on the memoirs. Actually, Rupp was quite taken with his first New South Wales appointment: 'There is some magnificent scenery on the way up here, but I think Tamworth itself is as fine as anything on this side of the Hawkesbury . . . The town . . . is lit by electricity, and the streets are quite handsome. All around is undulating country, exceedingly fertile, and covered with magnificent fields of lucerne, wheat and grass . . . The whole district is a revelation to me; I had no idea it was so good. Everything betokens prosperity and progress.' (*GGSQ,* October 1903, pp. 17–18.) Rupp also notified his old school that his duties as senior curate included 'superintendence of the Chinese Mission', and that as well as the two churches in Tamworth, there were 'ten other places to work'.

29 *Tamworth News,* 4 January 1905.

30. 'Rachel' after Rupp's stepmother, and 'Mary' after his wife's mother.

31. Archdeacon R. J. Moxon, Letterbooks 1908–9. AD Arm.

32. Bishop Cooper to Rupp, 18 March 1909. Cooper Letterbook, 1909, AD Arm.

33. *ABMR,* 1 September 1914, p. 118.

34. *Op. cit.,* 1 November 1914, p. 149; 1 December 1914, p. 165.

35. *Op. cit.,* 1 February 1915, p. 210.

36. *Glen Innes Guardian,* 29 March 1915.

37. Rev. George P. M. Ware to Bishop Cooper, 21 July 1915. Cooper Papers, courtesy of Mr David Stammer.

38. *ABMR,* 7 December 1919, p. 135; 7 January 1920, p. 155.

39. *Op. cit.* 7 March 1920, p. 2. Rupp summarized this service as 'five years . . . on deputation work which took me over most of N.S.W. east of the Darling, once to Queensland, once to Vic., and twice to Tasmania. Any spare day or hour that cropped up I was off in the bush.' Rupp to G. V. Scammell, 17 April 1926. RBGS.

40. Rupp to T. E. Hunt, 19 and 27 July 1947. RBGS.

41. Rupp to T. E. Hunt, 20 September 1944. RBGS.

42. Now *Rhizanthella slateri* (Rupp) M. Clements et Cribb.

43. Rupp to G. V. Scammell, 17 April 1926. The period was 1923–4.

44. For H. L. Kesteven, DSc, MD (1881–1964), physician and naturalist, see *ADB,* 9, pp. 579–80, the source of this quotation; see also J. N. Rentoul in *AOR,* June 1965. Rupp derived great pleasure from his old HMV recordings of G&S., more especially *The Mikado* and *HMS Pinafore.*

45. Rupp to Bishop Long, 15, 21 and 24 May, 27 July 1929. Bishop's Letters, AB6568. AD Newc. See also *NDC,* 1 November 1924, p. 32; 1 January 1925, p. 33; 2 March 1925, p. 30.

46. Rupp to H. G. Geissmann, 8 June and 19 August 1926. RBGS. The plant was a Greenhood Orchid, *Pterostylis truncata* Fitzgerald.

47. Rupp to G. V. Scammell, 23 July 1926. RBGS. See *Newcastle Herald,* 21 July 1926, for the accident at the Prince Street level crossing, Paterson, involving Joseph Andrew Mate, aged thirty-two.

48. See H. M. R. Rupp, 'Treasures of an Orchid Herbarium', *AOR,* March 1940, pp. 24–5, and Rupp to H. G. Geissmann, 14 November 1925 and 6 January 1926. RBGS.

49. Rupp to Dr J. C. V. Behan, 20 November 1925. Trinity College Archives.

50. J. C. V. Behan to Rupp, 25 November 1925. Trinity College Archives. For Sir John Behan (1881–1957) see *ADB,* 7, pp. 247–8.

51. Rupp to J. C. V. Behan, 28 November 1925. Trinity College Archives. For J. H. Maiden (1859–1925), director of Sydney Botanic Gardens, 1896–1924, and his herbarium boxes, see L. A. Gilbert, *The Royal Botanic Gardens, Sydney: A History, 1816–1985,* Melbourne 1986, Chapter 6. See also *ADB,* 10, pp. 381–3.

52. For A. J. Ewart (1872–1937), foundation professor of Botany, University of Melbourne, see *ADB,* 8, pp. 448–50.

53. J. C. V. Behan to Rupp, 16 December 1925. Trinity College Archives.

54. Rupp to J. C. V. Behan, 19 December 1925. Trinity College Archives.

55. Melbourne *Herald,* 9 February 1926; Melbourne *Argus,* 10 February 1926; *SMH,* 6 February 1926; *Daily Telegraph,* 5 February 1926, with an appreciation of Rupp's work on 6 February perhaps by Alec Chisholm.

56. Melbourne *Herald,* 9 February 1926. See also Ian C. Clarke, 'The History of the Herbarium, School of Botany, University of Melbourne', in P. S. Short (ed.), *History of Systematic Botany in Australasia,* Melbourne 1990, pp. 13–21.

57. Rupp to A. J. Ewart, 16 January, 26 June and 2 November 1926. Department of Botany, University of Melbourne.

58. Ian Clarke, 'The History of the Herbarium at the Botany School, University of Melbourne', in *Botanical History Symposium: Development of Systematic Botany in Australia,* Melbourne, May 1988, and in *History of Systematic Botany in Australasia.*

59. Rupp to J. H. Willis [8 April] 1955. RBGS.

60. *SMH,* 28 August 1961. Like her brother, she was cremated at Northern Suburbs Crematorium following a service in St Thomas's, North Sydney.

61. Rupp to Bishop Long, 10 October, 9 November 1928, 23 January 1929. Bishop's Letters, AB6568. AD Newc. See also *NDC,* 1 October 1929, p. 32.

62. Bishop Long to Rupp, 28 October 1929, and reply 31 October 1929. Bishop's Letters, *loc. cit.*

63. Rupp to Bishop Long, 6 November 1929, Bishop's Letters, *loc. cit.*

64. Bishop Long to Rupp, 8 November 1929. Bishop's Letters, *loc. cit.* and memo of 7 November 1929. Parish Files, AB6905, AD Newc.
65. Rupp to Bishop Long, 16 January 1930. Bishop's Letters, AB6568, AD Newc. See also *NDC*, 1 May 1930, pp. 33–4.
66. Rupp to G. V. Scammell, 10 May 1926. RBGS.
67. Rupp to Bishop Long, 18 February 1930, and reply of 21 February 1930. Bishop's Letters, AB6568. AD Newc.
68. Rupp to A. H. Chisholm, 5 February 1930. RBGS. And to G. V. Scammell, 25 April 1930: 'we find the miners . . . delightful people to live amongst. The popular idea of them as a pack of ruffians might as well be applied to Potts Point.'
69. Rupp to A. H. Chisholm, *loc. cit.*
70. *Sunday Pictorial,* 19 January 1930.
71. Rupp to A. H. Chisholm, 20 January 1930. RBGS.
72. *Sunday Pictorial,* 26 January 1930.
73. Rupp to A. H. Chisholm, 20 September 1930. RBGS.
74. Rupp to A. H. Chisholm, 3 December 1930. RBGS.
75. H. K. C. Mair, destined to become Rupp's friend and the director of the Royal Botanic Gardens, Sydney, 1964–70, received his copy as a gift from an aunt on 8 April 1930. I have been privileged to use it during the preparation of this work.
76. See Bibliography.
77. *SM,* 30 April 1930, p. 16.
78. *Education Gazette* (N.S.W.), 2 June 1930, pp. 95–6. See also Rupp to A. H. Chisholm, 24 June 1930. RBGS.
79. *Aust.* 10 May 1930.
80. Rupp to A. H. Chisholm, 23 April 1930. RBGS.
81. Rupp to A. H. Chisholm, 19 August 1930. RBGS.
82. Rupp to F. Fordham, 28 October 1931. RBGS. See also Rupp to T. E. Hunt, 16 June 1941. RBGS.
83. Rupp to A. H. Chisholm, 3 September 1931. RBGS.
84. Rupp to H. G. Curtis, 14 September 1931. RBGS.
85. Rupp to A. H. Chisholm, 3 September 1931. RBGS.
86. Rupp to H. G. Curtis, 14 September 1931. RBGS.
87. Rupp to A. H. Chisholm, 3 September 1931. RBGS.
88. Rupp to Bishop Batty, 23 October 1931; Bishop Batty to Rupp, 13 February 1932. Bishop's Letters, AB6627. AD Newc.
89. Rupp to A. H. Chisholm, 3 February 1932. RBGS. The Rupps 'all fell violently in love with Port Macquarie . . . the loveliest place . . . so fresh and green and clean . . . brush forests, palm scrubs, heathlands'. Rupp to H. G. Curtis (née Geissmann), 27 January 1931. RBGS.
90. Rupp to A. H. Chisholm, 14 May 1932. RBGS. See also *NDC*, 1 June 1932, pp. 41–2.
91. Rupp to A. H. Chisholm, 3 February 1932. RBGS.
92. Memo of enquiry of 9 March 1932, Parish Files, AB6905. AD Newc. Rupp to Bishop Batty, 28 March 1932. Bishop's Letters, AB6627. AD Newc.
93. Rupp to Bishop Batty, 5 and 28 March, 2 and 27 April 1932. Bishop's Letters. AB6627, AD Newc.
94. Rupp to A. H. Chisholm, 3 February 1932. RBGS.
95. Rupp to A. H. Chisholm, 3 December 1930. RBGS.
96. Rupp to A. H. Chisholm, 3 February 1932. RBGS.
97. Rupp to A. H. Chisholm, 14 May 1932. RBGS.
98. Rupp to F. Fordham, 16 August 1932. See also Rupp to A. H. Chisholm, 6 April 1932. RBGS.
99. Rupp to A. H. Chisholm, 14 May 1932. RBGS.
100. Rupp to Bishop Batty, 26 and 30 July 1932. Bishop's Letters, AB6627. AD Newc.
101. Rupp to F. Fordham, 10 September 1932. RBGS.
102. Rupp to A. H. Chisholm, 17 September 1932. RBGS.
103. Rupp to Bishop Batty, 3 October 1932. Bishop's Letters, *loc. cit.*
104. Rupp to A. H. Chisholm, 21 February 1933. RBGS.
105. Rupp to Bishop Batty, 1 January 1933; Rupp to Archdeacon Woodd, 11 February 1933; Archdeacon Woodd to Rupp, 2 March 1933; Bishop Batty to Rupp, 7 March 1933 and reply of 8 March 1933. Bishop's Letters, *loc. cit.*
106. Rupp to A. H. Chisholm, 24 April 1933. RBGS. Rupp to Archdeacon Woodd, 5 May 1933. Bishop's Letters, AB 6627. AD Newc. *NDC,* 1 May 1933, p. 35.
107. Rupp to F. Fordham, 5 June 1933. RBGS.
108. Rupp to F. Fordham, 28 June 1933. RBGS.
109. Rupp to F. Fordham, 23 October 1933. RBGS.
110. Rupp to F. Fordham, 14 March 1936. RBGS.
111. Rupp to C. A. Brown, 25 April, 13 July 1933. Parish Files, AB6919, AD Newc.
112. Rupp to Archdeacon Woodd, 14 July 1933. Bishop's Letters, AB6627, AD Newc.
113. Letter to C. A. Brown, 27 March 1934. Parish Files, AB6919, AD Newc.
114. Bishop Batty to Rupp, 17 and 25 February 1936. Bishop's Letters, AB6628. AD Newc.
115. Rupp to Bishop Batty, 18 and 22 February 1936. Bishop's Letters, *loc. cit.*
116. Rupp to F. Fordham, 28 March 1936. RBGS. See also *NDC,* 1 July 1936, p. 30, where Rupp noted 'the extent of the dilapidations may be gauged by the fact that only the stone walls have escaped demolition!'
117. Bishop Batty to Rupp, 4 March 1936. Bishop's Letters, *loc. cit.*
118. Rupp to Bishop Batty, 11 December 1936. Bishop's Letters, *loc. cit.*
119. Rupp to F. Fordham, 14 October 1936. RBGS. *Maitland Mercury,* 17 October 1936.
120. Rupp to F. Fordham, 2 November 1936. RBGS. See also Bishop's Letters, October 1936, *loc. cit.*
121. Rupp to Bishop Batty, 18 March 1938. Bishop's Letters, AB6629, AD Newc.
122. Rupp to Bishop Batty, 2 September 1937. Bishop's Letters, AB6628, AD Newc.
123. Rupp to F. Fordham, 5 February 1938. RBGS.
124. Rupp to Bishop Batty, 15 October 1938 and reply of 17 October. Bishop's Letters, AB6629, AD Newc.
125. Bishop Batty to Rupp, 7 February 1939. Bishop's Letters, *loc. cit.*
126. *NDC,* 1 May 1939, p. 82.
127. *NDC,* 1 June 1939, Supplement, p. 1.
128. A. C. Cox and L. C. Cox, descendants of William Cox (1764–1837) of Clarendon, near Windsor.
129. Rupp to T. E. Hunt, 7 March 1949. RBGS.
130. Rupp to Bishop Batty, 23 October 1931. Bishop's Letters, AB6627, AD Newc.
131. Rupp to Bishop Batty, 19 December 1931. Bishop's Letters, *loc. cit.*
132. Bishop Batty to Rupp, 3 March 1932, Bishop's Letters, *loc. cit.*
133. Rupp to C. A. Brown, 20 September 1934. Parish Files, AB6919, AD Newc.
134. Rupp to C. A. Brown, 16 May 1934, Parish Files, *loc. cit.*

Chapter 8: Man of Letters, Man of Science

1. Rupp to G. W. Althofer, 19 September 1945. RBGS.
2. Rupp to L. A. Gilbert, 9 October 1947. RBGS.

3. Rupp to L. A. Gilbert, 16 October 1947. RBGS.

4. Rupp to G. V. Scammell, 17 April 1926. RBGS.

5. J.H. Willis in *Vic. Nat.*, November 1956, p. 110.

6. *AOR*, Summer 1986, pp. 8, 9. Rupp addressed the Society on the Ground Orchids of the South Maitland Coalfields, 31 January 1936 — see QVF 584/050/944R in RBGS.

7. Rupp to T. E. Hunt, 29 April 1947. RBGS. William Henry Nicholls (1885–1951), bookbinder, municipal gardener and orchidologist, was born at Ballarat. An accomplished cyclist, bushwalker and photographer, 'he developed an obsessive interest in native orchids' about 1920, and thereafter made superb paintings which were ultimately reproduced in *Orchids of Australia*, Melbourne 1969. See J. H. Willis in *Vic. Nat.*, April 1951, pp. 241–3 and in *ADB*, 11, pp. 23, 24.

8. Note by A. H. Chisholm, February 1948, with Rupp Correspondence (Willis Collection) RBGS. Alexander Hugh Chisholm, OBE, FRZS (1890–1977), naturalist and journalist, was born at Maryborough, Victoria. He was author of numerous nature books, and editor of the *Australian Encyclopaedia*, Sydney 1958 and 1963; and of several newspapers and journals, including *The Emu* and the *Victorian Naturalist*. See *WWA*, 1941–74.

9. Hilda Gladys Curtis (née Geissmann; 1890–1988) was born in Brisbane on 22 November 1890, the third of seven children of Swiss-born William Felix Geissmann and his wife Bertha Elfrieda Dorothea, who was of German extraction. In 1898 the family moved to Mount Tamborine (North) where they opened the Capo di Monte guesthouse in December. In the then extensive forests and in the Mountain school (conducted for some time by her mother), Hilda developed into a keen and competent naturalist, with special interests in orchids and birds — some (including Rupp) saw her love for cats as somewhat incongruous with that for birds. Seeking an even more paradisaic habitat, the father went to join the New Australia settlement in Paraguay, where he died. Both before and after Capo was converted into flats in 1914, Hilda helped her mother to run the establishment which became a mecca for bushwalkers and naturalists, as well as for political and vice-regal figures and visiting scientists. She relished scientific company and both taught and learnt much while accompanying visitors on their bush excursions. One regular visitor from about 1920 was Alec Chisholm, who nicknamed her 'Hildy-gei' ('Hildigei' or 'Hilligei'), by which she became known to her friends. Hilda Geissmann married Herbert Curtis, a local farmer, on 1 September 1926. After Capo di Monte was destroyed by fire on 31 October 1932, some Tamborine visitors found welcome accommodation at the Curtis home, Windermere, where Hilda lived until she was ninety. Long since widowed, Hilda Curtis died in Brisbane on 10 June 1988, survived by her son, Mr H. S. Curtis, who has kindly presented Rupp's letters to his mother to the Royal Botanic Gardens, Sydney. Her vast knowledge of wildlife and her technical skills with the camera were widely acclaimed — see, e.g., Charles Barrett, *Australia, My Country*, Melbourne 1941, p. 84, and *On the Wallaby*, Melbourne 1942, p. 166; Bernard O'Reilly, *Green Mountains*, Brisbane 1962, p. 155; H. J. Carter, *Gulliver in the Bush*, Sydney 1933, pp. 84–100; Francis Ratcliffe, *Flying Fox and Drifting Sand*, London 1938, pp. 16–21; Eve Curtis, *The Turning Years: A Tamborine Mountain History*, Beaudesert 1988, p. 41; F. S. Colliver, 'Hilda Curtis' (obituary), *The Queensland Naturalist*, Vol. 28, 1988, pp. 42–3. Alec Chisholm used some of her photographs in *Birds and Green Places*, London 1929, and in *Bird Wonders of Australia*, Sydney 1934. W. H. Nicholls named the Greenhood Orchid, *Pterostylis hildae*, in her honour, and

in 1989 Mrs Susan Cantrell and other Mt Tamborine residents resolved to erect a local monument to commemorate her life and work.

10. Edith Coleman (née Harms; 1874–1951), teacher and naturalist, was born at Woking, Surrey, 29 July 1874. She married pioneer Victorian motorist, James George Coleman, at South Yarra on 7 April 1898, and had two daughters. On 11 September 1922 she joined the Field Naturalists' Club of Victoria, 'and the *self same evening* read a paper — ''Some Autumn Orchids'' '. (See *Vic. Nat.*, December 1922 and September 1950.) Mrs Coleman died at Sorrento, Victoria, on 3 June 1951. See *WWA*, 1947, p. 236; H. M. R. Rupp in *AOR*, Vol. 16, 1951, p. 122; and *Biog. Reg.*, I, p. 136. Information also kindly supplied by Dr P. F. Thomson (grandson), Blackburn, 1987.

11. Charles Leslie Barrett, FRZS (1879–1959), naturalist and journalist, was born at Hawthorn, Victoria, 26 June 1879. He joined the Field Naturalists' Club of Victoria in September 1899, later holding various offices including that of editor of the *Victorian Naturalist*, 1925–40. The author of several books, he served on the Melbourne *Herald* for thirty-three years, and edited the celebrated 'Sun Nature Books' and 'Sun Travel Books'. See F. S. Colliver in *Vic. Nat.*, August 1959, pp. 110, 111; A. H. Chisholm in *ADB*, 7, p. 185; and *WWA*, 1941–55. In January 1926 Barrett called at the Paterson Rectory with specimens from Barrington Tops, including a new Double-tail Orchid which Rupp named *Diuris venosa*. Rupp to G. V. Scammell, 9 April 1926, RBGS; and H. M. R. Rupp, *The Orchids of New South Wales*, Sydney 1943, pp. 15, 16.

12. Rupp to G. V. Scammell, 5 April 1926. RBGS. Dr Richard Sanders Rogers, MA, ChM, MD, DSc (1861–1942), physician and orchidologist, was born in Adelaide on 2 December 1861. His wife Jean (née Paterson), also 'gifted with the ''orchid eye'' ', shared his botanical interests, and for a time maintained correspondence with Rupp after her husband's death. Rupp considered that Rogers became 'recognized as the leading authority on Australian Orchids'. See H. M. R. Rupp in *AOR*, June 1942, pp. 22–4; and Jean Gibberd in *ADB*, 11, p. 443.

13. Florence Perrin (née Dawson; 1883–1952), wife of George E. Perrin of Launceston, Tasmania, was a keen phycologist who worked with Professor A. H. S. Lucas (1853–1936) on the seaweeds of Tasmania, Queensland and Lord Howe Island. See Sophie C. Ducker on A. H. S. Lucas in *ADB*, 10, pp. 163, 164.

14. Rupp to H. G. Geissmann, 24 March 1925. RBGS.

15. Rupp to G. V. Scammell, 6 July 1926. RBGS.

16. Rupp to F. Fordham, 8 March 1935. RBGS.

17. Rupp to G. V. Scammell, 18 August 1926. RBGS.

18. Rupp to T. E. Hunt, 29 April 1947. RBGS.

19. Rupp to G. V. Scammell, 17 April and 27 May 1926; to T. E. Hunt, 19 January and 7 February 1945; to H. G. Geissmann, 2 and 14 April 1925; 8 May 1925, and 12 June 1925: 'you are a genius at these photographic studies'. Rupp mounted the photographs himself, taking care to acknowledge each one for the exhibition. Rupp to H. G. Geissmann, 25 August 1925. RBGS.

20. F. Ratcliffe: *Flying Fox and Drifting Sand*, London 1938, p. 20.

21. Rupp to H. G. Geissmann, 27 [March 1925?] RBGS.

22. Rupp to H. G. Geissmann, 29 May 1925. RBGS.

23. Rupp to H. G. Geissmann, 30 May and 14 October 1925. RBGS.

24. Rupp to H. G. Curtis (née Geissmann), 7 November 1930. RBGS.

25. Rupp to G. V. Scammell, 27 May and 5 June 1926. RBGS.

26. Rupp to W. H. Nicholls, 4 March 1927. RBGS.

27. Rupp to A. H. Chisholm, 22 March 1933. RBGS.

28. Rupp in *AOR,* December 1951, p. 122, where the date of first contact is given as 1923 instead of either late 1925 or early 1926. Unfortunately, no correspondence has so far been found.

29. Dr F. A. Rodway (1880-1956) was a son of Leonard Rodway (1853-1936), the dental surgeon who served as honorary government botanist of Tasmania, 1896-1932. A dedicated collector, he amassed a herbarium of some 10,000 specimens which were transferred as a gift to the National Herbarium of New South Wales between 1952 and 1956. As well as maintaining contact with Dr Rodway, Rupp also wrote to Mrs Olive Rodway (née Barnard), Leonard Rodway's second wife, concerning Tasmanian Caladenias. See Rupp to Olive Rodway, 18 and 30 March 1933, and Rupp to F. A. Rodway, 1 January and 21 November 1924. RBGS.

30. Adam Forster (1852-1928), naturalist and artist, came to Australia about 1875. He became secretary of the Pharmaceutical Association of New South Wales and registrar of the Pharmacy Board, 1897-1920. He took a botanical as well as artistic interest in native plants, and was also a keen philatelist. He died on 11 April 1928 after collapsing in a tram soon after leaving his home at Dulwich Hill. His wife, two daughters and a son survived him. In 1949 the National Library purchased about 800 of Forster's paintings from one of the daughters, Mrs E. McGregor. See *Biog. Reg.,* I, p. 327; *SMH,* 12 April 1928; Rupp to H. G. Geissmann, 19 May and 25 August 1925. RBGS.

31. Rupp to A. H. Chisholm 4 October 1924. RBGS.

32. See Rupp to A. Forster, 7, 23 and 31 January 1928; 7 and 9 February 1928; 9 March 1928. RBGS.

33. Thistle Yolette Stead (née Harris) AM, BSc, MEd, DLD, DipEd, was born in Sydney on 29 July 1902, the daughter of Charles Thomas Harris and Ilma (née Rokes). After teaching in N.S.W. secondary schools 1924-38, she lectured in Biological Science at Sydney Teachers' College, 1938-58. In August 1953 she married fellow-naturalist, David George Stead. Mrs Stead has taken a lead in many environmental movements and is the author of several books. See *WWA,* 1947 onwards, and *Who's Who of Australian Women,* 1982, pp. 222-3.

34. Erwin Nubling (1876-1953) was a German-born manager of the Poldi Steel Co. and an orchid enthusiast with a special interest in bower birds (thereby winning A. H. Chisholm's acclamation, see *Birds and Green Places,* London 1929, pp. 182, 186). Interned during World War I and strongly anti-Nazi during World War II, Nubling is remembered by Dr J. H. Willis as 'a diminutive, precise little man, full of old-world charm and courtesy'. He was an expert calligrapher who, from Rupp's experience, was capable of penning a 'quite Prussian letter' when the finer points of *Prasophyllum nublingii* Rogers were brought into question! But they remained good friends with deep respect for each other's work. Rupp to T. E. Hunt, 4 and 12 July 1946. RBGS.

35. F. A. Weinthal, a conveyancer, was a collector and cultivator of orchids, at least during the period 1903-39. F. M. Bailey named *Sarcochilus weinthalii* in 1903 in recognition of his work. Weinthal sent this and other species from southern Queensland to Rupp in September 1926. At that time he was working 'in the Govt. Savings Bank in Martin Place', Sydney. Rupp to G. V. Scammell, 16 September 1926. RBGS. By the late 1940s, Weinthal was living in Roseville.

36. A. G. Hamilton (1852-1941), teacher, naturalist and lecturer in Nature Study at Sydney Teachers' College, was born in County Cavan, Ireland, on 14 April 1852, and came to New South Wales with his parents in 1866. Rupp, who 'knew him well' was especially grateful for the 'wonderful lot' of W.A. specimens Hamilton collected for him in 1926. Rupp to A. G. Hamilton, 27 October 1926; to G. V. Scammell, 17 December 1926; to G. W. Althofer, 9 August 1945. RBGS. See L. A. Gilbert in *ADB,* 9, pp. 173, 174.

37. Rupp to A. H. Chisholm, 4 October 1924; Rupp to G. V. Scammell, 14 June 1926. RBGS.

38. O. D. Evans, BSc (1889-1975) was born at Shepherd's Darling Nursery, Bourke Street, Sydney. In 1924 he was appointed curator of the John Ray Herbarium at Sydney University, where he also served as Chief Laboratory Attendant. He later worked in the Botany Department of the University of New South Wales, and was part-time botanist at the National Herbarium of New South Wales, 1959-71. In addition to publishing several papers on monocotyledonous taxa, he was co-author with N. C. W. Beadle and R. C. Carolin of the *Handbook of the Vascular Plants of the Sydney District and the Blue Mountains,* Armidale 1962, and Sydney 1963. See R. C. Carolin and L. A. S. Johnson in *Telopea,* Vol. 1, No. 2, 1976.

39. Edwin Cheel (1872-1951) was born near Canterbury in Kent on 14 January 1872. Coming to Mackay, Queensland, in 1892, he worked on the sugar cane fields and in Sydney's Centennial Park before joining the staff of the Botanic Gardens in 1901. He became a botanical assistant in 1908, principal botanical assistant in charge of the Herbarium in 1913, then curator in 1924. He retired in 1933 and died at Ashfield in 1951. *WWA,* 1941-47; see Ann G. Smith in *ADB,* 7, pp. 628, 629.

40. Rupp to G. V. Scammell, 5 April 1926. RBGS.

41. G. V. Scammell, BSc, ARACI, FCS (1903-1986) was born at Glenelg, S.A., on 16 August 1903, but was educated in Sydney. Leaving Sydney University, he joined F. H. Faulding & Co., serving as a director, 1928-64; he also had a long association with the Red Cross Society. In May 1930 he married Lorna Jones, and lived at Clifton Gardens, Sydney. A competent artist, he executed most of the illustrations for Rupp's *Orchids of New South Wales.* Scammell died in September 1986, survived by his wife, a daughter and two sons. *WWA,* 1947-80; *SMH,* 4 September 1986.

42. Rupp to G. V. Scammell, 17 April 1926. RBGS.

43. Rupp to G. V. Scammell, 27 September 1927. RBGS.

44. Rupp to T. E. Hunt, 20 November 1946. RBGS.

45. Rupp to G. V. Scammell, 17 December 1926. RBGS.

46. Edward Edgar Pescott, FLS, FRHS (1872-1954), teacher and horticulturist, was born at Geelong on 11 December 1872. After teaching in several schools, he joined the Victorian Department of Agriculture. By the time he retired in December 1937, he had been a long-standing member of the Field Naturalists' Club of Victoria, and had developed special interests in orchids, the Wattle Day movement and the work of Baron von Mueller. *WWA,* 1941-50; I. F. McLaren in *ADB,* 11, pp. 206-8.

47. Alfred James Tadgell (1863-1949) was born at North Melbourne on 6 February 1863 and served the Clarke family as estate accountant for fifty-five years. He joined the Field Naturalists' Club of Victoria in 1920, and became a frequent contributor to the *Victorian Naturalist.* His large plant collection is now part of the National Herbarium of Victoria, in the building he worked so hard (if almost anonymously) to have constructed. He died on 6 September 1949. See P. F. Morris in *Vic. Nat.,* November 1949, p. 135.

48. Timothy Green (1860-1949), advertising officer, naturalist and photographer, was a Yorkshireman ('a quaintly old-fashioned English gentleman') who worked for John Danks and

Son for some twenty-five years, helping to produce the firm's illustrated catalogues. He established contact with the National Herbarium, Melbourne, about 1920, and applied his special talent for stereoscopic photography to making exquisite bushland and orchid studies, which were lodged in Kew and Melbourne. See P. F. Morris in *Vic. Nat.*, July 1950, pp. 55, 56. Rupp noted in a letter to A. H. Chisholm, 23 April 1930, that he had received 'some splendid stereo pictures' of a new *Microtis* 'from Mrs Eaves of Melbourne' who had learned her skill from Green, and who bade 'fair to excel her master'.

49. Rupp to G. V. Scammell, 3 February 1927. RBGS.

50. Rupp to A. G. Hamilton, 6 January 1928. RBGS. Patrick Francis Morris (1896–1974), the son of a South Yarra veterinarian, joined the staff of the National Herbarium of Victoria in 1913, at the age of seventeen, and remained for forty-eight years, retiring as Senior Botanical Officer. He became a member of the Field Naturalists' Club of Victoria in June 1918, and later served as president. See J. H. Willis in *Vic. Nat.* July 1974. pp. 205, 206.

51. Dr (later Sir) Arthur William Hill (1875–1941) was Director of Kew Gardens from 1 March 1922 until his death in a riding accident on 3 November 1941.

52. A. W. Hill to Rupp, 4 March 1930. Library of Kew Gardens, by kind permission of the Trustees.

53. Rupp to A. W. Hill, 8 April 1930. *Loc. cit.*

54. A. W. Hill to Rupp, 10 June 1930. *Loc. cit.*

55. Rupp to A. W. Hill, 18 July 1930 and reply of 29 August 1930. *Loc. cit.*

56. A. W. Hill to Rupp, 10 October 1930. *Loc. cit.*

57. Rupp to G. V. Scammell, 18 October 1930. RBGS. Rev. Ernest Norman McKie, BA (1882–1948) was born at Barraba, and worked in a bank before training for the Presbyterian ministry. After ordination, he served at Manilla, 1909–12, then at Guyra, 1912–48. He was elected Moderator in 1938. McKie took a special interest in eucalypts and grasses, but certainly corresponded with Rupp on orchids. His papers and herbarium were, it seems, most regrettably destroyed after his death on 19 May 1948. See Norman Hall, *Botanists of the Eucalypts,* Melbourne 1978, p. 89; *Proc. Linn. Soc. N.S.W.*, Vol. 74, 1949, p.v.

58. Rupp to F. Fordham, 15 May, 7 June, 17 September and 20 November 1934; 19 June 1935. RBGS.

59. Cyril Tenison White (1890–1950) was born in Brisbane. In 1905 he became pupil-assistant to his grandfather, Frederick Manson Bailey (1827–1915), colonial botanist of Queensland, 1881–1915. On the resignation in 1917 of his uncle, John Frederick Bailey, as government botanist and director of the Brisbane Botanic Gardens, White succeeded him and remained in office until his death on 16 August 1950.

60. Rupp to F. J. Rae, 7 September 1939. RBGM. For F. J. Rae (1883–1941) see *WWA,* 1947, pp. 557, 558.

61. Ernest William Slater, formerly of Bulahdelah, 1929–50, and Bathurst, now of Glenbrook, was born at Killingworth, N.S.W., on 21 August 1904, the son of William Frederick Slater and Sarah Jane, née Hoskins. Rupp knew him as 'a young bootmaker . . . a very intelligent youth' but he later worked as a wood supplier and road contractor, a saw miller and fuel-depot proprietor. When interviewed on 29 April 1988, he recalled quite clearly the occasion of discovering *Cryptanthemis.*

62. Frederick and Rose Fieldsend were very keen and successful horticulturists who maintained a fine greenhouse. F. J. Fieldsend was proprietor of a pottery at East Maitland.

63. Dr R. S. Rogers of Adelaide named the W.A. plant *Rhizanthella gardneri.* See *J. Roy. Soc. W.A.*, Vol XV, 1928.

64. *Proc. Linn. Soc. N.S.W.*, Vol. 57, 1932, pp. 57–61. Rupp's appropriate generic name was superseded when this species was placed under *Rhizanthella.*

65. Rupp to F. Fordham, 28 June 1933. RBGS.

66. Rupp to F. Fordham, 16 June 1933; to W. H. Nicholls, 15 October 1933. RBGS.

67. Rupp to A. W. Hill, 3 July and 12 September 1934, and replies. Library of Kew Gardens by kind permission of the Trustees.

68. *SMH,* 12 November 1985 for reference to a further discovery by Dr Mark Clements of the Australian National Botanic Gardens. See also D. L. Jones, *Native Orchids of Australia,* Sydney 1988, pp. 332–3, under *Rhizanthella.*

69. Fred Fordham (1890–1978) was born at Moss Vale, N.S.W., on 27 December 1890, Rupp's eighteenth birthday. He worked on the family farm as a boy, and later attended school at Bowral, 1904–7. Winning a scholarship to the old Training College at Blackfriars, he very likely came under the influence of A. G. Hamilton. His first appointment was to Milsons Point School in February 1911. After service in France during World War I, he rejoined the N.S.W. Department of Education in 1919, and taught in Sydney, Wauchope and the Young district before moving in August 1928 to Brunswick Heads, where he remained for nearly twenty years. Fordham retired from Martin's Gully School, Armidale, where he served 1948–50, and died at Port Macquarie on 7 September 1978. Rupp named the Helmet Orchid, *Corybas fordhamii,* in his honour. (Taped interview, 1975, from Mr Graeme Fordham, Armidale; information from daughters, Mrs Jean Trotter, Port Macquarie, and Mrs Helen Faint, Hillgrove.)

70. Rupp to F. Fordham, 28 October 1931. RBGS.

71. Rupp to F. Fordham, 21 February 1934. Similarly, Rupp to H. G. Geissmann, 8 May 1925: 'May I say that my "queer name" as some people dub it, is pronounced Rupe'. RBGS.

72. Rupp to F. Fordham, 10 April 1940. RBGS.

73. Rupp to F. Fordham, 26 April 1940. RBGS.

74. Rupp to L. A. Gilbert, 5 April 1949. RBGS.

75. Rupp to F. Fordham, 30 May 1936. RBGS.

76. *Australian Museum Magazine,* 20 November 1936, pp. 138–44.

77. Rupp to F. Fordham, 6 September and 7 October 1936. RBGS.

78. Rupp to F. Fordham, 25 September 1936. RBGS.

79. Rupp to F. Fordham, 13 April 1942. RBGS.

80. Rupp to F. Fordham, 5 April 1937. RBGS.

81. Rupp to F. Fordham, 1 and 6 December 1939. RBGS.

82. Rupp to T. E. Hunt, 26 March, 1953. RBGS. Rupp in *Vic. Nat.*, March 1947, p. 3, and July 1953, pp. 54, 55. *C. dockrillii* Rupp is now included under *C. trilabra* FitzG. and *P. uroglossum* Rupp is considered identical with *P. fuscum* R.Br.

83. T. E. Hunt (1913–70) spent much of his life in Ipswich, although he taught in schools near Tannymorel and at Mungindi before serving for many years at Silkstone Primary School, Ipswich. After a break on a pineapple farm in North Queensland, he returned to teaching and was at Yeronga High School at the time of his death on 17 June 1970. A competent artist as well as a keen botanist, Hunt led regular student excursions to Bribie and Stradbroke Islands, and taught pharmaceutical botany at the University of Queensland. He married Etta Mathieson, and they had two sons. (Information from Mrs Betty Mathieson, Ipswich, 1987.)

84. Dr C. P. Ledward, MB, BS, MRCS, LRCP (1903–1963) graduated from London University in 1923 and was a member of the Kingston-on-Thames Division of the British Medical Association early in 1925. In 1926, while at Canungra,

Queensland, he applied to join the Australian Medical Association. After an apparently short time at Cloncurry, Dr Ledward moved to Burleigh Heads when there were still swamps and light rainforest to be investigated. He also explored widely in the McPherson Ranges, winning a reputation as a competent botanist and entomologist, and publishing some papers on butterflies. Rupp named the rare Pixie Cap Orchid, *Acianthus ledwardii,* in his honour in 1938. Ledward is remembered as a quiet, reserved, amiable man given to long walks along beaches and bush tracks. (Information from Prof. John Pearn, Brisbane, and from Mrs Jean Harslett, Amiens, Queensland, who with her sister Dorothy, used to visit Ledward at Burleigh Heads.)

85. Dr Hugo Flecker, MB, ChM, FRGS (1884–1957), physician, radiologist and naturalist, was born at Prahan, Victoria, and educated at Prince Alfred College, Adelaide, and Sydney University. After service in Egypt and France in World War I he practised in Melbourne as a pioneer radiologist. He moved to Cairns in 1932 and became foundation president of the North Queensland Naturalists' Club, 1932–45. He died in Cairns on 25 June 1957. *WWA,* 1947–55; *Biog. Reg.,* I, p. 231. (Information from Dr Pat Flecker, Townsville, 1987.) As early as the autumn of 1937 (when he hoped that Bernard O'Reilly's 'courage and splendid efforts' in finding the Stinson aeroplane which crashed in the McPherson Ranges would 'be properly recognised'), Rupp had noted that most of his current orchid work was associated with North Queensland. Rupp to Hilda Curtis, 3 and 31 March 1937. RBGS.

86. Rupp to T. E. Hunt, 16 June 1941. RBGS.

87. Rupp to T. E. Hunt, 4 December 1942. RBGS.

88. Rupp to T. E. Hunt, 22 September 1955. RBGS. The paper was published in *Vic. Nat.,* November 1955.

89. Rupp to A. H. Chisholm, 9 July 1946. RBGS.

90. N. A. Wakefield, MSc (1919–72), teacher and naturalist, joined the Field Naturalists' Club of Victoria in May 1938 and shortly contributed a paper on the orchids of the Orbost District as the first of 126 articles in the *Victorian Naturalist,* of which he became editor. He died on 23 September 1972 following an accident at his Sherbrooke home. See J. H. Willis in *Vic. Nat.,* April 1973, pp. 103–5.

91. Rupp to T. E. Hunt, 1 December 1943. RBGS.

92. Rupp to T. E. Hunt, 6 April, 25 May, 4 July, 15 August, 20 September and 31 December 1946; 15 April and 5 September 1953. RBGS.

93. T. S. Hart, MA, BCE (1871–1960) was born at Caulfield, Melbourne. Like Rupp, he was a graduate of Melbourne University, a member of the Field Naturalists' Club of Victoria and a correspondent of Baron von Mueller. A man of encyclopaedic interests and accomplishments, he taught in several institutions, the last being the Melbourne Correspondence School. See L. Farrall in *ADB,* 9, pp. 221, 222; J. H. Willis in *Vic. Nat.,* August 1960, pp. 111–14; and Rupp's letters to Hart, 30 October 1946 to 11 January 1947. RBGS.

94. Rupp to T. E. Hunt, 23 August 1947. RBGS. The plant was *Rimacola elliptica* (R.Br.) Rupp, the Green Rock Orchid.

95. See letters of Rupp to L. A. Gilbert, 23 June 1947 to 16 June 1956. RBGS.

96. Rupp to J. Scrivener, 20 November 1941. RBGS.

97. Rupp to F. Fordham, 21 February 1944. RBGS.

98. The Scrivener sisters — Joyce (who married Dr Archibald Telfer) and Gwen (who married Mr Ed Artlett) — were two of the six daughters of Charles P. Scrivener and Edith, née Lethbridge, through whom there was an association with the King family. There were also two sons. The family ran a guest house and orchard at Taihoa, Mt Irvine (near Mt Wilson) where grandfather Charles Robert Scrivener (1855–1923), a noted surveyor, built an unusual and interesting home of his own design, with hexagonal rooms arranged around a central skylight. C. P. Scrivener was a friend of E. J. Gregson (1882–1955) and George Valder (1861–1950) who shared keen interest in natural history and agriculture, and much of their knowledge was passed on to the Scrivener daughters. Predeceased by her husband, Joyce Telfer died in 1988. Mrs Gwen Artlett and her husband still live at Taihoa. In November 1988 she graciously presented to the Royal Botanic Gardens, Sydney, a collection of one hundred letters written by Rupp to the sisters between October 1941 and December 1948 and over thirty of his original botanical drawings. In May 1989 Mrs Artlett made a further presentation of nine more letters and nineteen botanical drawings found by Mrs Telfer's son and daughter in their mother's papers.

99. Rupp to J. Scrivener, 7 December 1941. RBGS.

100. Rupp to G Scrivener, 26 November 1941. RBGS.

101. Rupp to J. and G. Scrivener, 20 April 1942. RBGS.

102. Rupp to T. E. Hunt, 3 August, 1946. RBGS.

103. Dorothy Curtis and Morwenna Jean, daughters of Alec Gemmell and Morwenna, née Curtis, of Braemar, Glen Aplin, were born in Stanthorpe. Like their English mother, Dorothy was a talented artist, with a flair for natural subjects. She became an art teacher, and in 1952 married David Morrice Gordon, a keen naturalist with a special talent for propagating native plants. George Althofer (best man at the wedding) and his brother Peter (both Rupp's friends, also) frequently visited the Gordons at Myall Park, where the household ultimately included three daughters and a son. Tragically, Dorothy Gordon died in a vehicle accident on 5 September 1985. Jean Gemmell, five years Dorothy's senior, tended to follow their Scottish father's interest in natural history, more especially in entomology and botany. She has discovered several new species of beetles, some of which were named in her honour, and has found many species of orchids hitherto unknown to occur in Queensland. She married Robert Harslett in 1951 and has two daughters and a son. Mrs Jean Harslett, BEM, of Mountain View, Amiens, Queensland, is well-known as a community worker and author of historical works on Stanthorpe and local state schools. As a tribute to Dorothy Gordon and her fine artwork, Mrs Harslett prepared the Introduction and text for her sister's *Australian Wildflower Paintings,* Sydney 1988. In 1989 she kindly presented a dozen of Rupp's letters to the Royal Botanic Gardens, Sydney.

104. Rupp to T. E. Hunt, 8 October 1945; see also letter of 25 October 1945. RBGS.

105. Rupp to J. Telfer (née Scrivener) 21 February 1948. RBGS.

106. E.g., Rupp to T. E. Hunt, 3 August 1946; to G. W. Althofer, 16 August 1949. RBGS.

107. Rupp to G. W. Althofer, 16 August 1949. RBGS.

108. Isobel Kendall Bowden, OAM (1908–86), naturalist, teacher and authoress, was born at Woodford, the daughter of Eric Kendall Bowden (1871–1931), solicitor and MP for Nepean, and Reinetta May (née Murphy). A pioneer Blue Mountains conservationist with an abiding interest in bush fire control, Miss Bowden taught kindergarten at Frensham and Blue Mountains Grammar Schools, and wrote children's stories. Shortly before her death, she sorted and indexed her papers and photographs, naturalist's diaries and botanical watercolours for transfer to the Blue Mountains and Mitchell Libraries. Her forty-two letters from Rupp, June 1947 to May 1956, are in the Royal Botanic Gardens, Sydney. See *Blue*

Mountains Gazette, 4 June 1986; *ADB,* 7, p. 360.

109. I. K. Bowden, Diary, Book I, pp. 185-7. Bowden Papers, Blue Mountains City Library, Springwood.

110. Rupp to I. K. Bowden, 15 July 1947. RBGS.

111. Rupp to I. K. Bowden, 13 January 1948. RBGS.

112. Rupp to I. K. Bowden, 29 January 1948. RBGS.

113. Rupp to I. K. Bowden, 5 February 1948. RBGS.

114. Rupp to I. K. Bowden, 25 February 1948. RBGS.

115. Rupp to T. E. Hunt, 11 February 1948. RBGS.

116. Rupp to G. Scrivener, 18 February 1948. RBGS.

117. Rupp to T. E. Hunt, 13 October 1948; 7 March and 10 April 1949; 10 March 1951. RBGS.

118. Rupp to J. H. Willis, 7 February 1951. RBGS.

119. Rupp to T. E. Hunt, 3 October and 3 November 1952. RBGS.

120. Isobel Bowden's notes on Rupp's letter of 15 July 1948. RBGS, and on her letter to 'Helen' of 16 March 1983, Bowden Papers, Blue Mountains City Library, Springwood.

121. K. Mair to I. K. Bowden, 12 November 1968. Bowden Papers, *loc. cit.*

122. James Hamlyn Willis, DSc, DipFor, was born at Oakleigh, Victoria in January 1910. He worked as a Field Officer with the Victorian Forests Commission, 1931-9, before serving first as botanist, then as senior botanist at the National Herbarium of Victoria, 1939-61. Dr Willis was assistant government botanist of Victoria and deputy director of the Royal Botanic Gardens, Melbourne, 1961-72. Like many of Rupp's correspondents, he had a long and distinguished association with the Field Naturalists' Club of Victoria. In 1987, Dr Willis very kindly lent his collection of 284 original letters from Rupp to four recipients so that copies could be lodged with similar material in the library of the Royal Botanic Gardens, Sydney. His *Handbook to Plants in Victoria,* Melbourne, 2 vols, 1962, 1973, was widely acclaimed. See *WWA,* 1947-74.

123. Rupp to J. H. Willis, 5 April 1949. RBGS.

124. Rupp to T. E. Hunt, 7 October 1948. RBGS.

125. Rupp to J. H. Willis, 8 April 1949. RBGS.

126. Rupp to J. H. Willis, 28 September 1949, RBGS, and similarly on 18 April 1951: 'I've always been an incorrigible inflicter of my screeds on other people'.

127. Rupp to W. Hunter, 12 April 1943. RBGM. William Hunter (1893-1971), surveyor and botanist, was born at Tatura, Victoria, and had completed his surveyor's apprenticeship when he joined the AIF in 1915. After the war he spent most of his life in work and retirement in East Gippsland. He corresponded with fellow members of the Field Naturalists' Club of Victoria, W. H. Nicholls, T. S. Hart and H. M. R. Rupp. See J. H. Willis in *Vic. Nat.,* April 1971, pp. 88-91.

128. See Rupp to H. D. Gordon, 23 March, 27 June, 14 July 1946, and a reply of 21 June 1946; Rupp to W. M. Curtis, 5 February 1947, where he noted 'my chief contact with Tasmanian orchids now is through a Launceston lad named Neil Burrows, who is very keen, very observant, and something of an artist'.

129. Briefly interviewed by telephone in January 1989.

130. See Rupp to M. W. Nichols, 26 October 1940 to 5 February 1946. RBGS.

131. Rupp to T. E. Hunt, 17 June and 27 July 1945; 15 March 1946; 29 January 1947. RBGS.

132. Rupp to T. E. Hunt, 27 July 1945. RBGS.

133. Rupp to T. E. Hunt, 6 and 21 October 1944. RBGS.

134. Rupp to T. E. Hunt, 18 June and 4 December 1942. RBGS.

135. Copies of letters, F. W. Schmidt to Rupp, 19 July 1948, and Rupp to F. W. Schmidt, 2 April 1949; 14 August 1950; 26 February and 14 March 1951, in RBGS by courtesy of Mr Alec Dickins, Balgowlah.

136. Rupp to T. E. Hunt, 19 January 1947. RBGS.

137. Rupp to T. E. Hunt, 13 May 1943. RBGS.

138. G. W. Althofer, OAM, was born in January 1903. Having established himself as an orchardist and nurseryman, he founded the Burrandong Arboretum in 1935. Shortly after, he established the native plant nursery, Nindethana, at Dripstone where his grandfather had settled ninety years before. Althofer and Rupp shared a keen interest in things literary and botanical. See *Wild Life and Outdoors,* April 1953, pp. 355-9; *Debrett's Handbook of Australia,* 1987, p. 17.

139. Rupp to G. W. Althofer, 27 July 1945. RBGS.

140. Rupp to G. W. Althofer, 27 July 1945 to 13 August 1955. RBGS. In 1963, Mr C. K. Ingram, then of Bathurst, now of Mt Tomah, produced about a hundred copies of extracts of Rupp's letters to George Althofer for the interest of like-minded naturalists.

141. Rupp to T. E. Hunt, 27 July 1945. RBGS.

142. Rupp to G. W. Althofer, 17 September 1945. RBGS.

143. Rupp to G. W. Althofer, 21 September 1945. RBGS.

144 *Proc. Linn. Soc. N.S.W.,* Vol. 73, 1948, pp. 133-5.

145. Rupp to Dorothy Gemmell, 10 October 1945. RBGS.

146. Rupp to T. E. Hunt, 8 October 1945. RBGS.

147. Rupp to G. W. Althofer, 24 September 1947. RBGS.

148. Rupp to G. Scrivener, 1 October 1947. RBGS.

149. Rupp to G. W. Althofer, 12 and 19 November 1947. RBGS.

150. See A. W. Dockrill in *AOR,* December, 1956, p. 157; and *Vic. Nat.,* September, 1964, pp. 128-38. See also Jones, *Native Orchids,* pp. 222-44.

151. Rupp to G. W. Althofer, 4 October 1950. RBGS. See H. M. R. Rupp, 'In the Althofer Country', *Vic. Nat.,* August 1951, pp. 67-8, and 'The Orchid Flora of the Central Western Slopes of New South Wales', *Proc. Linn. Soc. N.S.W.,* Vol. 73, 1948, pp. 130-6.

152. Edwin Daniel Hatch, FLS, was born in London in 1919 and went to New Zealand with his parents in 1922. While stationed at a munitions store at Waiouru in the North Island, 1941-5, he spent his leisure time studying accountancy and botanizing widely. Subsequently he has published twenty formal papers on orchids and over sixty other articles on natural history. Now living in retirement at Laingholm, Auckland, Mr Hatch was elected a fellow of the Linnean Society of London in May 1988. Unfortunately, only one letter from Rupp remains in Mr Hatch's possession — it is dated Willoughby, 17 May 1954. See E. D. Hatch, 'The Small Green Orchid', *Journal of the Auckland Botanical Society,* February 1988, pp. 2-11.

153. Rupp to T. E. Hunt, 6 July and 12 August 1947. RBGS.

154. Rupp to G. W. Althofer, 10 September 1951; to T. E. Hunt, 2 September 1952 and 25 June 1953; to F. Fordham, 14 May 1953; to J. H. Willis, 11 April 1953. RBGS.

155. Rupp to F. Fordham, 26 April 1940. RBGS.

156. Rupp to T. E. Hunt, 28 July 1944 and 10 January 1947. RBGS.

157. Rupp to G. V. Scammell, 20 October 1926. RBGS. Without visiting Australia, George Bentham (with Baron von Mueller's assistance) produced the seven volumes of the classic *Flora Australiensis* (London 1863-78), using descriptions prepared from herbarium specimens — a fact which makes his work, despite its shortcomings, even more remarkable.

158. Rupp to G. V. Scammell, 20 October 1926. RBGS.

159. Rupp to T. E. Hunt, 9 July 1943 (and to G. W. Althofer, 26 September 1947). RBGS.

160. Rupp to G. W. Althofer, 31 October 1945. RBGS.
161. Rupp to G. W. Althofer, 19 November 1947. Rupp's friends, Mrs Edith Coleman and George Scammell, had long espoused similar views. 'I agree with you as regards splitting of species. I take my seat among the lumpers quite cheerfully. In the case of my own species I waited three seasons in each case and saw much material — quite enough to feel sure they were fixed — and entitled to rank as species. Moreover I based my separation on more than three definite distinctions which are fully apparent in dried material. This splitting will react on those who succeed us — who will, in many instances have only dried material to work on. And what tangles they will find . . . The more I see, the more I admire those old botanists who[se] species today seem as well defined as when they were created.' Edith Coleman to G. V. Scammell, 14 January 1931. RBGS.
162. *Vic. Nat.*, January 1944, pp. 135–8.
163. Rupp to T. E. Hunt, 22 March 1946. RBGS.
164. Rupp to T. E. Hunt, 26 January 1947. RBGS.
165. Rupp to T. E. Hunt, 19 February 1950. RBGS.
166. Rupp to G. V. Scammell, 15 April 1926. RBGS.
167. Rupp to G. V. Scammell, 7 May 1926. RBGS.
168. Rupp to G. V. Scammell, 5 June 1926. RBGS.
169. Rupp to F. Fordham, 16 August 1932. RBGS.
170. Rupp to T. E. Hunt, 6 August 1943 and 8 October 1945. RBGS.
171. Rupp to T. E. Hunt, 26 January 1943. RBGS.
172. Rupp to T. E. Hunt, 26 January 1943 and 14 September 1944. RBGS.
173. Rupp to G. V. Scammell, 27 September 1927. RBGS. The orchid in question, *Corybas dilatatus* (Rupp et Nicholls) Rupp, is now referred to *C. diemenicus* (Lindl.) H. G. Reichb. Rupp described the same incident for Mrs H. G. Curtis (née Geissmann) in his letter of 26 August 1927. RBGS.
174. Rupp to G. V. Scammell, 6 July 1926. RBGS.
175. Rupp to T. E. Hunt, 21 February 1948. RBGS.
176. Rupp to W. H. Nicholls, 7 April 1948. RBGS.
177. Rupp to F. Fordham, 26 April 1940. RBGS.
178. Rupp to T. E. Hunt, 15 March 1944. RBGS.
179. Rupp to J. H. Willis, 13 December 1945. RBGS.
180. Rupp to G. V. Scammell, 17 April 1927. RBGS.
181. Rupp to G. V. Scammell, 13 July 1926. RBGS. By mid-1925, Rupp had made '5 or 6 dozen' glass projection slides from his photographs. 'It isn't difficult: just like printing off bromides or any other dark-room papers, only you have the glass slide instead of the paper'. Rupp to H. G. Geissmann, 19 June 1925. RBGS.
182. Rupp to G. V. Scammell, 23 July 1926. RBGS.
183. Rupp to G. V. Scammell, 26 July 1926. RBGS.
184. Rupp to G. V. Scammell, 28 July 1926. RBGS.
185. Rupp to G. V. Scammell, 12 November 1926. RBGS. The *Sydney Mail,* 27 October 1926, depicted two species of *Diuris* and five species and one variety of *Dendrobium* to illustrate Rupp's article.
186. Rupp to T. E. Hunt, 14 July 1941; 25 March 1945. RBGS.
187. Rupp to T. E. Hunt, 28 February 1945. RBGS.
188. Rupp to G. V. Scammell, 16 August 1928. RBGS.
189. Marjory Loader Collection.
190. Rupp to G. V. Scammell, 16 August 1928. RBGS.
191. Note by Rupp headed 'Fitzgerald's Plates', Marjory Loader Collection. Rupp always spelt R.D.F.'s name thus, without a capital 'g', holding that otherwise one would presume that the surveyor-botanist did not know how to spell his own name!

192. Rupp to G. V. Scammell, 25 July 1926. RBGS. The 'Maitland man' was the solicitor, W. J. Enright (1874–1949).
193. Rupp to F. Fordham, 24 September 1937. RBGS.
194. Rupp to F. Fordham, 8 December 1937. RBGS.
195. Rupp to F. Fordham, 2 May 1935. RBGS.
196. Rupp to F. Fordham, 17 August 1935. RBGS.
197. Rupp to T. E. Hunt, 23 September 1942. RBGS.
198. Rupp to F. Fordham, 28 October 1936. RBGS. A decade earlier Rupp described his despair during a summer drought, with century temperatures (Fahrenheit) 'every day' in February, and 'Death & destruction everywhere! (lost about 50 per cent of my orchids)'. Rupp to H. G. Geissmann, 22 March 1926. RBGS.
199. Rupp to F. Fordham, 8 November 1941. RBGS.
200. Rupp to T. E. Hunt, 28 December 1944. RBGS.
201. Rupp to T. E. Hunt, 9 August 1946. RBGS.
202. Rupp to T. E. Hunt, 25 February 1952. RBGS.
203. Rupp to F. Fordham, 2, 26 February 1937. RBGS.
204. Rupp to F. Fordham, 14 March 1942. RBGS.
205. Rupp to H. G. Geissmann, 16 June 1925. RBGS.
206. Rupp to H. G. Geissmann, 4 September 1925. RBGS.
207. Rupp to H. G. Geissmann, 1 February 1929. RBGS.
208. Rupp to H. G. Geissmann, 23 March 1931. RBGS.
209. Rupp to T. E. Hunt, 22 April 1941. RBGS.
210. Rupp to T. E. Hunt, 10 March 1951. RBGS.
211. Rupp to F. Fordham, 5 November 1953. RBGS.
212. Rupp to F. Fordham, 4 February 1947. RBGS.
213. Rupp to F. Fordham, 15 December 1953. RBGS.
214. Rupp to T. E. Hunt, 16 January 1946; also to A. H. Chisholm, 13 January 1946. RBGS.
215. Rupp to T. E. Hunt, 19 December 1946. RBGS.
216. Rupp to T. E. Hunt, 26 January 1947. RBGS.
217. Rupp to T. E. Hunt, 25 February 1947. RBGS.
218. Rupp to T. E. Hunt, 3 July 1947. RBGS.
219. Rupp to T. E. Hunt, 19 July 1947. RBGS.
220. Rupp to T. E. Hunt, 15 August 1946. RBGS. From Alfred, Lord Tennyson's 'Maud: A Monodrama'.
221. Rupp to G. W. Althofer, 10 October 1945. RBGS.
222. Rupp to G. W. Althofer, 13 June 1946. RBGS.
223. Rupp to G. W. Althofer, 1 July 1947. RBGS.
224. Rupp to F. Fordham, 19 June 1949. RBGS.
225. Rupp to T. E. Hunt, 15 July 1949. RBGS.
226. Rupp to F. Fordham, 15 and 20 October 1941; 6 February 1952; to T. E. Hunt, 23 August 1944; 8 February and 27 November 1951; 31 January and 13 February 1952. RBGS.
227. Rupp to T. E. Hunt, 27 April 1946. RBGS.
228. Rupp to T. E. Hunt, 6 December 1946. RBGS.
229. Rupp to T. E. Hunt, 24 February 1947. RBGS.
230. Rupp to T. E. Hunt, 9 and 11 March 1947. RBGS.
231. Rupp to T. E. Hunt, 20 November 1946; also 6 September 1945: 'my sight is colour-defective'. RBGS.
232. Rupp to T. E. Hunt, 27 February 1947. RBGS.
233. Rupp to F. Fordham, 13 September 1945. RBGS. Similarly Rupp to H. G. Geissmann 7 July 1925, confessing that caterpillars, grubs, snails and slugs were regarded 'all alike as my enemies' for they 'eat my beloved plants'. RBGS.
234. Rupp to F. Fordham, 10 January 1934. RBGS.
235. Rupp to F. Fordham, 2 November 1936. RBGS.
236. Rupp to T. E. Hunt, 11 July 1946. RBGS. This species is now referred to *D. nindii* W. Hill.
237. Rupp to F. Fordham, 6 October 1933. RBGS.
238. Rupp to F. Fordham, 6 January 1937. RBGS.
239. Rupp to T. E. Hunt, 5 January 1946. RBGS.
240. Rupp to T. E. Hunt, 26 March 1953. RBGS.
241. Rupp to J. and G. Scrivener, 30 August 1942. RBGS.

242. Rupp to T. E. Hunt, 6 October 1944. RBGS.
243. Rupp to G. W. Althofer, 26 September 1945. RBGS.
244. Rupp to T. E. Hunt, 3 April 1946. RBGS.
245. Rupp to J. and G. Scrivener, 25 March 1943. RBGS.
246. Rupp to Bishop Batty, 4 December 1930. AB6627. AD Newc.
247. Rupp to H. A. Woodd, 11 February 1933. AB6627. AD Newc.
248. Bishops' Letterbooks and Parish Files, *passim*. AD Newc.
249. Letterbooks (replies only) AD Arm.
250. Rupp to J. Scrivener, 18 September 1942. RBGS.
251. Rupp to J. H. Willis, 31 January 1942. RBGS.

Chapter 9: Retirement and Recognition

1. Rupp to F. Fordham, 7 May 1939. RBGS.
2. Rupp to F. Fordham, 7 May and 3 August 1939. RBGS.
3. Rupp to T. E. Hunt, 1 July 1947. RBGS.
4. Rupp to F. Fordham, 31 January 1940. RBGS.
5. Rupp to F. Fordham, 13 December 1940.RBGS.
6. Rupp to T. E. Hunt, 18 April 1947. RBGS. Rupp conducted most Sunday services in St Luke's during April 1944, and on Christmas Day, 1945, he assisted at four services between 6 a.m. and noon to cater for 784 communicants. Rupp to T. E. Hunt, 1 May 1944; Rupp to G. W. Althofer, 26 December 1945. RBGS.
7. Rupp to F. Fordham, 6 December 1939. RBGS.
8. Rupp to G. W. Althofer, 1 July 1947. RBGS.
9. Rupp to F. Fordham, 4 March 1946. RBGS.
10. Rupp to T. E. Hunt, 7 March 1949. RBGS. Reminiscences of Mr H. K. C. Mair, 1987.
11. Rupp to T. E. Hunt, 28 December 1944. RBGS. See also Rupp to J. and G. Scrivener, 2 October 1942. RBGS.
12. Rupp to F. Fordham, 3 September 1939. RBGS.
13. Rupp to F. Fordham, 30 May 1940. RBGS.
14. Rupp to F. Fordham, 29 September 1940. RBGS.
15. Rupp to F. Fordham, 17 June 1940. RBGS.
16. See L. A. Gilbert, *The Royal Botanic Gardens, Sydney: A History, 1816-1985*, Melbourne 1986, p. 148.
17. Rupp to F. Fordham, 30 December 1941. RBGS.
18. Rupp to A. H. Chisholm, 3 March 1942. RBGS.
19. Rupp to F. Fordham, 14 March 1942. RBGS.
20. Rupp to J. Scrivener, 3 June 1942. RBGS.
21. *Ibid.* A torpedo intended for the USS *Chicago* hit a stone wall on the eastern side of Garden Island. The resulting explosion sank a naval depot vessel, the ferry *Kuttabul*, with the loss of nineteen ratings.
22. Rupp to F. Fordham, 9 June 1942. RBGS.
23. Rupp to J. Scrivener, 20 May 1942. RBGS.
24. Rupp to J. Scrivener, 9 May 1942. RBGS.
25. Rupp to F. Fordham, 19 April 1945. RBGS.
26. Rupp was variously described the complaint as 'arthritis', 'arthritic neuritis' or 'muscular rheumatism'.
27. Rupp to T. E. Hunt, 14 December 1942. Similarly to A. H. Chisholm, 13 October 1942. RBGS.
28. Rupp to J. and G. Scrivener, 16 December 1942. RBGS.
29. Rupp to J. Scrivener, 5 August 1943. RBGS. R. H. Anderson, BScAgr (1899-1969), director and chief botanist, Sydney Botanic Gardens, 1945-64, joined the Gardens staff in 1921, and succeeded Edwin Cheel as botanist and curator of the Herbarium in 1936. He had keen interests in native trees and in agricultural botany. *WWA*, 1962, and Gilbert, *Royal Botanic Gardens*, Chapters 6 and 7. Joyce Winifred Vickery,

MBE, DSc (1908-79) was appointed Assistant Botanist in August 1936, 'the first woman ever appointed a professional officer in a scientific capacity in the Public Service'. She became an acknowledged authority on grasses, and won unwanted publicity, but merited praise, in presenting forensic evidence in the Graeme Thorne kidnapping and murder case. This and other services led to the award of an MBE in 1962. See Alma Lee in *Telopea*, Vol. 2(1), 1980.
30. Rupp to F. Fordham, 19 August 1943. Similarly to T. E. Hunt, 31 August 1943. RBGS.
31. Rupp to J. and G. Scrivener, 31 August 1943. RBGS.
32. Rupp to T. E. Hunt, 3 October 1943. RBGS.
33. Rupp to T. E. Hunt, 29 April 1947; similarly to G. H. Curtis (née Geissmann) 19 May 1944 RBGS.
34. H. M. R. Rupp, *The Orchids of New South Wales*, Sydney 1943, pp. xiv-xv.
35. Rupp to A. C. Telfer, 7 and 22 October 1943. RBGS.
36. Rupp to J. Scrivener, 21 January 1944, and to T. E. Hunt, 27 February 1944. RBGS. See also D. F. Blaxell in *The Orchadian*, December 1978, p. 42.
37. Rupp to A. H. Chisholm, 20 March 1944. RBGS.
38. Rupp to A. H. Chisholm, 18 March 1944. RBGS.
39. Rupp gave his 'original sketch' to Gwen Scrivener. Rupp to G. Scrivener, 23 December 1943. See also Rupp to F. Fordham, 21 February 1944; and to H. G. Curtis (née Geissmann), 19 May 1944. RBGS.
40. Rupp to G. Scrivener, 23 December 1943. RBGS.
41. Rupp to T. E. Hunt, 18 May 1944; Rupp to H. G. Curtis (née Geissmann) 19 May 1944. RBGS.
42. Rupp to F. Fordham, 18 May 1944. RBGS.
43. Rupp to T. E. Hunt, 27 February 1944. RBGS. Rupp wrote similarly to Hilda Curtis (née Geissmann) on 19 May 1944, declaring 'My indebtedness to Joyce Vickery goes far beyond what is said in my acknowledgements . . . that is all she would allow me to put in'. RBGS.
44. Rupp to T. E. Hunt, 29 April 1947. RBGS.
45. Rupp to A. H. Chisholm, 5 August 1944; to H. G. Curtis (née Geissmann), 19 May 1944. RBGS.
46. Rupp to A. H. Chisholm, 31 October 1945. RBGS.
47. Rupp to F. Fordham, 29 January 1944. RBGS.
48. Rupp to F. Fordham, 24 July 1944. RBGS.
49. Rupp to T. E. Hunt, 15 August 1944. RBGS.
50. Rupp to A. H. Chisholm, 13 August 1944. RBGS.
51. Rupp to T. E. Hunt, 20 August 1945. RBGS.
52. Rupp to F. Fordham, 14 September 1945. RBGS.
53. Rupp to F. Fordham, 17 October 1945. RBGS.
54. Rupp to A. H. Chisholm, 31 October 1945. RBGS.
55. Rupp to J. H. Willis, 23 November 1945. RBGS.
56. Rupp to T. E. Hunt, 24 November 1945; to F. Fordham, 29 November 1945. RBGS.
57. Rupp to G. Scrivener, 2 April 1946. RBGS.
58. Rupp to G. W. Althofer, 26 December 1945. RBGS.
59. *Ibid.* Also Rupp to G. Scrivener, 2 April 1946; to A. H. Chisholm, 15 April 1946. RBGS.
60. D. L. Jones, *Native Orchids of Australia*, Sydney 1988, pp. 415-6.
61. Rupp to J. and G. Scrivener, 19 August 1943. RBGS.
62. Rupp to F. Fordham, 31 May 1945. RBGS.
63. Rupp to F. Fordham, 14 September 1945. RBGS. This Sun Orchid is now recognized as a natural hybrid.
64. Rupp to F. Fordham, 8 January 1946, and to A. H. Chisholm, 29 March 1947. RBGS.
65. Rupp to May Garrard, 22 November 1947. Original in possession of Mrs Eileen Cox (née Rupp). Copy in RBGS.
66. Rupp to A. H. Chisholm, 18 February 1946. RBGS.

67. *SMH,* 22 February 1946.
68. *SMH,* 12 March 1946.
69. *SMH,* 19 March 1946.
70. Rupp to T. E. Hunt, 15 March 1946; to A. H. Chisholm, 23 March 1946. RBGS.
71. Rupp to G. Scrivener, 19 March 1946; to T. E. Hunt, 11 March 1947. RBGS.
72. Rupp to F. Fordham, 12 April 1947. RBGS.
73. Rupp to T. E. Hunt, 15 and 22 March, 3 April 1946. RBGS.
74. Rupp to T. E. Hunt, 15 November 1946; to J. H. Willis, 16 November 1946. RBGS.
75. Rupp to Phyllis Mander Jones, 20 November 1946, and The Unpublished Plates of R. D. Fitzgerald's 'Australian Orchids' in the Mitchell Library, Sydney. Typescript. ML D248.
76. Rupp to T. E. Hunt, 19 June 1947. RBGS. Mrs Campbell also wrote to L. A. Gilbert, 8 March 1949, about this trunk, but the result of any investigation is not known. Rupp published an article about five such plates in the Herbarium. See *AOR,* September 1939, pp. 65-6.
77. Rupp to T. E. Hunt, 11 February, 11 and 28 March 1947. RBGS.
78. Rupp to J. H. Willis, 21 April 1947. RBGS. Dr Joyce Vickery has already been noticed in reference 29; Dr Mary Tindale, well-known for her special interests in ferns and wattles, is now an honorary research associate at the Herbarium; Miss Joy Gardiner-Garden became Mrs Thompson, and was also elected an honorary research associate.
79. Rupp to T. E. Hunt, 25 June and 1 July 1947. RBGS. See also A. P. Elkin, *The Diocese of Newcastle,* Sydney 1955, pp. 692-4.
80. Rupp to T. E. Hunt, 31 July and 2 August 1947. RBGS.
81. Rupp to A. H. Chisholm, 3 December 1930 and 22 March 1933. RBGS.
82. See Bibliography.
83. Rupp to A. H. Chisholm, 22 August 1947. RBGS. Also Rupp to T. E. Hunt, 16 August 1947. RBGS.
84. Rupp to A. H. Chisholm, 10 October 1947. RBGS.
85. Rupp to T. E. Hunt, 10 September 1947. RBGS.
86. Rupp to T. E. Hunt, 10 April and 16 July 1948. RBGS.
87. Rupp to T. E. Hunt, 13 June 1948. RBGS.
88. Rupp to T. E. Hunt, 28 June 1948. RBGS. Rupp began a revision of this catalogue, 'then woefully out of date', as late as March 1954. Rupp to F. Fordham, 13 March 1954. RBGS.
89. Rupp to T. E. Hunt, 12 November 1952. RBGS.
90. Rupp to T. E. Hunt, 15 August 1948; also to J. H. Willis, 21 August 1948, and to F. Fordham, 23 August 1948. RBGS.
91. Rupp to F. Fordham, 23 August 1948. RBGS.
92. *Colac Herald,* 10 September 1948.
93. Rupp to F. Fordham, 12 October 1948. RBGS.
94. Rupp to T. E. Hunt, 7 October 1948. RBGS.
95. Rupp to J. H. Willis, 10 October 1948; to F. Fordham, 12 October 1948. RBGS.
96. Rupp to F. Fordham, 4 November 1948; also to G. Scrivener, 5 November 1948, to T. E. Hunt, 7 November 1948, and to J. H. Willis, 9 November 1948. RBGS.
97. Rupp to J. H. Willis, 13 November 1948. RBGS.
98. Rupp to G. Scrivener, 26 December 1948. RBGS. Rupp had begun writing by November.
99. Rupp to T. E. Hunt, 17 January 1949. RBGS. Frederick Sefton Delmer, MA (1864-1931) was a native of Hobart and graduate of the University of Melbourne. He became a journalist and professor of English at Berlin. *Biog. Reg.* I, p. 176.
100. *SMH,* 8 August 1949.

101. Rupp to J. H. Willis, 8 April 1949; also to W. H. Nicholls, 7 April 1949. RBGS.
102. Rupp attributed this honour to J. H. Willis's influence. See Rupp to J. H. Willis, 3 February 1953; to T. E. Hunt, 24 February 1953. RBGS.
103. Rupp to G. W. Althofer, 7 June 1955. See also Rupp to J. H. Willis, 6 June 1955; to I. K. Bowden, 29 June 1955. RBGS.
104. *AOR,* September 1955, p. 132.
105. Rupp to J. H. Willis, 8 July 1955. RBGS.
106. *Acacia ruppii* Maiden et Betche. *Proc. Linn. Soc. N.S.W.,* 1912, pp. 244-5; *Boronia ruppii* Cheel. *Proc. Roy. Soc. N.S.W.,* 1927, pp. 404-5; *Prasophyllum ruppii* Rogers. *Proc. Roy. Soc. S.A.,* 1927, pp. 292-3; *Cadetia ruppii* St. Cloud. *N.Q. Nat.,* XXIII, 110, pp. 2-3 (1 January 1955). This name was withdrawn as the species was found to be identical with that then known as *Dendrobium glabrum* J. J. Smith, now known as *Diplocaulobium glabrum* (J. J. Smith) Kraenzlin. *N.Q. Nat.,* XXIV, p. 112 (1 September 1955). Hawkes's name was published in the *Orchid Weekly,* 2, p. 129 (1960). See also *AOR,* March 1964, p. 40. Previously known as *D. fusiforme* Bail., the species is now referred to *D. jonesii* Rendle.
107. Rupp to J. Scrivener, 27 February 1942. RBGS.
108. Rupp to F. Fordham, 13 April 1942; to T. E. Hunt, 22 April 1942. RBGS.
109. Rupp to T. E. Hunt, 17 August 1950. RBGS.
110. Rupp to F. Fordham, 18 September 1950. RBGS.
111. Rupp to T. E. Hunt, 13 March 1951. RBGS.
112. Rupp to F. Fordham, 17 March 1951. RBGS.
113. Rupp to J. H. Willis, 25 March 1951. RBGS.
114. Rupp to J. Gemmell, 26 March 1951. RBGS.
115. Rupp to F. Fordham, 6 May 1951. RBGS.
116. Rupp to T. E. Hunt, 13 May 1951. RBGS.
117. Rupp to T. E. Hunt, 12 June 1951. RBGS.
118. Rupp to J. H. Willis, 3 January 1952. RBGS.
119. Rupp to T. E. Hunt, 1 May 1953. RBGS.
120. Rupp to F. Fordham, 6 May 1951. RBGS.
121. Rupp to T. E. Hunt, 31 January 1953; also to J. H. Willis, 17 January 1953. RBGS.
122. Rupp to J. H. Willis, 20 February 1953. RBGS.
123. Rupp to J. H. Willis, 13 January 1953. RBGS.
124. Rupp to J. H. Willis, 17 January 1953. RBGS.
125. Rupp to J. H. Willis, 2, 14 and 26 July 1949. RBGS.
126. Rupp to J. H. Willis, 17 November 1949. RBGS.
127. Rupp to J. H. Willis, 11 February 1950. RBGS. Rupp mentioned Mrs Henriette Sinclair's death 'at Rockhampton' in Rupp to J. H. Willis, 27 August 1953. RBGS.
128. Rupp to T. E. Hunt, 1 June, 1 July and 17 September 1951; to J. H. Willis, 14 October 1951. RBGS.
129. Rupp to T. E. Hunt, 31 May 1951. RBGS.
130. Rupp to J. H. Willis, 10 December 1951. RBGS.
131. Rupp to T. E. Hunt, 12 November 1952, 28 September and 15 December 1953. Also Fr B. Lowery to L. A. Gilbert, 14 April 1989.
132. Rupp to T. E. Hunt, 17 October 1953. RBGS.
133. Rupp to F. Fordham, 19 November 1948; 8 November 1949; 6 May 1951. RBGS.
134. Rupp to T. E. Hunt, 17 October 1953. RBGS.
135. Rupp to J. H. Willis, 6 July, 28 July and 20 August 1952. RBGS.
136. Rupp to J. H. Willis, 5 February 1954. RBGS.
137. Rupp to J. H. Willis, 8 July 1954. RBGS.
138. Rupp to J. H. Willis, 12 August 1954. RBGS. These two species, long considered synonymous, are now regarded as distinct.

139. Rupp to J. H. Willis, 13 January 1955. RBGS. Fr B. Lowery to L. A. Gilbert, 14 April 1989.

140. Rupp to J. H. Willis, 20 November 1955. See Bibliography. Rupp referred to 'a few Diuris descriptions' as his 'swan song' in letters to F. Fordham, 8 August 1955, and to T. E. Hunt, 22 September 1955. RBGS.

141. Rupp to G. W. Althofer, 13 August 1955, and to I. K. Bowden, 21 July 1954. RBGS.

142. Rupp to I. K. Bowden, 29 June 1955. RBGS.

143. Rupp to J. H. Willis, March 1956. RBGS.

144. Rupp to F. Fordham, 14 June 1956. RBGS.

145. Rupp to I. K. Bowden, 17 May 1956. Original in Blue Mountains City Library, Springwood.

146. *Transactions of the Royal Society of New South Wales*, 1867, Vol. I, p. 27. Also *SMH*, 10 July 1867.

Part II Recollections of an Amateur Botanist

Chapter 1: Early Days

1. I.e., stepmother, Rachel.

2. Rev. Charles Ludwig Herman Rupp (1839–1917).

3. Armstrong Bay.

4. *Clematis microphylla* DC.

5. Musk Daisy-bush, *Olearia argophylla* (Labill.) Benth.

6. Soft Tree-fern, *Dicksonia antarctica* Labill.

7. January 1884.

8. John Bracebridge Wilson (1828–95), naturalist and educationist, married Oriana Maria Rowcroft, H. M. R. Rupp's aunt, who died in 1911. *ADB*, 6, pp. 417–18.

9. Baron Sir Ferdinand Jakob Heinrich von Mueller (1825–96), government botanist of Victoria, 1853–96, director of Melbourne Botanic Gardens, 1857–73. *ADB*, 5, pp. 306–8.

10. Sir Charles Frederic Belcher, OBE, MA, LLB (1876–1970), who served widely in the Colonial Service as legal adviser and judge, published books on the birds of Geelong (1914) and Nyasaland (1930). *WWA*, (1968); *Biog. Reg.*, I, p. 47.

11. Helmet Orchids are now known as *Corybas*. The species would have been *C. diemenicus* (Lindl.) Reichb.f.

12. I.e., 1932/3.

13. *Prostanthera nivea* A. Cunn. ex Benth.

14. *Dendrobium speciosum* Sm. and *D. striolatum* Reichb.f., *Sarcochilus australis* (Lindl.) Reichb.f. and *S. falcatus* R.Br. A fifth Victorian epiphytic orchid was recorded later — *Thrixspermum tridentatum* (Lindl.) T. E. Hunt, now *Plectorrhiza tridentata* (Lindley) Dockrill.

15. Ferdinand von Mueller, *Key to the System of Victorian Plants*, 2 vols, Melbourne 1885–8.

16. John George Robertson (1803–62) arrived in Victoria from Van Diemen's Land in 1840, and took up Wando Vale. Before retiring to his native Scotland in 1854, he had sent some 4,000 plant specimens to Kew. *Biog. Reg.*, II, p. 219.

Chapter 2: The Hobby Progresses

1. Alexander Leeper, BA, LLD (1884–1934), classicist, was appointed principal of Trinity College in 1876, and warden, 1881–1918. *ADB*, 10, pp. 54–7.

2. Typhoid fever.

3. Sir Walter Baldwin Spencer, KCMG, MA, LittD, FRS (1860–1929), biologist and anthropologist, was Professor of Biology, University of Melbourne, 1887–1920. He joined the Royal Society and the Field Naturalists' Club of Victoria in 1887, and in 1899 succeeded Sir Frederick McCoy as honorary director of the National Museum of Victoria. See D. J. Mulvaney and J. H. Calaby, *'So Much that is New': Baldwin Spencer, 1860–1929: A Biography*, Melbourne 1985.

4. Thomas Sergeant Hall, MA, DSc (1858–1915) was appointed lecturer in Biology, University of Melbourne, in December 1893. He joined the Field Naturalists' Club of Victoria in 1888 and took a leading role in its affairs. *ADB*, 9, pp. 166–7.

5. William Austin Horn (1841–1922), pastoralist and promoter of mining and exploration, sponsored a scientific expedition to central Australia in 1894. *ADB*, 9, pp. 367–9.

6. Sir Frederick McCoy, KCMG, MA, DSc, FRS (1817–99) was appointed professor of Natural Science, University of Melbourne, in 1854, and worked virtually until his death in May 1899. *ADB*, 5, pp. 134–6.

7. See the portrait reproduced nearby.

8. Wilson died at Geelong on 22 October 1895; von Mueller died at South Yarra on 10 October 1896.

9. *Diuris punctata* var. *albo-violacea* Rupp ex Dockrill (now *D. fragrantissima* Jones et Clements).

10. E.g., the Manna or Ribbon Gum, *E. viminalis* Labill.; Messmate Stringybark, *E. obliqua* L'Hérit.; Mountain Grey Gum, *E. cypellocarpa* L.A.S. Johnson.

11. *S. parviflorus* Lindl., long considered identical to *S. australis* (Lindl.) Reichb.f.

12. In August 1895.

13. December 1894 to February 1895. Florence was the wife of Arthur Augustus Monypenny.

14. Issues of 31 December 1894; 4, 11, 15 and 18 January 1895; 19 February 1895.

15. Both the yellow-flowering Wilcannia Lily, *Calostemma luteum* Sims, and the pink, or reddish-purple Garland Lily, *C. purpureum* R.Br., occur in the Riverina. For some time the species were combined under the latter name, but now are regarded as separate species once more.

16. R. S. Rogers, in *Proc. Roy. Soc. Vic.* 1915. Previously this species had been confused with the Leafy Greenhood, *Pterostylis cucullata* R.Br. It is now referred to *P. furcata* Lindl.

17. Rt Rev. Samuel Thornton (1835–1917), first bishop of Ballarat, 1875–1900.

18. Rt Rev. Henry Edward Cooper, MA, DD (1845–1916), former curate to Rev. Samuel Thornton in Birmingham; vicar

of Hamilton, Victoria, 1884–93; archdeacon of Ballarat, 1894–95; coadjutor bishop to Thornton at Ballarat, 1895–1901; bishop of Grafton and Armidale, 1901–14; and first bishop of Armidale, 1914–16.

Chapter 3: Tamworth and Warialda

1. Thomas Kingsmill Abbott, MA (Oxon.) (1866–1912) was born 18 February 1866 near Murrurundi, N.S.W. He was appointed archdeacon of Tamworth in 1902, after service in London; St James's, Sydney; St Jude's, Randwick; St Paul's College, Sydney; and St Hilda's, Katoomba. From 1910 to 1912 he was headmaster of The Armidale School, where on 6 December 1912 he died from a heart attack while 'playing vigorously at the net' on the school tennis court.
2. Florence Mabel Dowe (1879–1956) was born at Tamworth, N.S.W., on 18 October 1879, the daughter of Richard Andrew Dowe (1848–1915), solicitor, and Mary, née Bloomfield (1849–1930). Richard Dowe was the son of Dr Joshua Dowe (1812–75), resident physician at Windsor, N.S.W., 1841–67, and Sarah, née Loder. The Loder and Howe families were related. George Loder accompanied the expeditions led by his father-in-law John Howe from Windsor to the Hunter Valley, 1818–20. The families are commemorated by Howe's Park, Singleton, Howe's Valley, Loder Creek and Loder House, Windsor.
3. Joseph Henry Maiden (1859–1925), director of Sydney Botanic Gardens and government botanist, 1896–1924. *ADB*, 10, pp. 381–3.
4. *Cymbidium canaliculatum* R.Br.
5. *Actinotus helianthi* Labill.
6. White Cypress, *Callitris glaucophylla* Thompson et Johnson; Black Cypress, *C. endlicheri* (Parl.) F.M. Bail.
7. *Casuarina cristata* Miq.
8. *Santalum lanceolatum* R.Br.
9. *Eremophila mitchellii* Benth.
10. Now *A. floribunda* (Sm.) Sweet.
11. E.g. *Pterostylis mutica* R.Br.
12. 1908.

Chapter 4: The Clarence

1. *Dendrobium beckleri* F. Muell.; *D. teretifolium* R.Br.
2. *D. linguiforme* Sw.
3. *Calanthe triplicata* (Willem.) Ames, formerly *C. veratrifolia* (Willd.) R.Br.
4. Giant Stinging Tree or Gympie, now known as *Dendrocnide excelsa* (Wedd.) Chew.
5. *Acacia ruppii* Maiden et Betche. *Proc. Linn. Soc. N.S.W.*, 1912.
6. The Lemon-scented Tea Tree, *Leptospermum citratum* (J.F. Bailey et C.T. White) Challinor, Cheel et Penfold, is now referred to *L. petersonii* F.M. Bailey (1905).
7. John Luke Boorman (1864–1938), an Englishman, was trained at Kew Gardens. He was appointed gardener at the Sydney Botanic Gardens in January 1887, and collector in October 1901. He retired in 1924 after a career which J. H. Maiden noted had shown him to be a 'a hardworking, unselfish, most intelligent servant of the public'.
8. Now Mimosaceae.

Chapter 5: Barraba and the Nandewars; and Pickings Here and There

1. Mt Kaputar is correct.
2. Bed-bug, *Cimex lectularius.*
3. *Dendrobium linguiforme* Sw.
4. *Cymbidium canaliculatum* R.Br.
5. *Boronia ruppii* E. Cheel. *Proc. Roy. Soc. N.S.W., 1927.*
6. *Coprosma hirtella* Labill.
7. 1914–20.
8. Henry Bruné Atkinson (1874–1960), appointed rector of Holy Trinity, Hobart, 1916, collated archdeacon, 1924. *ADB*, 7, pp. 120–1.
9. Rev. Henry Dresser Atkinson.
10. Rt Rev. Robert Snowdon Hay (1867–1943), bishop of Tasmania, 1919–43.
11. These species, then known as *Prasophyllum baueri* (R.Br.) Poir and *P. deaneanum* FitzG., have both been referred by some authors to *Genoplesium baueri*. See H. M. R. Rupp, 'Robert Brown's *Genoplesium baueri*', *Vic. Nat.*, August 1949, pp. 75–9.
12. I.e., during World War I.
13. Robert David FitzGerald (1830–92), surveyor and orchidologist, lived at Adraville, Hunter's Hill. His magnificent folio work, *Australian Orchids*, was published in parts between 1875 and 1894. *ADB*, 4, pp. 178–9. Rupp maintained that the orchidologist himself did not use a capital 'G' in his name.
14. Pearl R. Messmer (née Finckh).
15. Richard Sanders Rogers, MA, MD, DSc (1862–1942), medical practitioner, consulting physician and orchidologist. See H. M. R. Rupp, *AOR*, June 1942.

Chapter 6: Tasmania

1. Leonard Rodway (1853–1936), dentist and member of honorary staff, Hobart General Hospital; honorary botanist to Tasmanian government, 1896–1932. Author of *Tasmanian Flora*, Hobart 1903.
2. For Dr F. A. Rodway, see Part I, Chap. 8, ref. 29.
3. St Aidan's parish, East Launceston, was formed in 1921 from the existing parishes of St John and Holy Trinity. See W. R. Barrett, *History of the Church of England in Tasmania*, Hobart 1942, p. 78.
4. For Mrs F. Perrin, see Part I, Chap. 8, ref. 13.
5. Arthur Henry Shakespeare Lucas (1853–1936), schoolmaster and phycologist, was foundation editor of the *Victorian Naturalist*, 1884–92. In retirement, he retained his wide diversity of interests, becoming acting professor of Mathematics in the University of Tasmania and honorary curator of algae in the Herbarium of Sydney Botanic Gardens. *ADB*, 10, pp. 163–4. See also *A. H. S. Lucas, Scientist: His Own Story*, Sydney 1937.
6. Myrtle Beech, *Nothofagus cunninghamii* (Hook.) Oerst.
7. Tasmanian Christmas Bush, *Prostanthera lasianthos* Labill.; New South Wales Christmas Bush, *Ceratopetalum gummiferum* Sm.
8. *Telopea truncata* (Labill.) R.Br.
9. *Pterostylis falcata* Rogers. *Proc. Roy. Soc. Vic.*, 1915. Now referred to *P. furcata* Lindl.
10. Rt Rev. Reginald Stephen (1860–1956), fifth bishop of Newcastle, 1919–28, bishop of Tasmania, 1914–19, dean of Melbourne, 1910–14, a native of Geelong and, like Rupp, educated at Geelong Grammar and Trinity College, University of Melbourne.

Chapter 7: Bulahdelah

1. At Branxton.
2. See Rupp in *Proc. Linn. Soc. N.S.W.*, 1932, pp. 57–61. Now known as *Rhizanthella slateri* (Rupp) M. Clements et Cribb.
3. For E. W. Slater, see Part I, Chap. 8, ref. 61.
4. *Dipodium punctatum* (Sm.) R.Br.
5. *Corybas undulatus* (A. Cunn.) Rupp et Nicholls. See R. S. Rogers in *Proc. Roy. Soc. S.A.*, 1927.
6. For Dr H. L. Kesteven, see Part I, Chap. 7, ref. 44.
7. *Dendrobium kestevenii* Rupp. *Proc. Linn. Soc. N.S.W.*, 1931, p. 137. This is now considered to be a hybrid between *D. speciosum* var. *speciosum* and *D. kingianum.*
8. Dr Rogers served as president of the Botany Section at the 1932 meeting of the Australian and New Zealand Association for the Advancement of Science.
9. For Edith Coleman, see Part I, Chap. 8, ref. 10.
10. For W. H. Nicholls, see Part I, Chap. 8, ref. 7.
11. *Proc. Linn. Soc. N.S.W.*, 1928. *Corysanthes* is now known as *Corybas.*
12. For E. Cheel, see Part I, Chap. 8, ref. 39.
13. For A. H. Chisholm, see Part I, Chap. 8, ref. 8.
14. For H. G. Curtis, née Geissmann, see Part I, Chap. 8, ref. 9.
15. For C. T. White, see Part I, Chap. 8, ref. 59.
16. For A. Forster, see Part I, Chap. 8, ref. 30.
17. Herman E. Finckh, father of Mrs Pearl R. Messmer.
18. For A. G. Hamilton, see Part I, Chap. 8, ref. 36.

Chapter 8: Paterson and Barrington Tops

1. Negrohead or Antarctic Beech.
2. Sir John Clifford Valentine Behan (1881–1957), who succeeded Dr Alexander Leeper as Warden of Trinity College, 1918–46. *ADB*, 7, pp. 247–8.
3. See Part I, Chap. 7.
4. For C. L. Barrett, see Part I, Chap. 8, ref. 11.
5. Edward Edgar Pescott, FLS, FRHS (1872–1954), teacher, botanist, historian and author, government pomologist, Victorian Department Agriculture, 1917–37, long associated with Wattle Day League. President, Field Naturalists' Club of Victoria, 1926–7.
6. For G. V. Scammell, see Part I, Chap. 8, ref. 41.
7. 16 November 1925.
8. *Prasophyllum ruppii* R.S. Rogers. *Proc. Roy. Soc. S.A.*, 1927.
9. John Hopson (1867–1928), a farmer with special interests in birds and insects, lived all his life at Eccleston where for over forty years he was a leader of the Congregational Church. He guided and otherwise assisted many scientific excursions to the Barrington Tops from 1915, and joined the Linnean Society of New South Wales in 1918. His name is commemorated by some insect species he discovered. *Proc. Linn. Soc. N.S.W.*, 1929, pp. vi–vii.
10. *Diuris venosa* Rupp. *Vic. Nat.*, 1926, p. 153; *Proc. Linn. Soc. N.S.W.*, 1926, p. 313, and 1928, p. 336.
11. *Prasophyllum rogersii* Rupp and *P. hopsonii* Rupp. *Proc. Linn. Soc. N.S.W.*, 1928. The latter is now referred to *P. nudum* Hook.f.
12. *Pterostylis falcata* Rogers, now referred to *P. furcata* Lindl.
13. 17 June 1928.
14. Walter John Enright (1874–1949), solicitor and 'amateur scientist of distinctly high calibre', with special interests in local government, anthropolgy, geology and modern languages. *ADB*, 8, pp. 439–40.
15. Robert David FitzGerald, *Australian Orchids*, Sydney. Vol. 1 (Parts 1–7) 1875–82; Vol. II (Parts 1–5) 1884–94.

Chapter 9: The 'Guide'; the Coalfields and Pilliga Scrub

1. Charles Moore and Ernst Betche, *Handbook of the Flora of New South Wales . . .*, Sydney 1893.
2. The only edition.
3. Florence Sulman, *A Popular Guide to the Wild Flowers of New South Wales*, 2 vols, London/Sydney 1913–14.
4. William A. Dixon, *The Plants of New South Wales*, Sydney 1906.
5. Sir Wilfrid Russell Grimwade, CBE, BSc, FACI (1879–1955), Melbourne businessman, published *An Anthography of the Eucalypts* (illustrated with his own photographs) in 1920; 2nd ed., 1930. *ADB*, 9, pp. 126–8.
6. George Robertson (1869–1933) entered into partnership in January 1886 with David Angus (1855–1901) to form the well-known firm of booksellers and publishers.
7. 'Hope deferred maketh the heart sick: but when the desire cometh, it is a tree of life'.
8. *Pterostylis furcillata* Rupp. *Proc. Linn. Soc. N.S.W.*, 1930.
9. Rt Rev. John Stoward Moyes (1884–1972), bishop of Armidale, 1929–64.
10. *Cymbidium canaliculatum* R.Br., forma *aureolum* Rupp. *Proc. Linn. Soc. N.S.W.*, 1934.
11. Hector Macquarrie, *We and the Baby*, Sydney 1929. The 'baby' was a Baby Austin.
12. Bede Theodoric Goadby (1862–1945), lieutenant-colonel, Royal Australian Engineers, came to W.A. about 1895 and went to New Britain with the First Contingent in 1914. He later joined the Western Australian Naturalists' Club and the Field Naturalists' Club of Victoria. Many of his specimens are preserved in the herbaria of Perth, Melbourne and Kew. See S. S. Mackenzie, *The Australians at Rabaul (Official History of Australia in the War of 1914–18*. Vol. X); and Edith Coleman in *Vict. Nat.*, June 1945.

Part III: Retrospect

Chapter 1: Early Days

1. Actually, Rupp's mother died a fortnight after he was born.
2. December 1948 to January 1949. Rupp took the MS to Angus and Robertson on 17 January 1949.
3. Carl Ludwig Rupp
4. Or a little later.
5. It is not clear whether the parents and baby Paul died before or after embarkation at Hamburg in October 1848.
6. The College was then (1862) located at Liverpool. It was moved to Newtown in 1891.
7. The family left Koroit towards the end of 1883. Rupp originally wrote '66 years ago', changing it to '72' in a subsequent revision, presumably in 1955.
8. Rupp would have been delighted with the recent efforts to restore some of 'those glories'.
9. Dugald Graeme MacDougall (1873-1947), a Bendigo-born newspaperman, became publicity officer for the opening of Sydney Harbour Bridge in 1932 and general organizing secretary for the Sesquicentenary Celebrations, 1938. *WWA*, 1947; *Biog. Reg.*, II, p. 50.
10. 4 Macleay Street, Potts Point.
11. Respectively of the genera *Caladenia* and *Diuris*.
12. J. Bracebridge Wilson, MA, FLS, *Florula Corioensis; or Excursions near Geelong in Search of Plants, with a List of those Collected*, Geelong 1889.
13. J. Bracebridge Wilson, MA, FLS, *List of Algae from Port Phillip Heads and Western Port Collected while Dredging for Polyzoa and Sponges*, Geelong 1889.
14. J. L. Cuthbertson, BA (1851-1910), classics teacher and poet, taught at Geelong Grammar 1875-8 and 1885-96. Retaining in retirement his keen interest in the school, he sought Rupp's reminiscences in 1905. *ADB*, 3, pp. 514-15 and Geelong Grammar School Archives. For a new assessment of the influential and beloved, if alcoholic, Cuthbertson, see Weston Bate, *Light Blue Down Under: The History of Geelong Grammar School*, Melbourne 1990, pp. 41-53, and elsewhere.
15. Albert Finchett Garrard (1857-1947) married Oriana Mary (May) Bracebridge Wilson, the headmaster's elder daughter, Rupp's cousin. Their daughter, Edith, Mrs Adair, Rupp's first cousin once removed, now in her ninety-fourth year, has greatly enriched the school's archives by her presentations.
16. Arthur Alfred Lynch, MA, MRCS, LRCP (1861-1934), universal scholar and prolific author, had a remarkably varied career in teaching, engineering, journalism, politics and medicine, and in subversive movements which led to his being formally sentenced to death for treason in 1903. A Melbourne graduate, Lynch was studying in the University of Berlin by 1888 and did not return to Australia. *ADB*, 10, pp. 176-7.
17. Rev. George Goodman, MA (1821-1908) arrived in Melbourne in 1853, and served as vicar of Christ Church, Geelong, from January 1855 until his retirement in September 1906. He lectured in theology at Trinity College 1877-99 and was examining chaplain to Bishop Perry and his successors for fifty years. *ADB*, 4, pp. 264-5.
18. Although later changed to '50', the original figure was apparently forty-four or forty-six. Rupp was married in December 1904, and he probably wrote this in December 1948.

19. 'The Fields of Coleraine' was first published in 1866 as part of 'Whiffs from the Pipe' or 'Hippodromania', concerning the chances of various entrants in the Great Western Stakes.
20. Ada Cambridge (1844-1926) published a novel and some devotional works before coming to Australia with her husband, the Rev. George Frederick Cross. She was a prolific writer of novels, reminiscences and verse during her long residence in Victoria.
21. Samuel Winter Cooke, BA (1847-1929), grazier, was member for Western Province in the Victorian Legislative Council, 1888-1901, and for Wannon in the first Federal parliament, 1901-3. *ADB*, 8, pp. 101-2. The magnificence of Murndal, near Hamilton, is vividly described by Weston Bate in *Historic Homesteads of Australia*, Sydney 1969.
22. 27 December 1893.
23. John George Robertson (1803-62), pastoralist and botanist.
24. Also J. G. Robertson, appointed Professor of German Literature, University of London, 1903.
25. Ronald Campbell Gunn (1808-81), public servant and botanist, arrived in Van Diemen's Land in 1830. He was an enthusiastic collector for W. J. and J. D. Hooker. *ADB*, 1, pp. 492-3. Rupp became very interested in Gunn's work and began to gather material for a biography.
26. Abram Louis Buvelot (1814-88), Swiss-born landscape painter and photographer, painted 'Waterpool at Coleraine' in 1869. See Joan Kerr, *Dictionary of Australian Artists*, Sydney 1984, I, pp. 119-120.

Chapter 2: University Days

1. This crisis apparently occurred in 1890, when Leeper's effigy was burnt and two expelled students and their supporters left in an impressive procession of hansom cabs, 'leaving only twelve depressed survivors in residence'. J. R. Poynter in *ADB*, 10, p. 54.
2. A delightful allusion to the Thirty-nine Articles of Religion 'agreed upon — in the Convocation holden at London in the Year 1562'. The basic tenets of belief, containing 'the true Doctrine of the Church of England agreeable to God's Word'.
3. George Upward (1855-1915), rower and parliamentary officer.
4. Typhoid fever, 1893.
5. Conferred 23 December 1897.
6. Thomas Slaney Poole, MA LLB (1873-1927) made a long-enduring contribution to South Australian law and has been described as 'the best judge South Australia has ever had'. *ADB*, 11, pp. 256-7.
7. Hereward Humfry Henchman, MA, LLB (1874-1939). See *Biog. Reg.* 1, p. 321.
8. David John Davies Bevan, MA, LLB (1873-1954) became the first judge of the Supreme Court of the Northern Territory, 1912-21. See *Biog. Reg.*, I, p. 56.
9. Sir Charles Frederic Belcher, OBE, MA, LLB (1876-1970). See *WWA*, 1968; *Biog. Reg.* I, p. 47.
10. Sir Augustus Andrewes Uthwatt, Baron Uthwatt of Lathbury (1879-1949), Judge of Chancery, Lord of Appeal in Ordinary. Rupp later noted, 'He died rather suddenly in 1949'.

11. St Andrew's Church, Dennington, was opened in December 1914 as a memorial to the Venerable Andrew Edward Peacock (1863-1912), archdeacon of the Otway, 1906-12. He is also commemorated in Trinity College Chapel by a brass plaque with a most moving inscription.

12. Sir Alexander James Peacock (1861-1933), thrice premier of Victoria and later speaker of the Legislative Assembly, 1928-33.

13. Rt Rev. George Merrick Long, CBE, MA, DD, LLD (1874-1930), bishop of Bathurst, 1911-27, and of Newcastle, 1928-30. Director of Education for the Australian forces, 1917-19. *ADB*, 10, pp. 134-5.

14. Rupp later noted, 'Alas! he died in 1950'.

15. William Lionel Russell Clarke, MA, MLC (1876-1954), pastoralist, politician and philanthropist; Sir Francis Grenville Clarke, KBE, MLC (1879-1955), pastoralist and businessman, was president of the Legislative Council of Victoria, 1923-43. *ADB*, 8, pp. 16-18.

16. John Beacham Kiddle, OBE, LLB (1878-1950), Melbourne solicitor and pastoralist, father of the authoress Margaret Loch Kiddle (1914-58). Rupp made a marginal note, 'Died in 1950'.

17. Archdeacon Forster died in Armidale, N.S.W., in 1946. Of the others, Austine Bithray Rowed was ordained priest in 1899; Francis Lynch, priest, 1896; Leonard Townsend, priest, 1899; Herbert Thomas Fowler, priest, 1899; John Stanley Wells, priest, 1899; John Henry Frewin, priest, 1898; Alfred Ernest Jones Ross, deacon, 1894; Charlton George Brazier, priest, 1899. Other clergy noticed include Henry Thomas Langley, MA, priest, 1901, who became dean of Melbourne in 1942; George Ambrose Kitchen, MA, ThL, priest, 1904, who served as archdeacon of Hay, 1923-34; Thomas Keyran Pitt, priest, 1899; Edward Nowill Wilton, BA, priest, 1902, one-time assistant bishop of Melanesia, who served at St Thomas's, 1931-48.

18. Robert Browning (1812-89) published *Strafford* in 1837.

19. 28 Arnold Street, South Yarra, near Melbourne Grammar School and the Botanic Gardens.

20. It will be noted that such an attempt was made, to good effect, in the 'Recollections'.

21. For the remarkable Baron Sir Ferdinand Jakob Heinrich von Mueller, KCMG, PhD, MD, FLS, FRS, etc, government botanist of Victoria, 1853-96, and director of the Melbourne Botanic Gardens, 1857-73, see Margaret Willis, *By Their Fruits*, Sydney 1949; Edward Kynaston, *A Man on Edge*, London 1981; *ADB*, 5, pp. 306-8.

22. Edward Ellis Morris, MA, LittD (1843-1902), professor of Modern Languages and Literatures, 1884-1902; chairman of the Professorial Board, 1888, 1890-3. *ADB*, 5, pp. 293-4.

23. Thomas George Tucker, CMG, MA, LittD (1859-1946), professor of Classical Philology, 1886-1919.

24. Henry Laurie, MA, LLD (1837-1922), professor of Mental and Moral Philosophy, 1886-1911.

25. John Simeon Elkington, MA, LLB (1841-1922), professor of History and Political Economy, 1876-1912. *ADB*, 8, p. 425.

26. Sir Frederick McCoy, KCMG, MA, DSc, FRS (1817-99), foundation professor of Natural Science, 1855-1899; first (honorary) director of the National Museum of Victoria, 1857-99. *ADB*, 5, pp. 134-6.

27. Sir Walter Baldwin Spencer, KCMG, MA, DSc, LittD, FRS (1860-1929), professor of Biology, 1887-1920.

28. This date of birth has been widely recorded, but 1817 is now considered more likely.

29. Rev. Canon Robert Potter (1832-1908).

30. Rt Rev. Reginald Stephen, MA, FACT (1860-1956), dean of Melbourne, 1910-14; bishop of Tasmania, 1914-19, and of Newcastle, 1919-28.

31. I.e., about 1925.

32. Actually Rupp made his 'debut in print' while still at Geelong Grammar, and the newspaper articles consisted of more than a mere 'list of the native plants' of the district. See Bibliography.

33. Venerable Peter Teulon Beamish, MA, DD, LLD (1824-1914) of Warrnambool had baptized H. M. R. Rupp in January 1873.

34. *Pterostylis baptistii* FitzG.

35. *Thelymitra antennifera* (Gunn ex Lindl.) Hook.f.

Chapter 3: Ordination — and After

1. Rt Rev. Samuel Thornton, MA, DD (1835-1917), first bishop of Ballarat, 1875-1900.

2. Rev. Henry Samuel Robinson Thornton (1871-c.1950) was not unmindful of Rupp's difficulties. In presenting Rupp with a couple of Geelong Grammar book prizes as a farewell gift, Thornton inscribed them 'In Memory of a happy association Colac Beeac 1898-1901' and 'To H. M. R. Rupp, An acknowledgement of my many shortcomings in the Vicarial relation 1898-1901'.

3. For Bishop Cooper, see Part II, Chap. 2, ref. 19.

4. In 1902.

5. Ven. William George Hindley (1853-1936), one-time archdeacon of Melbourne. *ADB*, 9, pp. 304-5.

6. 1911.

7. Rt Rev. Arthur Vincent Green, MA, LLD (1857-1944), archdeacon of Ballarat, 1890-4; bishop of Grafton and Armidale, 1894-1900, and of Ballarat 1900-15. *ADB*, 9 pp. 90-1.

8. Thomas Kingsmill Abbott, MA (1866-1912), archdeacon of Tamworth, 1902-10; headmaster of The Armidale School, 1910-12. A fine brass plaque with a Latin tribute is in the school's chapel. See also Part II, Chap. 3, ref. 1.

9. John Howe (1774-1852), settler and explorer, made two journeys with his son-in-law, George Loder, from Windsor to the Hunter Valley, 1819-20, and finally settled near Morpeth. *ADB*, 1, p. 560.

10. I.e, 1904-48.

11. Joseph Henry Maiden, ISO, FLS, FRS (1859-1925), director of Sydney Botanic Gardens 1896-1924. See L. A. Gilbert, *The Royal Botanic Gardens, Sydney: A History, 1816-1985*, Melbourne 1986, Chapter 6; *ADB*, 10, pp. 381-3.

12. Rupp later noted: 'My lay reader at Warialda became the Rev. A. M. S. Wilson, who recently retired from St Paul's, Malvern, Victoria'. He was the Rev. Albert Montagu Stephen Wilson (1881-1963).

Chapter 4: 1905-1914

1. *Actinotus helianthi* Labill.

2. *Olearia ramosissima* (DC.) Benth.

3. Western or Pink Wedding-bush, *Ricinocarpos bowmanii* F. Muell.

4. *Cymbidium canaliculatum* R.Br.

5. Stringybark Sheoak, now known as *Allocasuarina inophloia* (F. Muell. et Bailey) L. Johnson.

6. Belah, *Casuarina cristata* Miq.

7. Buddah, *Eremophila mitchellii* Benth.

8. Kangaroo Grass, *Themeda australis* (R.Br.) Stapf. and/or Oat

Kangaroo (Tall Oat) Grass, *T. avenacea* (F. Muell.) Maiden et Betche.

9. Darling or 'Cranky' Pea, *Swainsona galegifolia* (Andr.) R.Br. can cause stock to become 'pea-struck' when a craving for the plant develops.

10. Darling, Murray or Macquarie Lily, *Crinum flaccidum* Herb.

11. Wilcannia or Garland Lily, *Calostemma*, may have purplish or yellow flowers. Both forms have been included in *C. purpureum* R.Br. but Rupp's suggestion of 'Native Jonquil' would indicate the yellow form, now again known as *C. luteum* Sims.

12. Respectively, *Acacia harpophylla* F. Muell. ex Benth.; *A. homalophylla* A. Cunn. ex Benth.; and *A. pendula* A. Cunn. ex G. Don (also known as Boree).

13. Rt Rev. Thomas Henry Armstrong (1857–1930).

14. Sir Earle Christmas Grafton Page, GCMG, CH, MB, ChM, DSc, FRACS (1880–1961) was elected MHR for Cowper in 1919; first chancellor of the University of New England, 1955; one of the founders of the Federal Country Party, 1920, and deputy prime minister in the Bruce-Page Government in the 1920s.

15. Jessie Mary Grey Lillingston, BA (1889–1970) married (February 1916) the Hon. Sir Kenneth Whistler Street, one-time lieutenant-governor of New South Wales and chief justice.

16. Charles Frederick Tindal (1857–1938), eldest son of Charles Grant Tindal (1823–1914), founder of Ramornie station (1852), and the Australian Meat Co. (1865) based at Ramornie on the Clarence River. *ADB*, 6, pp. 277–8.

17. I.e., rainforests.

18. Giant Stinging Tree or Gympie, *Dendrocnide excelsa* (Wedd.) Chew.

19. This Lemon-scented Tea Tree is now referred to *L. petersonii* Bail.

20. *Proc. Linn. Soc. N.S.W.* 1912, pp. 244–5.

21. Rupp noted: 'Our sons-in-law [Leslie Clarendon Cox and Ashley Clarendon Cox] are direct descendants of William Cox of Clarendon, near Windsor, N.S.W. who built the first road across the Blue Mountains in 1815'. For William Cox (1764–1837) see *ADB*, 1, pp. 258–9, and for the family see P. C. Mowle, *A Genealogical History of Pioneer Families of Australia*, Sydney 1939.

22. Rev. George Peppin Maitland Ware (1877–1947).

23. Dr Gwenda L. Davis became associate professor of Botany at the University of New England. She now lives in retirement at Port Macquarie.

24. Rough Tree-fern, *Cyathea australis* (R.Br.) Domin.

25. *Boronia ruppii* Cheel. *Proc. Linn. Soc. N.S.W.*, 1927, pp. 404–5.

26. Edwin Cheel (1872–1951) worked in Queensland on arriving from England in 1892. In December 1897 he became a gardener in Centennial Park. Transferring to the Botanic Gardens in January 1902, Cheel became a botanical assistant in 1908, and five years later succeeded Ernst Betche as senior botanical assistant. He was curator of the Herbarium 1924–36. *ADB*, 7, pp. 628–9.

27. Rev. John Jones, BA (1875–1942), a former vicar of Thursday Island, was general secretary of the Australian Board of Missions, 1912–17, and chairman, 1917–22. *Biog. Reg.* I, pp. 377–8.

Chapter 5: The First World War — and After

1. In 1917.
2. 8 January 1915.

3. 28 September 1917.

4. Rt Rev. Montagu John Stone-Wigg (1861–1918), bishop of New Guinea, 1898–1908.

5. Rt Rev. Gerald Sharp (1865–1933), bishop of New Guinea, 1910–21; archbishop of Brisbane, 1921–3.

6. Rev. Copland King (1863–1918), great-grandson of Governor Philip Gidley King and grand-nephew of John Macarthur, worked in the New Guinea mission, 1891–1918.

7. Rev. Samuel Tomlinson (1856–1937) and his wife Elizabeth first went to the New Guinea mission in 1891.

8. Rt Rev. Gilbert White (1859–1933), bishop of Carpentaria, 1900–15, and of Willochra, 1915–25.

9. Rev. Ernest Richard Bulmer Gribble (1869–1957) was in charge of the Yarrabah Mission, 1894–1910, the Forrest River Mission, 1914–28, and chaplain of Palm Island Aboriginal Settlement, 1931–41. *ADB*, 4, p. 299.

10. Most Rev. St Clair George Alfred Donaldson (1863–1935), bishop of Brisbane, 1904, and archbishop, 1905–21. *ADB*, 8, pp. 319–20.

11. Rt Rev Francis de Witt Batty (1879–1961), bishop of Newcastle, 1931–58. *ADB*, 7, pp. 210–12.

12. Rt Rev. Ernest Augustus Anderson (1859–1945), bishop of Riverina, 1895–1925. *ADB*, 7, p. 53.

13. Rt Rev. Henry Edward Cooper died at Armidale on 1 July 1916.

14. Rt Rev. Cecil Henry Druitt (1875–1921), coadjutor bishop of Grafton and Armidale, 1911, and first bishop of Grafton, 1914–21.

15. Rt Rev. Lewis Bostock Radford (1869–1937), bishop of Goulburn, 1915–33.

16. Ven. Henry Bruné Atkinson (1874–1960), appointed rector of Holy Trinity, Hobart, 1916; collated archdeacon, 1924. *ADB*, 7, pp. 120–1.

17. 1942.

18. Streaked Rock Orchid, *Dendrobium striolatum* Reichb.f.

19. Butterfly Orchid, *Sarcochilus australis* (Lindl.) Reichb.f.

20. Most probably it was John Smith, MP, chairman of meetings held in London in April 1824 to promote formation of the Australian Agricultural Company which took up an extensive area near Port Stephens. The lake was so known by the mid-1820s.

21. *Aust.*, 3 April 1926.

22. For Dr R. S. Rogers see Part II, Chap. 5, ref. 15.

23. Ferdinand von Mueller, *Fragmenta Phytographiae Australiae*, 11 vols, Melbourne 1858–81.

24. For Edith Coleman, see Part I, Chap. 8, ref. 10.

25. For W. H. Nicholls, see Part I, Chap. 8, ref. 7.

26. After an initial project in 1950, Nicholls's splendid illustrations were published by Nelson in *Orchids of Australia*, Melbourne, in 1969, with text edited by D. L. Jones and T. B. Muir.

27. H. L. Kesteven, DSc, MD (1881–1964), physician and naturalist. *ADB*, 9, pp. 579–80.

28. Now *Rhizanthella slateri* (Rupp) M. Clements et Cribb.

Chapter 6: Paterson and the South Maitland Coalfields

1. Sadly while this work was being compiled, Mrs Eileen Cox died on 12 August 1989 and Mr Arthur Rupp died on 30 March 1990.

2. A constitution finally came into force on 1 January 1962.

3. For George Robertson, see Part II, Chap. 9, ref. 6.

4. For Bishop J. S. Moyes, see Part II, Chap. 9, ref. 9.

Chapter 7: Botany at Paterson and Weston

1. Sir John Clifford Valentine Behan, BA, BCL, LLD (1881–1957), warden of Trinity College, 1918–46, in succession to Dr Alexander Leeper. *ADB*, 7, pp. 247–8.
2. Alfred James Ewart, DSc, PhD, FLS, FRS (1872–1937), foundation professor of Botany, University of Melbourne, 1906–37. *ADB*, 8, pp. 448–50.
3. For John Hopson see Part II, Chap. 8, ref. 9.
4. *Dendrobium tenuissimum* Rupp. *Proc. Linn. Soc. N.S.W.*, 1927, p. 570. This species is now referred to *D. mortii* F. Muell.
5. *Nothofagus moorei* (F. Muell.) Krasser, also known as Antarctic Beech.
6. *Dendrobium speciosum* Sm.: 'Australia's most widely distributed epiphytic orchid', according to David L. Jones.
7. *Dysoxylum fraseranum* (A. Juss.) Benth.
8. I.e., Soft Tree-fern, *Dicksonia antarctica* Labill.; the other species are the Rough or Hill Tree-fern, *Cyathea australis* (R.Br.) Domin, and Prickly Tree-fern, *C. leichhardtiana* (F. Muell.) Copeland.
9. *Dendrobium falcorostrum* FitzG. which R. D. FitzGerald somewhat unconventionally described and named in the *Sydney Morning Herald* (18 November 1876) instead of in one of the accepted scientific journals.
10. Sickle Greenhood, *Pterostylis falcata* Rogers, now referred to *P. furcata* Lindley.
11. *Prasophyllum rogersii* Rupp. *Proc. Linn. Soc. N.S.W.*, 1928, p. 340; and *P. hopsonii* Rupp, *op. cit.*, p. 341.
12. *Diuris venosa* Rupp., *Vic. Nat.*, 1926, p. 153; *Proc. Linn. Soc. N.S.W.*, 1926 and 1928. It has subsequently been recorded from the New England National Park.
13. *Pterostylis furcillata* Rupp. *Proc. Linn. Soc. N.S.W.*, 1930, p. 415. Now considered to be a natural hybrid between *P. ophioglossa* R.Br. and *P. obtusa* R.Br.

Chapter 8: From Pilliga to Raymond Terrace

1. This list would have included the Narrow-leaved Ironbark, *Eucalyptus crebra* F. Muell.; Pilliga Grey Box, *E. pilligaensis* Maiden; Tumble-down Gum, *E. dealbata* A. Cunn.; Mugga, *E. sideroxylon* A. Cunn. ex Woolls; Carbeen, *E. tessellaris* F. Muell.; Rough-barked Apple, *Angophora floribunda* (Sm.) Sweet; Coolibah Apple, *A. melanoxylon* R. T. Baker; Buddah, *Eremophila mitchellii* Benth.; Belah, *Casuarina cristata* Miq.; Bull Oak, *Allocasuarina luehmannii* (R. T. Baker) L. Johnson; Brigalow, *Acacia harpophylla* F. Muell. ex Benth.; Cooba, *A. salicina* Lindley; Ironwood, *A. excelsa* Benth; Yarran, *A. homalophylla* A. Cunn ex Benth.; White Cypress, *Callitris glaucophylla* Thompson et Johnson; Black Cypress, *C. endlicheri* (Parl.) F. M. Bail.; Quandong, *Santalum acuminatum* (R.Br.) DC.; Whitewood, *Attalaya hemiglauca* (F. Muell.) F. Muell. ex Benth.; Leopard Tree, *Flindersia maculosa* (Lindl.) Benth.; Wilga, *Geijera parviflora* Lindl.; Gruey (Gruie) or Colane, *Owenia acidula* F. Muell.; Beefwood, *Grevillea striata* R.Br.; Native Pomegranate or Wild Orange, *Capparis mitchellii* Lindl.
2. Fringe-myrtle, *Calytrix tetragona* Labill.
3. Rupp distinguished this form as *Cymbidium canaliculatum* forma *aureolum*. *Proc. Linn. Soc. N.S.W.*, 1934.

4. Rupp turned sixty in December 1932.
5. The bridge was commenced in 1946 and opened in December 1952. A second bridge was opened in April 1987.
6. The church has been modified since Rupp's time.
7. Venerable Henry Alexander Woodd, BA, was made deacon in 1888. He was archdeacon of Newcastle, 1921–49.

Chapter 9: And So to the End

1. Two cities of Philistia mentioned in the Old Testament. See II Samuel i. 20.
2. At 24 Kameruka Road.
3. Robert Henry Anderson, BScAgr (1899–1969) succeeded Edwin Cheel as botanist and curator of the Herbarium, Sydney Botanic Gardens, in 1936, and was director, 1945–64.
4. The present Herbarium complex was opened in November 1982.
5. Henry Deane, MA, FLS (1847–1924), engineer and botanist, was associated with the survey and construction of the Transcontinental Railway between Port Augusta and Kalgoorlie. He was well-known to R. D. FitzGerald, Baron von Mueller and Sir Baldwin Spencer and in 1894 he saw the final published part of FitzGerald's *Australian Orchids* through the press. *ADB*, 4, pp. 178–9.
6. Dr Rudolf Schlechter (1872–1925), German botanist.
7. For E. D. Hatch, see Part I, Chap. 8, ref. 53.
8. For T. E. Hunt, see Part I, Chap. 8, ref. 83.
9. For C. L. Barrett, see Part I, Chap. 8, ref. 11.
10. For A. H. Chisholm, see Part I, Chap. 8, ref. 8.
11. For J. H. Willis, see Part I, Chap. 8, ref. 23.
12. For H. Flecker, see Part I, Chap. 8, ref. 85.
13. For C. T. White, see Part I, Chap. 8, ref. 59.
14. For A. G. Hamilton, see Part I, Chap. 8, ref. 36.
15. For E. N. McKie, see Part I, Chap. 8, ref. 57.
16. William Faris Blakely (1875–1941), author of *A Key to the Eucalypts*, Sydney 1934.
17. For the Scrivener sisters, see Part I, Chap. 8, ref. 98.
18. Dr Friedlander.
19. Dr Anderson.
20. With the encouragement of Mr Knowles Mair, director of the Royal Botanic Gardens, Sydney, 1964–70, and a friend and admirer of H. M. R. Rupp, *The Orchids of New South Wales* facsimile edition appeared in 1969 with a Supplement by Donald J. McGillivray.
21. First and last verses of 'Life' by Anna Laetitia Barbauld (1743–1825) in her *Works*, London 1825, Vol. I, pp. 261–2.

Bibliography of an Orchidologist

1. Rupp to T. E. Hunt, 24 February 1947. RBGS.
2. Mair to J. H. Willis, 25 October 1956. Original in possession of J. H. Willis; copy in RBGS.
3. *AWM*, 19 June 1934, p. 10.
4. *AWM*, 31 July 1934, p. 10.

Botanical Index

General Index

THE GREENHOOD ORCHIDS OF THE SOUTH MAITLAND COALFIELDS.

1, Pterostylis rufa. 2 , P. Mitchellii. 3 , P. nutans. 4 , P. curta. 5 , P. parviflora. 6 , P. truncata. 7 , P. concinna. 8 , P. longifolia. 9 , P. mutica. 10 , P. pedunculata. 11 , P. furcillata. 12 , P. ophioglossa. 13 , P. acuminata. 14 , P. reflexa. 15 , P. revoluta. 16 , P. pusilla var. prominens. 17 , P. obtusa. 18 , P. grandiflora.